Physical Processes

A VISUAL APPROACH

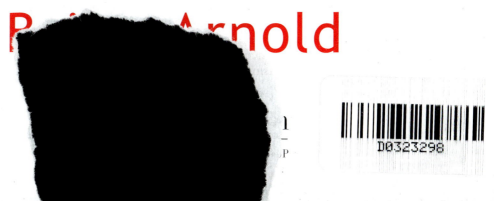

R[illegible]rnold

The publishers would like to thank the following individuals, institutions and companies for permission to reproduce photographs in this book. Every effort has been made to trace ownership of copyright. The publishers are happy to make arrangements with any copyright holder whom it has not been possible to contact.

© Allsport 66 (right); Andrew Lambert 98, 213, 214, 216; The Brett Weston Archive/ CORBIS 112; David A Hardy/ Science Photo Library 252 (top two); © Ecoscene/CORBIS 66 (left); European Space Agency/ Science Photo Library 254 (bottom right); Frank Zullo/ Science Photo Library 251; Geoffrey Taunton, Cordaiy Photo Library Ltd/ CORBIS 192; Kent Wood/ Science Photo Library 140; Michael S Yamashita/CORBIS 177; NASA 255; NASA/Roger Reessmeyer/CORBIS 254 (bottom right); NASA/Science Photo Library 242 (bottom), 256; Pekka Parviainen/Science Photo Library 254 (top); PLI/Science Photo Library 238; Ronald Rover/Science Photo Library 242 (top); Tony Hallas/Science Photo Library 252 (bottom right); Wendy Brown 215; Yves Baulieu, Publiphoto Diffusion/ Science Photo Library 107

The illustrations were drawn by Ian Foulis & Associates.

Orders: please contact Bookpoint Ltd, 130 Milton Park, Abingdon, Oxon OX14 4SB. Telephone: (44) 01235 827720, Fax: (44) 01235 400454. Lines are open from 9.00–6.00, Monday to Saturday, with a 24 hour message answering service. Email address: orders@bookpoint.co.uk

A catalogue record for this title is available from The British Library

ISBN 0 340 77295 6

First published 2001
Impression number 10 9 8 7 6 5 4 3 2 1
Year 2007 2006 2005 2004 2003 2002 2001

Cover photo from Chris Bell, Teleg[illegible]
Printed in Spain for Hodder & Sto[illegible] er
Headline Plc, 338 Euston Road, L[illegible]

Contents

Preface

Traditionally Physics has been regarded as a mathematical and difficult science, to be studied and understood by only the more able student. After many years of teaching pupils of all abilities, I have come to the conclusion that this is absolutely not the case. An understanding of the ideas behind most concepts in physics and their application in everyday life is available to all, regardless of academic ability. In my view it is the manner in which these ideas are presented to students which opens or closes doors to their understanding. It is the aim of this book to strip away unnecessary scientific jargon, to use language understood by most students and reduce text to a minimum by presenting ideas and information in as visual a way as possible. Where scientific terminology is important it has been explained simply and then opportunities provided for students to reinforce their understanding of the word

Physics is an exciting subject with an ever-increasing relevance to the real world. I hope this book opens the door to students who previously thought that this was a world to which they had no access. Physical Processes covers most of the core material in GCSE physics specifications and the physics component of GCSE science specifications.

My thanks go to all those folks who have supported me in the writing of this book. In particular Stephen Halder at Hodder and Stoughton, friends and colleagues at Roade School and of course my family who continually provided the support I needed.

Brian Arnold
April 2001

1 Energy and Energy Transfer

We all need energy to live, to grow and to be active. If we are very active and do lots of things we might be described as being **energetic.** We obtain the energy we need from the food we eat.

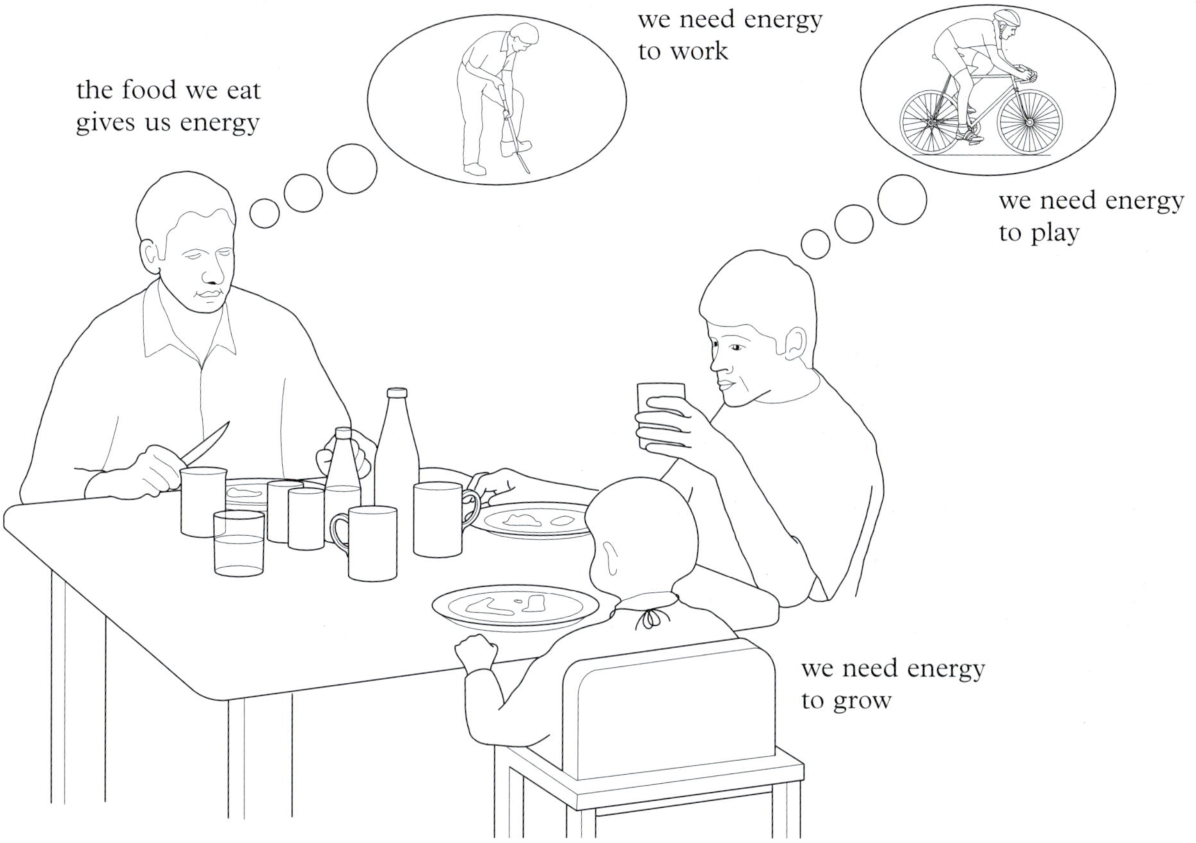
the food we eat
gives us energy
we need energy
to work
we need energy
to play
we need energy
to grow

Different types of energy

Food is not the only source of energy. There are many different kinds and sources of energy.

We obtain light energy from the Sun, from fires and from light bulbs.

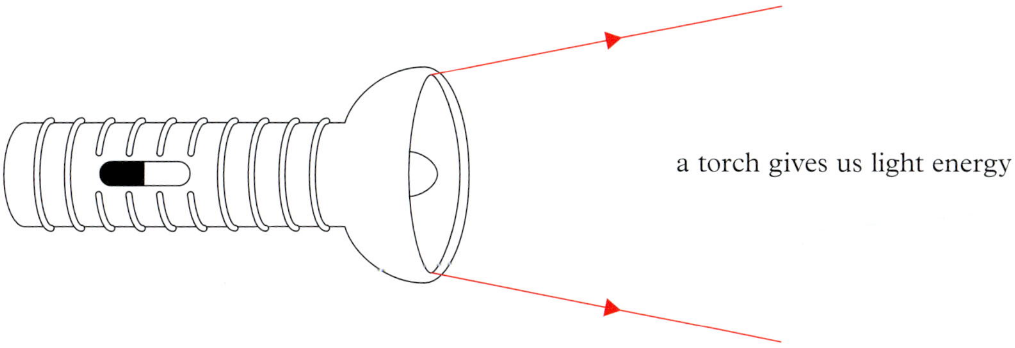
a torch gives us light energy

We obtain heat energy (sometimes called thermal energy) from objects that are hot.

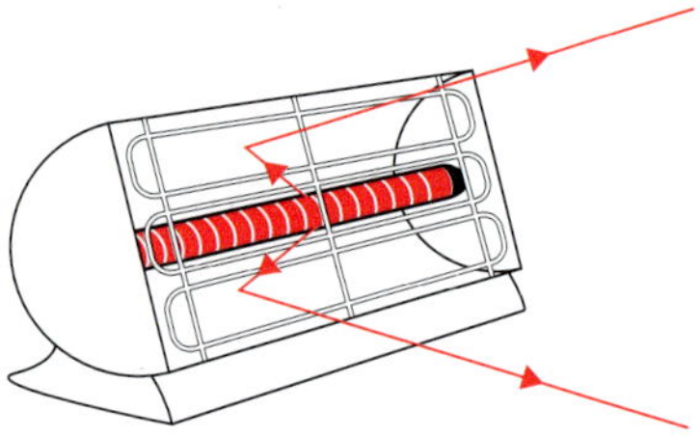

an electric fire gives out heat energy

We obtain sound energy from vibrating objects such as musical instruments and loudspeakers.

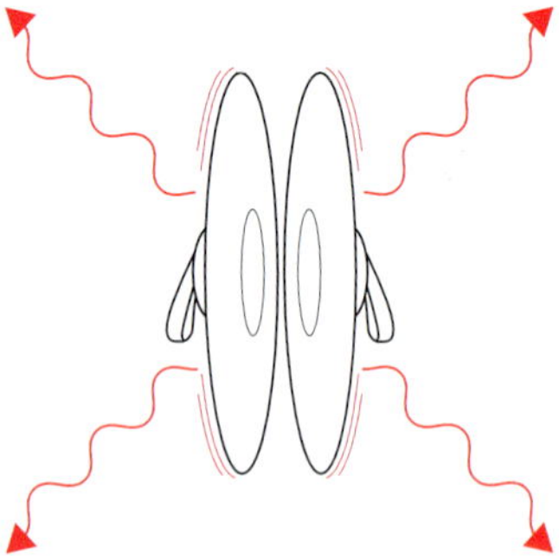

a vibrating cymbal is a source of sound energy

Chemical energy is stored in foods, fuels and batteries.

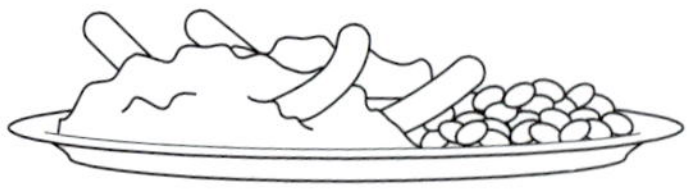

food and batteries are both sources of chemical energy

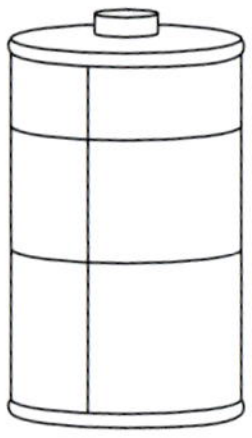

We obtain nuclear energy from the centre of atoms when nuclear reactions take place.

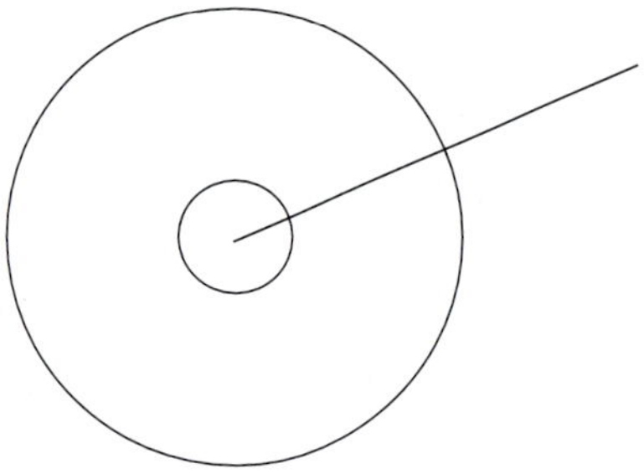

the centre of an atom is called the nucleus

reactions within the nucleus release energy

We obtain electrical energy whenever a current flows.

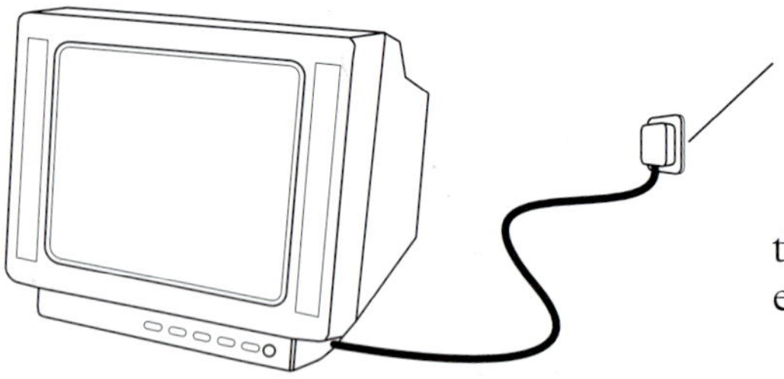

a current flows into the TV

the current supplies electrical energy to make the TV work

Objects which are moving have movement or kinetic energy.

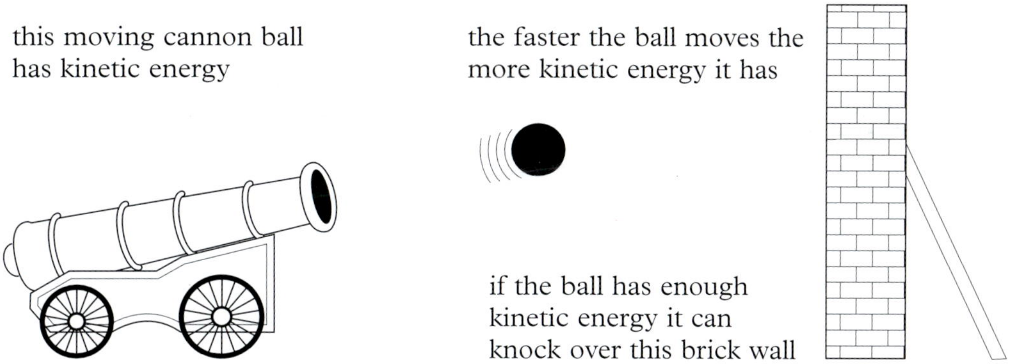
this moving cannon ball
has kinetic energy
the faster the ball moves the
more kinetic energy it has
if the ball has enough
kinetic energy it can
knock over this brick wall

Objects which are up high have gravitational potential energy.

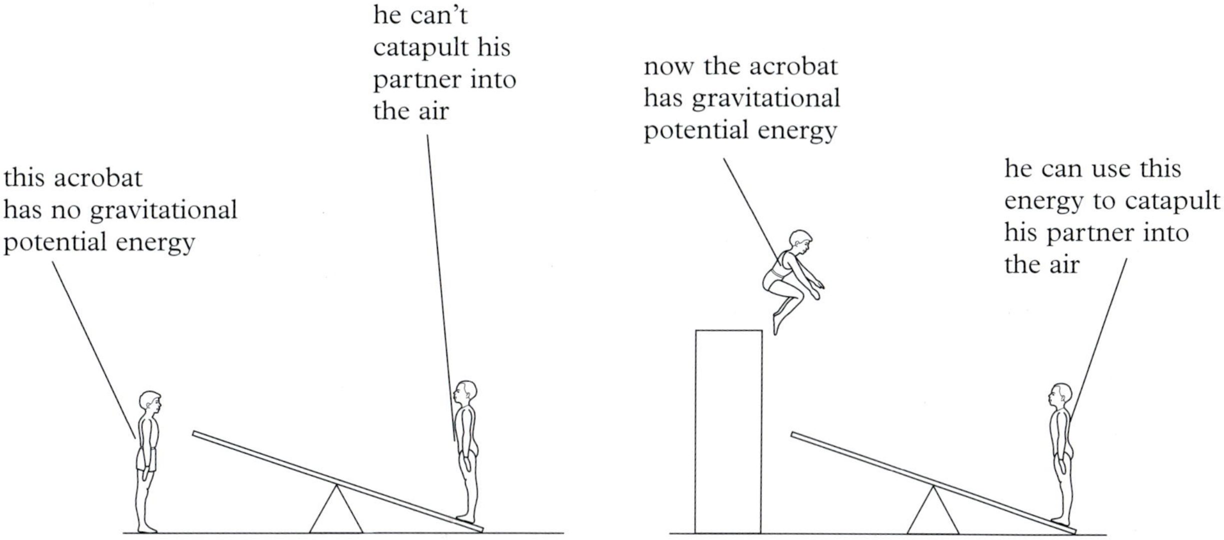
he can't
catapult his
partner into
the air
this acrobat
has no gravitational
potential energy
now the acrobat
has gravitational
potential energy
he can use this
energy to catapult
his partner into
the air

The higher an object, the more gravitational potential energy it has.

Objects which are stretched or squashed or twisted out of shape have elastic or strain potential energy.

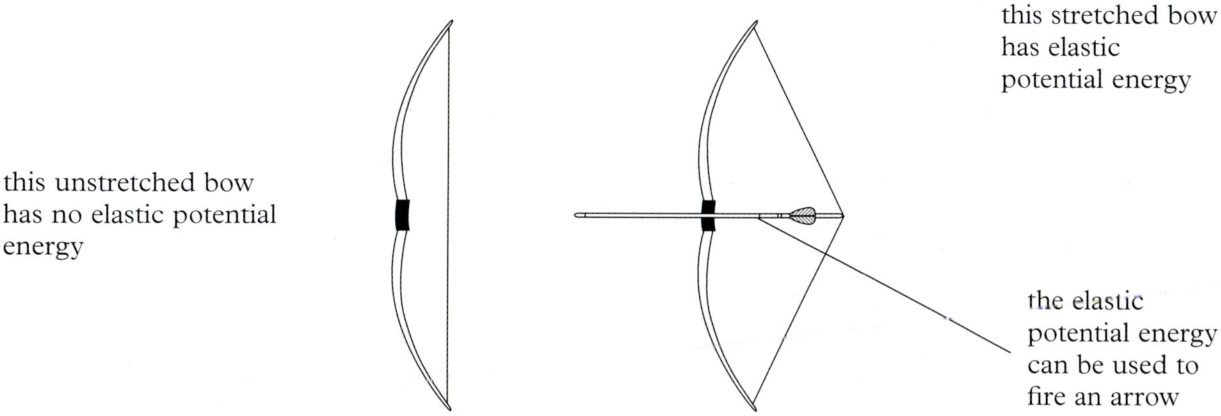
this stretched bow
has elastic
potential energy
this unstretched bow
has no elastic potential
energy
the elastic
potential energy
can be used to
fire an arrow

Questions

1 What forms of energy do the following have?
a) a stretched catapult
b) a can of petrol
c) a beaker of boiling water
d) water at the top of a waterfall
e) a sprinter during a 100 m race.

Energy transfers

When we use energy it changes into a new form.

This loudspeaker is changing electrical energy into sound energy.

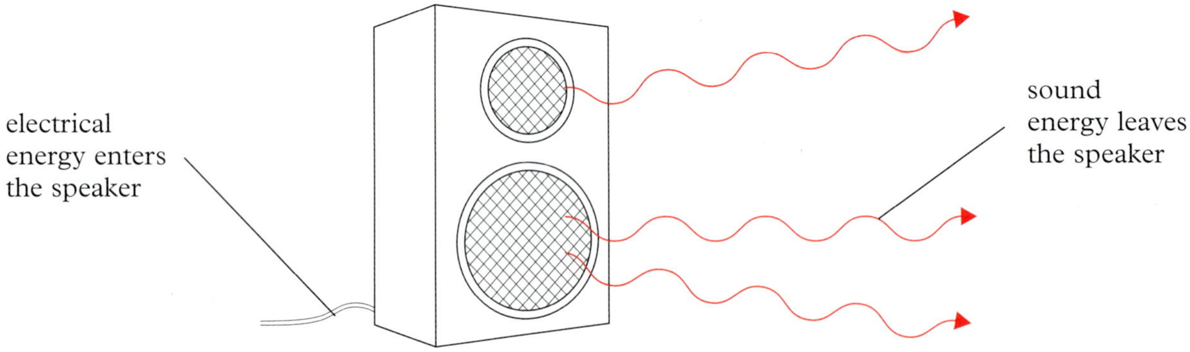
electrical
energy enters
the speaker
sound
energy leaves
the speaker

This electric motor is changing electrical energy into kinetic energy.

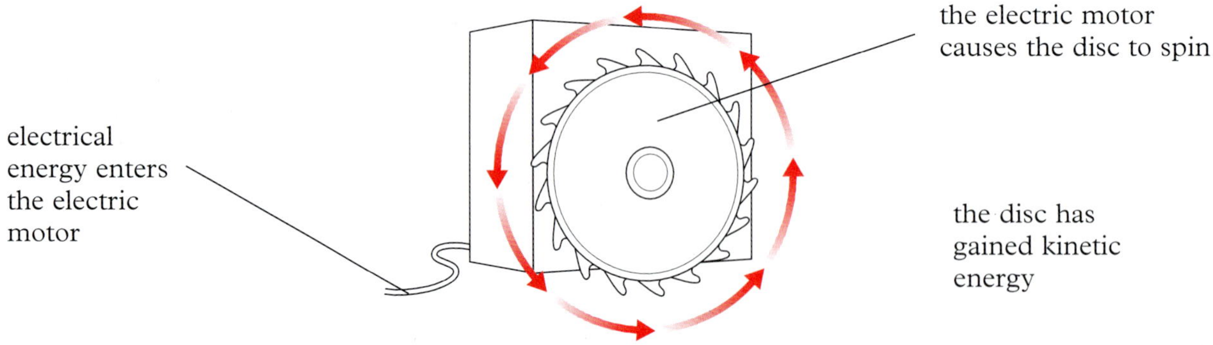
the electric motor
causes the disc to spin
electrical
energy enters
the electric
motor
the disc has
gained kinetic
energy

This crane changes electrical energy into gravitational potential energy.

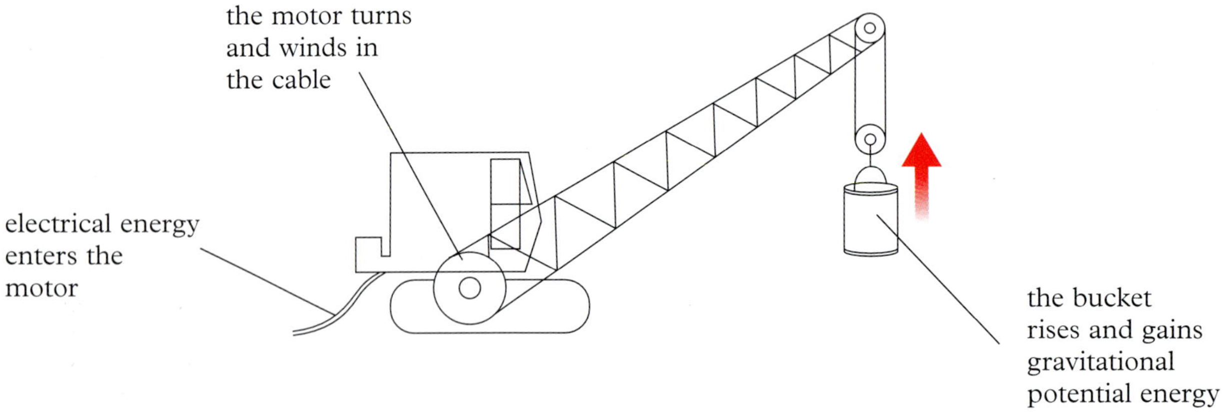
the motor turns
and winds in
the cable
electrical energy
enters the
motor
the bucket
rises and gains
gravitational
potential energy

This wind turbine is changing the kinetic energy of the wind into electrical energy.

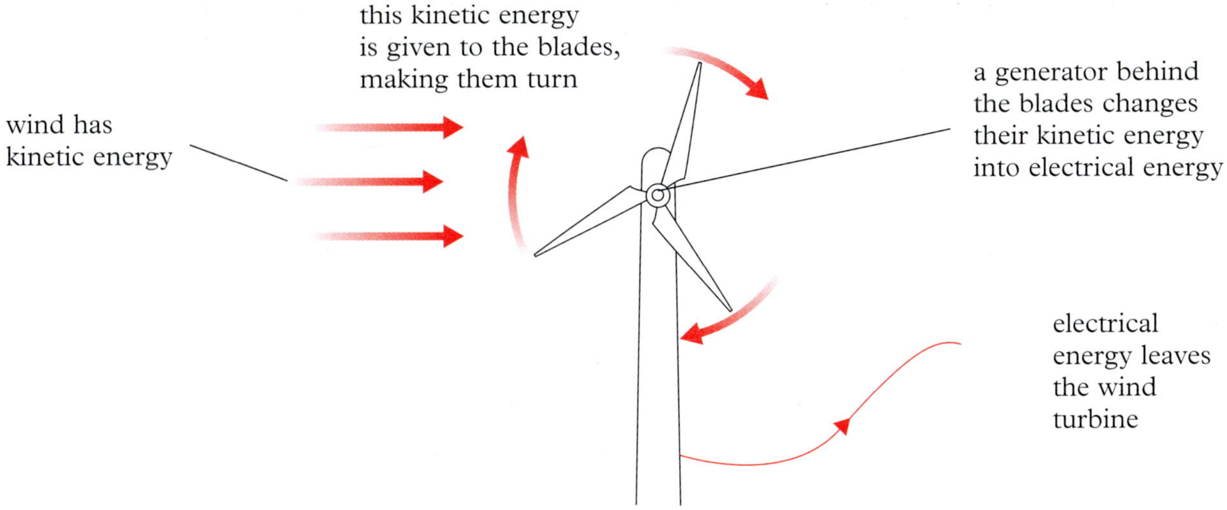
this kinetic energy is given to the blades, making them turn
wind has kinetic energy
a generator behind the blades changes their kinetic energy into electrical energy
electrical energy leaves the wind turbine

This light bulb is changing electrical energy into heat and light energy.

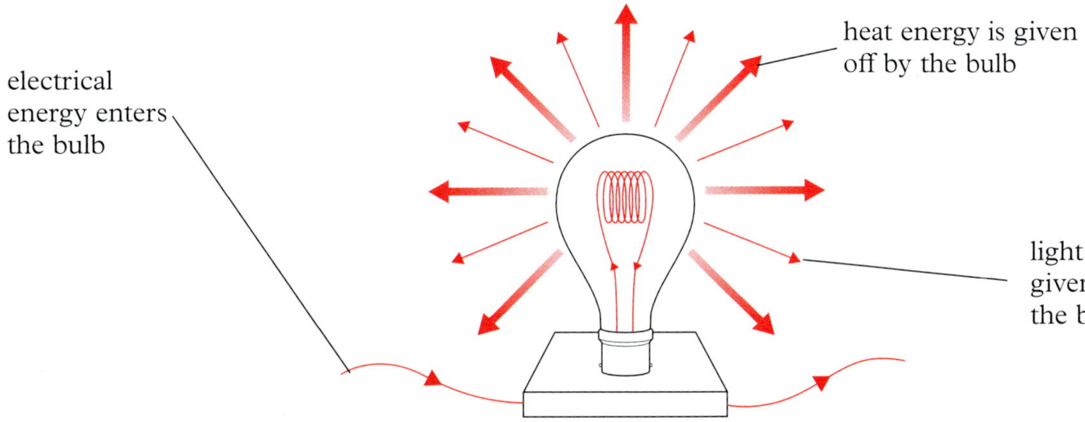
electrical energy enters the bulb
heat energy is given off by the bulb
light energy is given off by the bulb

This candle is changing chemical energy into heat and light energy.

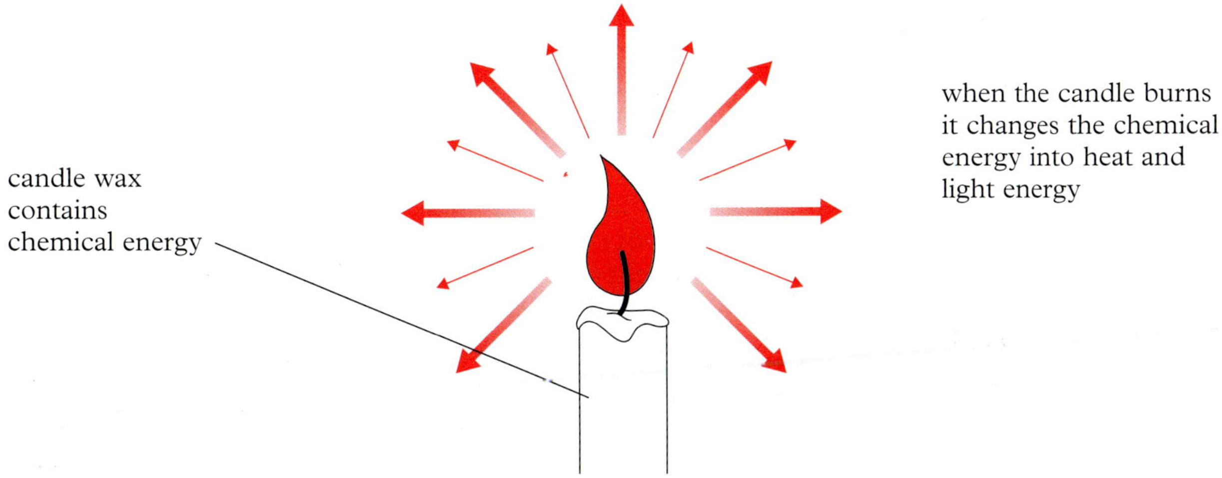
candle wax contains chemical energy
when the candle burns it changes the chemical energy into heat and light energy

This solar cell is changing light energy into electrical energy.

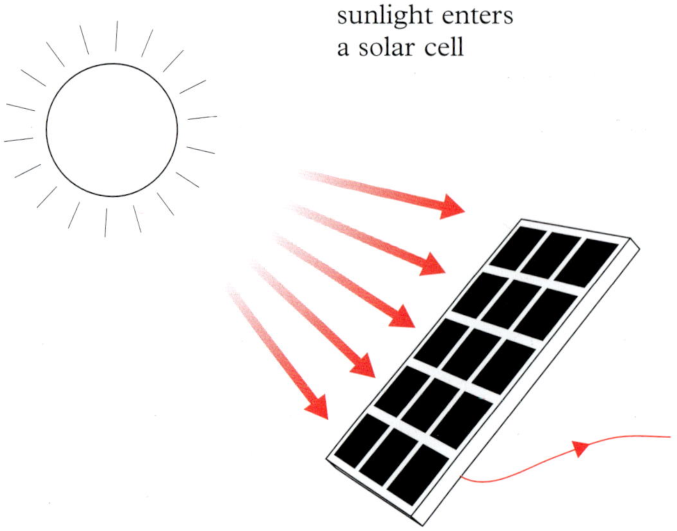
sunlight enters
a solar cell
electrical energy
leaves the solar cell

Questions

1 Fill in the missing words:

a) A burning candle changes _________ energy into _________ and _________ energy.

b) An electric fire changes _________ energy into _________ energy and _________ energy.

c) A loudspeaker changes _________ energy into _________ energy.

d) A microphone changes _________ energy into _________ energy.

e) A dynamo changes _________ energy into _________ energy.

f) An electric motor changes _________ energy into _________ energy.

2 What piece of apparatus would you use to change:

a) Electrical energy into light energy?

b) Light energy into electrical energy?

c) Chemical energy into kinetic energy?

d) Strain potential energy into kinetic energy?

Summary

Energy has many different forms. Heat (thermal energy), light and sound are forms energy may take when it is moving from place to place. Chemical energy, gravitational potential energy and strain energy are forms of energy which can be stored until they are needed. Kinetic or movement energy is the energy an object has because it is moving. Electrical energy is one of the most convenient forms of energy as it is easily changed into other forms.

Whenever energy is used it changes into other forms of energy. This is called energy transfer.

Key words

energetic Describes someone who is active and able to do lots of work.

energy This is needed to make things happen.

energy transfer The changing of one form of energy into a different form.

End of Chapter 1 Questions

1 Complete the following sentences using the words from the list below. Each word in the list may be used once, more than once or not at all.

heat light sound chemical electrical kinetic

a) In an electric motor the electrical energy is transferred mainly into _________ energy.

b) In a radio the electrical energy is transferred mainly into _________ energy.

c) In a torch the_________ energy stored inside the battery is transferred into electrical energy. This electrical energy is then transferred by the bulb into_________ energy and_________ energy.

d) In a coal fire_________ energy is being transferred into_________ energy and_________ energy.

2 What forms of energy are stored in the following?

a) water at the top of a waterfall
b) a slice of bread
c) a stretched catapult
d) a travelling bullet
e) a wound up clock.

3 Choose a device from the list below which could be used to make the following energy transfers.

solarcell electric bell television radio microphone motor car

a) electrical energy to light and sound energy
b) chemical energy to kinetic energy
c) sound energy to electrical energy
d) electrical energy to sound energy
e) light energy to electrical energy.

2 Measuring Energy

We measure energy in joules (J) and kilojoules (kJ). 1 kJ is 1000 J.

a man who does physical work like digging needs a lot of energy

this man needs around 15 000 kJ of energy from the food he eats every day

a young child needs less energy than an adult

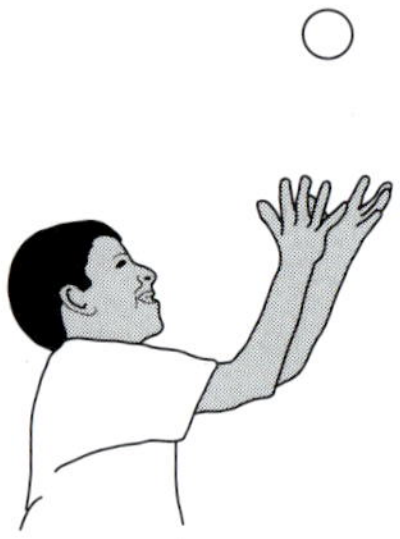

this boy needs around 8000 kJ of energy from the food he eats every day

The graph below shows approximate energy needs for one day. Boys and men tend to need a little more energy than girls and women.

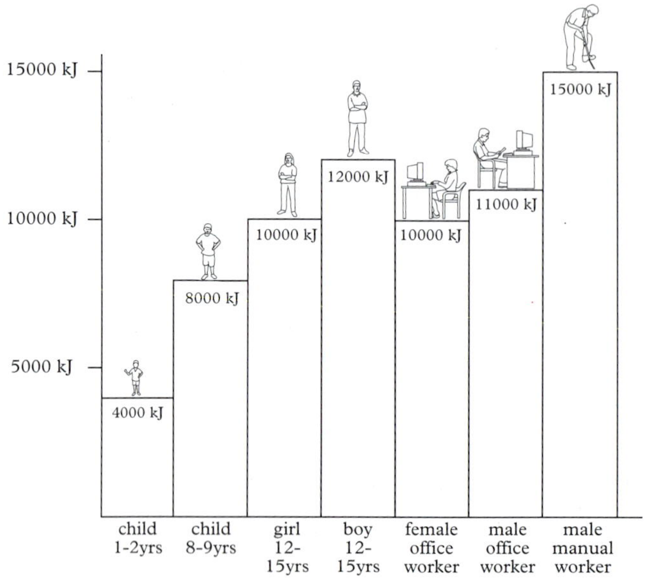

Different types of food contain different amounts of energy. The table below shows how much energy is contained in an average portion of different foods.

Food	Energy content in kJ	Food	Energy content in kJ
Apple	200	Cabbage	80
Banana	300	Carrots	80
Orange	200	Lettuce	40
Boiled potatoes	400	Bread and butter (1 slice)	400
Chips	1000	Ice cream	500
Boiled rice	500	Crisps	600
Spaghetti	500	Cake	700
Pizza	1200	Chocolate	1500
Peas	300	Cup of tea	200
Baked beans	400	Lemonade	700

Questions

1 Using the bar chart on page 8 write down the daily energy requirement for each of the following:
a) a very young child
b) an active teenage boy
c) a man who does heavy manual work.

2 Make a list of everything you have eaten during the last 24 hours. Now try to work out the amount of energy you have taken in.

3 If you ate an average portion of the following foods in one day how much energy would you obtain from it?
- A cup of tea and slice of toast with butter for breakfast.
- A chocolate bar at morning break.
- Pizza and an apple for lunch.
- A bag of crisps in the afternoon.
- Chips and beefburger (100 kJ) followed by a piece of cake and two more cups of tea for your dinner.

Conservation of energy

If 50 J of electrical energy flow into a light bulb, 50 J of heat and light energy must flow out of the bulb. The energy input must equal the energy output. This is true for all energy transfers and is expressed as a law.

The Law of Conservation of Energy says that during an energy transfer no energy is gained or lost

We can show the different amounts of energies in a simple **energy transfer diagram**.

The energy entering the bulb is 50 J. The energy leaving the bulb is 20 J + 30 J = 50 J. The Law of Conservation of Energy is being obeyed.

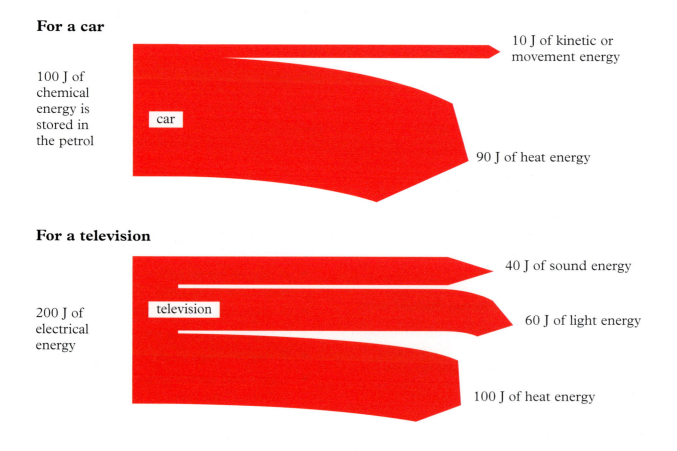

For a car

For a television

Questions

1 a) Write down the Law of Conservation of Energy.

b) Explain in your own words what this law means.

2 Draw an energy transfer diagram for each of the following:

a) a solar cell when 100 J of light energy is changed into 85 J of electrical energy and 15 J of heat energy.

b) a jack-in-a-box when 50 J of elastic potential energy is changed into 35 J of kinetic energy, 10 J of sound energy and 5 J of heat energy.

Efficiency

We use light bulbs to change electrical energy into light energy. However, as we have already seen some of the electrical energy is changed into heat. Only 20 J of the input energy is changed into light, which is what we want it to do. This light bulb is not very efficient.

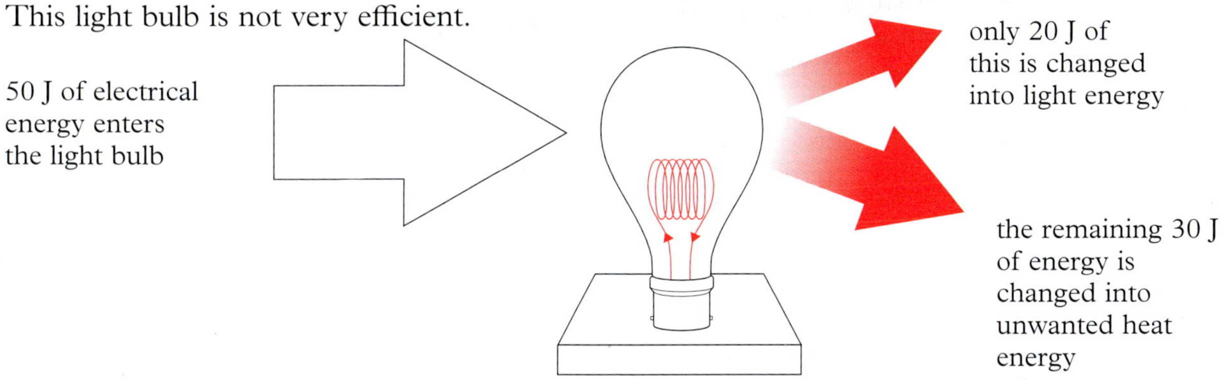

To calculate how efficient the bulb is we use the equation

$$\textbf{efficiency} = \frac{\textbf{useful energy out}}{\textbf{total energy in}} \times 100\%$$

For this bulb, efficiency $= \frac{20\text{ J}}{50\text{ J}} \times 100\%$

efficiency = 40%

This means that 60% of the energy entering the bulb is wasted. It is changed into an 'unwanted' form of energy.

Efficiency of a radio

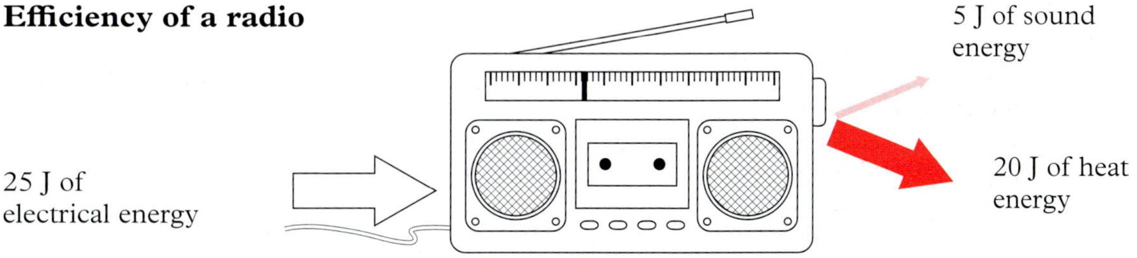

This radio converts 25 J of electrical energy into 5 J of sound energy and 20 J of heat. The efficiency of the radio is $\frac{5\text{ J}}{25\text{ J}} \times 100\% = 20\%$

Efficiency of a car

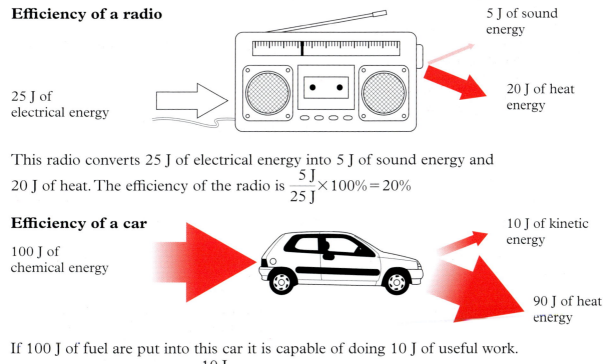

If 100 J of fuel are put into this car it is capable of doing 10 J of useful work.

The efficiency of the car is $\frac{10\text{ J}}{100\text{ J}} \times 100\% = 10\%$

Questions

1 a) Calculate the efficiency of an electric fire which changes 100 J of electrical energy into 80 J of heat energy and 20 J of light energy.

b) Calculate the efficiency of a steam engine which changes 50 J of chemical energy into 40 J of heat energy and 10 J of kinetic energy.

c) Calculate the efficiency of a dynamo which changes 200 J of kinetic energy into 150 J of electrical energy, 40 J of heat energy and 10 J of sound energy.

Summary

We obtain the energy we need to live and grow from the food we eat. Some people need lots of energy whilst others need far less. Our age, sex and lifestyle determine how much food we ought to eat each day.

When an energy transfer takes place the total amount of energy before and after the change must be the same. This is known as the Law of Conservation of Energy.

Often during an energy transfer some of the energy will change into a form which is not wanted. If most of the energy is changed into the required form we may describe the transfer as efficient. If much of it is changed into an unwanted form the transfer is inefficient and energy is wasted. We can calculate the efficiency of a transfer using the equation:

efficiency = useful energy out / total energy in × 100%

Key words

efficiency A measure of how much of the energy input was converted into the form of energy which was required. An efficiency of 100% means that all the input energy was converted into the energy wanted.

energy transfer diagram A diagram which shows the forms of energy before and after the transfer.

joule (J) The unit we use to measure energy. A thousand joules is one kilojoule (kJ).

End of Chapter 2 Questions

1 The energy transfer diagrams below show the forms of energy before and after each transfer. Write the correct amount of energy produced in each of the empty spaces.

a)

b)

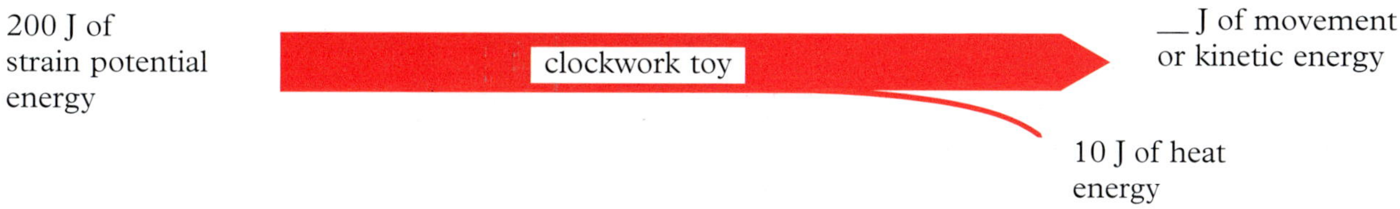

c)

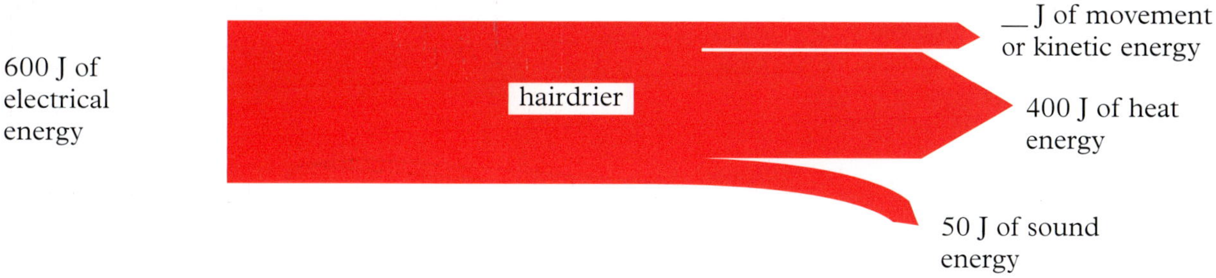

2 For every 200 J of electrical energy which enters this fire, 150 J of heat are produced. Calculate the efficiency of the fire.

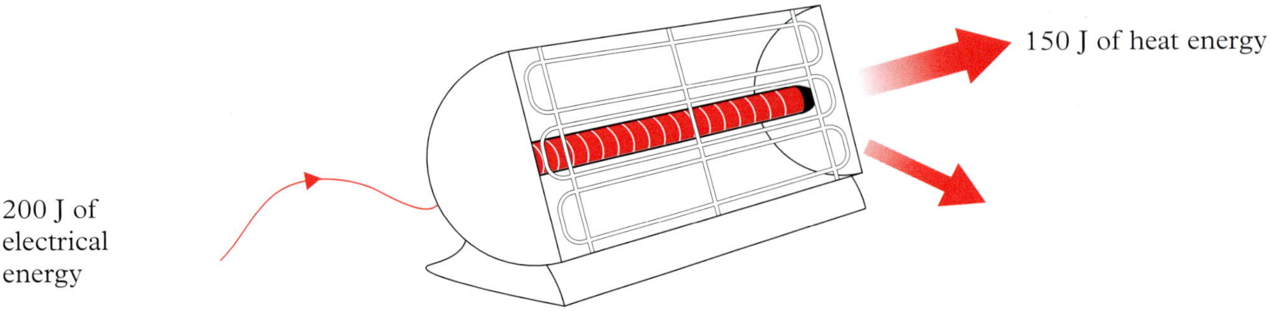

3 For every 500 J of fuel it burns this paraffin lamp produces 50 J of light energy. Calculate the efficiency of the lamp.

4 For every 100 J of electrical energy which enters this television set 10 J of useful light energy is produced and 5 J of useful sound energy.
a) Calculate the efficiency of the television set.
b) Suggest what form the remaining energy might have changed into.

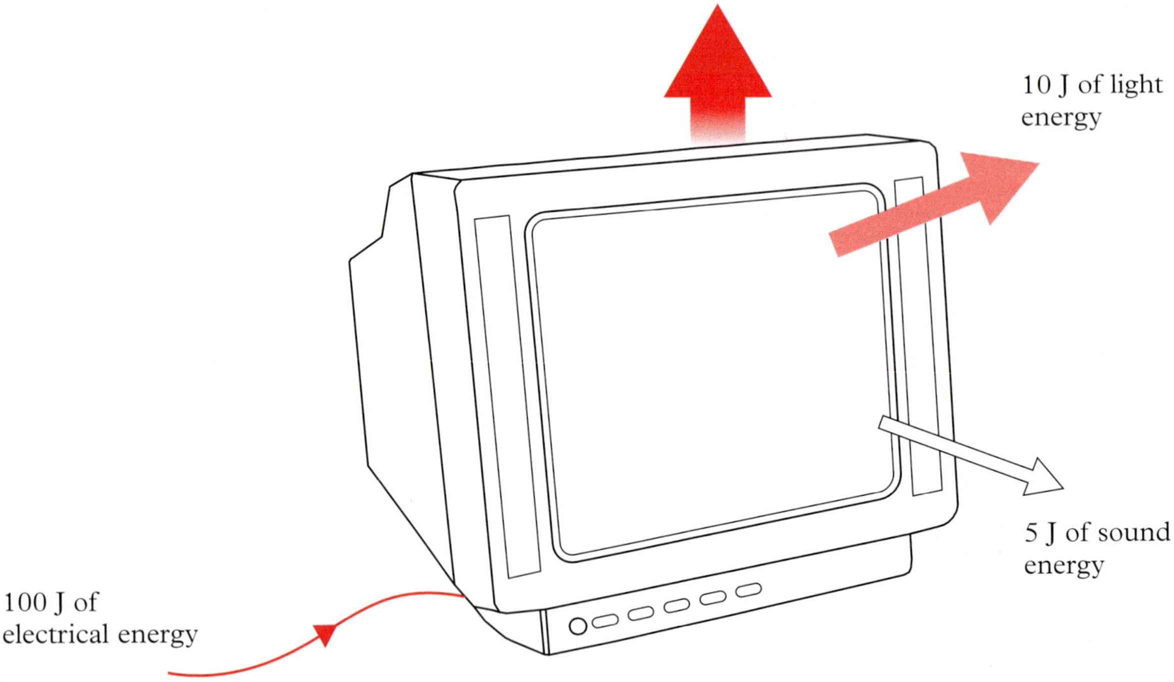

3 Energy Resources

As we have already seen, when energy is used it does not disappear. It simply changes into other forms of energy. Often however the change may produce **dilute** forms of energy which are not so easily reused.

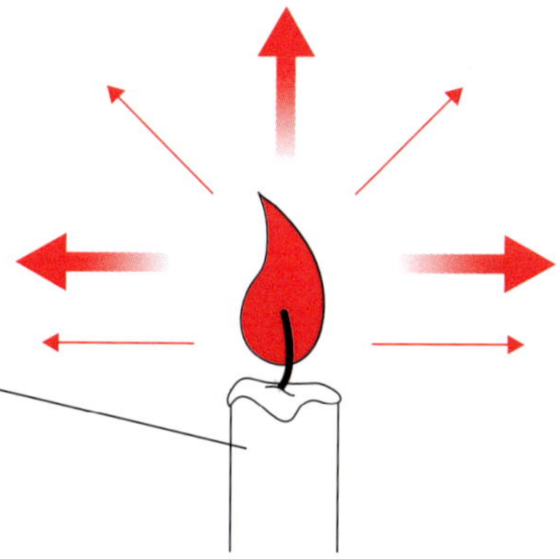
the heat and light energy the candle releases spreads out and becomes diluted
the wax this candle is burning is a concentrated form of chemical energy
energy that is dilute is difficult to reuse

It is important that we use sources of concentrated energy wisely. In the UK the fossil fuels, coal, oil and gas provide almost 90% of our energy needs. But if we continue to use them at this rate they will soon be gone. Coal, oil and gas are examples of **non-renewable** sources of energy. Once they have been used up they cannot be replaced.

the energy consumption in the UK can be shown in a pie chart

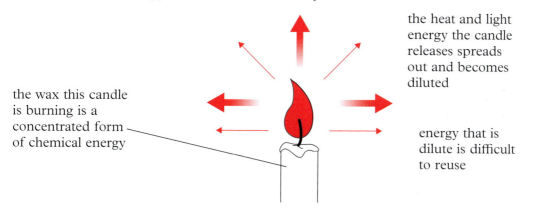

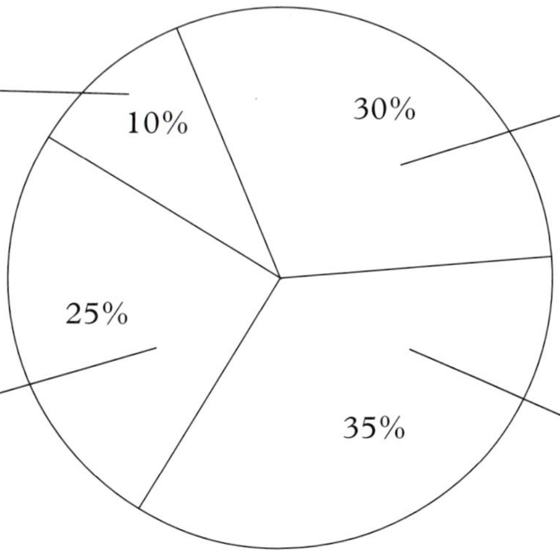
other energy sources
10%
30%
coal
25%
natural gas
35%
oil

How fossil fuels are formed

Fossil fuels are formed from dead plants and animals which lived on the Earth millions of years ago.

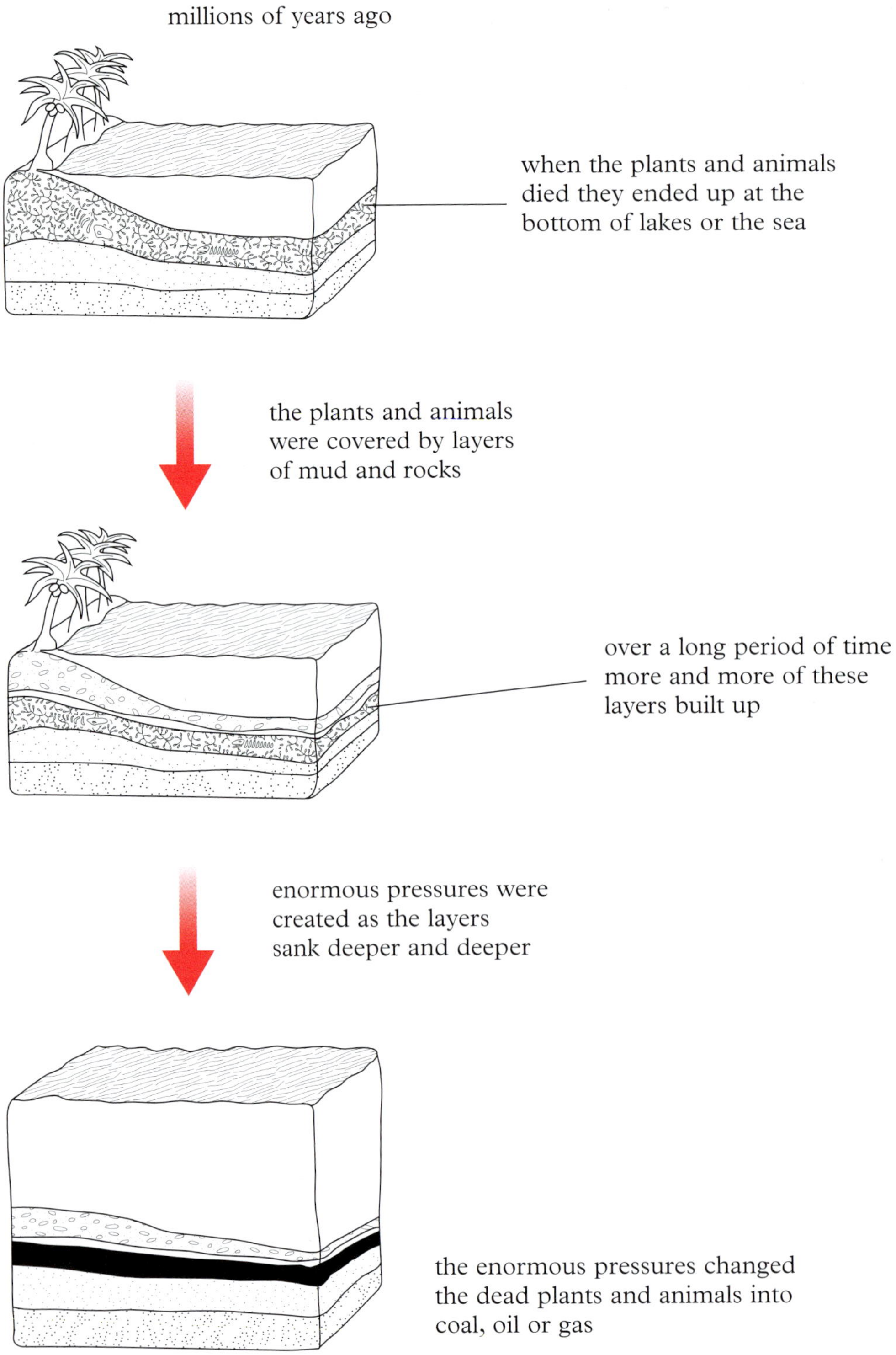
millions of years ago
when the plants and animals died they ended up at the bottom of lakes or the sea
the plants and animals were covered by layers of mud and rocks
over a long period of time more and more of these layers built up
enormous pressures were created as the layers sank deeper and deeper
the enormous pressures changed the dead plants and animals into coal, oil or gas

A fuel is a substance which releases energy when it is burned. We need fuel in order to generate electricity.

To avoid using up precious energy resources, like coal, oil and gas, we must look for alternative sources of energy which are **renewable**. Renewable energy resources can be continually replaced and won't ever run out.

Renewable sources of energy

There are many different possible sources of renewable energy.

Wind energy
This has been used for many hundreds of years to pump water or to turn millstones which ground wheat to produce flour. Nowadays we use wind turbines to produce electricity.

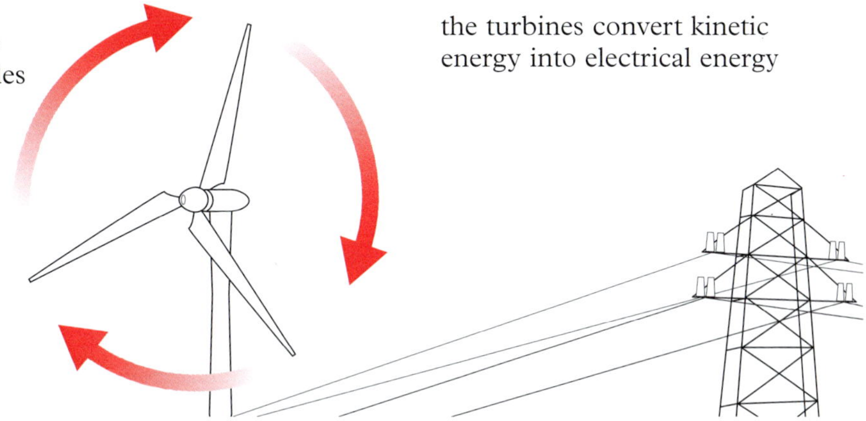
the wind turns round the blades of the wind turbines
the turbines convert kinetic energy into electrical energy
electrical energy is sent to the homes, factories, shops and so on

This is a very useful source of energy for isolated communities with no national electrical supply. But it is very dependant upon the weather. If there is no wind there is no electricity.

Hydroelectric energy
Rain water stored behind a tall dam has lots of potential energy.

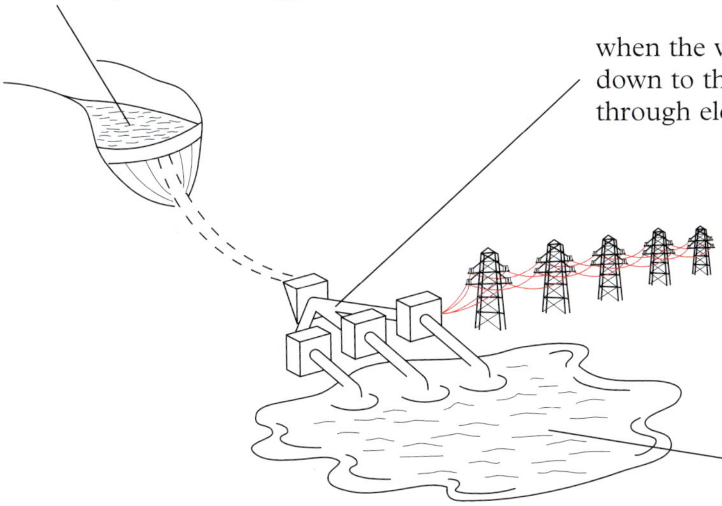
water in this upper lake has lots of potential energy
when the water is allowed to flow down to the lower lake it flows through electrical generators
the generators convert the kinetic energy of the moving water into electrical energy
the water collects in the lower lake at the bottom of the hill

This source of energy is very clean. It creates no pollution but the construction of the dam and its generators can be very expensive. Constructing the dam can have a high impact on the environment because it involves the flooding of valleys which destroys wildlife and their habitats.

Tidal energy

The Moon and the Sun cause the waters around the world to rise and fall. If the water in a river estuary is held behind a barrier when the tide is high and is released when the tide is low, its stored potential energy can be can be used to produce electricity.

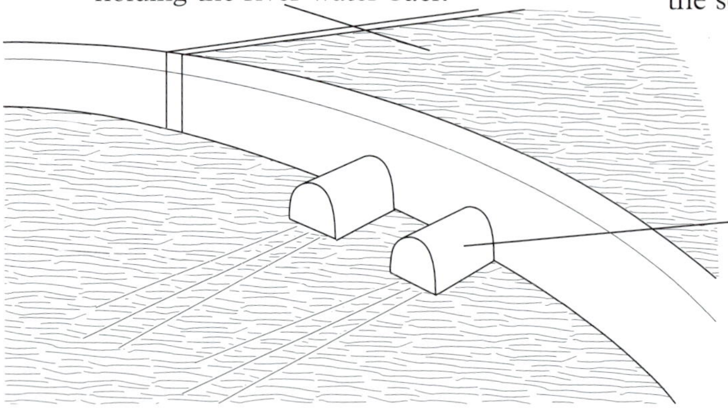
an artificial lake is created by holding the river water back
the water is released when the sea is at low tide
the water flows through electrical generators to create electricity

Although this energy source is constantly renewable the initial cost of a suitable installation is very high.

Solar energy

Most of the Earth's energy comes from the Sun. This solar energy can be converted directly into electricity using devices called solar cells.

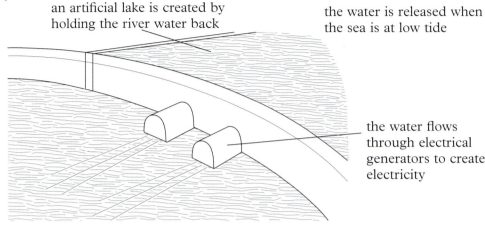

0.
MC
MR
M-
M+
C
7
8
9
÷
CE
4
5
6
X
√
1
2
3
—
+/–
0
·
%
+
=
solar cells like these convert solar energy into electrical energy

Solar energy can also be collected by solar panels and perhaps be used to heat water in the home. This energy source is renewable but if there is no sunshine there is no energy.

Biomass

Biomass is used to refer to any material that is or has been alive. Plants and trees are biomass. The Sun's energy may be captured by plants and trees.

the sun shines on the leaves of plants and trees
plants and trees can change this solar energy into concentrated chemical energy
they do this using a process called **photosynthesis**
if plants and trees are cut down and burnt, they release this energy
plants and trees can store this chemical energy

New plants and trees can be planted to replace those that have been cut down. This is a cheap source of energy particularly for Third World countries. However, the burning of the fuel may cause pollution and over a long period of time it may damage the Earth's atmosphere and cause changes in the Earth's weather.

Wave energy

The constant up and down movement of the surface of the sea can be used to generate electricity.

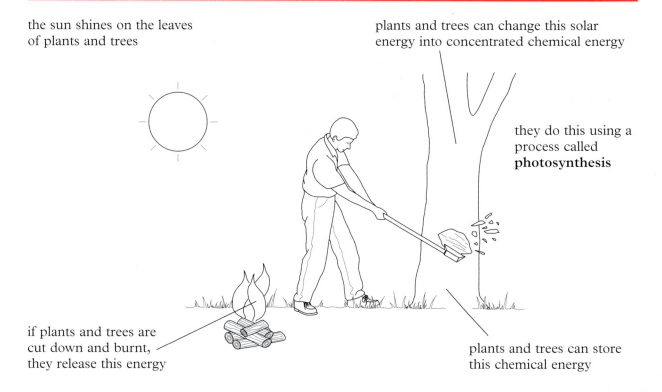

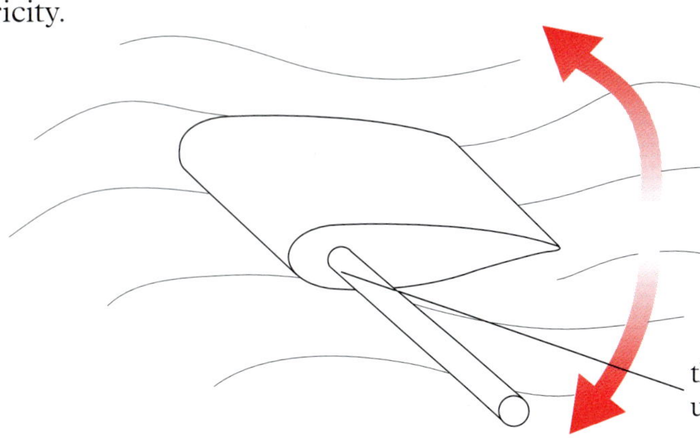
up and down motion is created by the waves
the movement of this axle is used to generate electricity

This source of energy produces no pollution and is renewable but very large areas of water are needed to collect enough energy to justify the cost of construction.

Geothermal energy

Deep inside the Earth nuclear reactions are taking place which release large quantities of energy. This energy increases the temperature of the surrounding rocks.

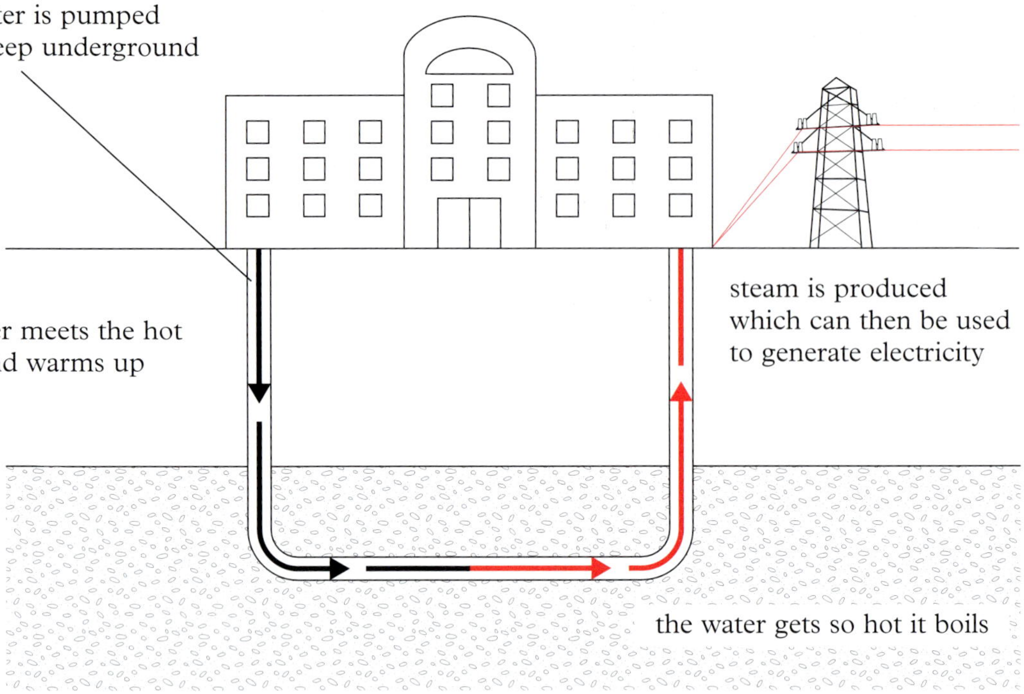
cold water is pumped down deep underground
the water meets the hot rocks and warms up
steam is produced which can then be used to generate electricity
the water gets so hot it boils

It is possible to obtain large amounts of energy from geothermal stations but deep drilling is expensive and there are not many sites where the hot rocks are not too deep.

Questions

1 a) Explain the difference between a renewable source of energy and a non-renewable source of energy.
b) Give two examples of each.

2 Name two renewable sources of energy which could be used by people who:
a) live in mountainous regions
b) live in the desert
c) live by the coast.

3 How do we obtain the energy which is stored in a fuel?

4 What is a fossil fuel and how is it formed?

Energy from the Sun

The Sun is our main source of energy.

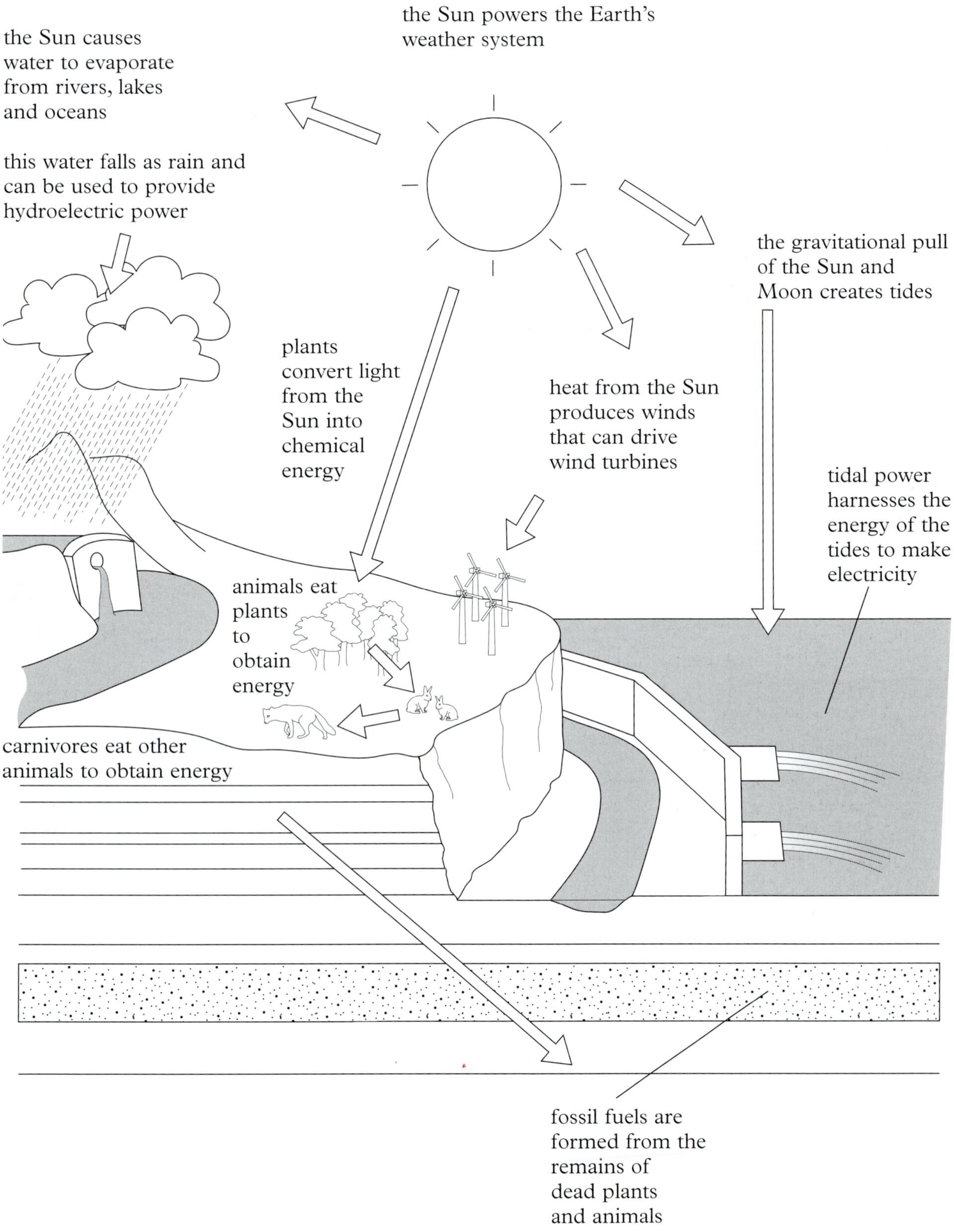
the Sun powers the Earth's weather system
the Sun causes water to evaporate from rivers, lakes and oceans
this water falls as rain and can be used to provide hydroelectric power
the gravitational pull of the Sun and Moon creates tides
plants convert light from the Sun into chemical energy
heat from the Sun produces winds that can drive wind turbines
tidal power harnesses the energy of the tides to make electricity
animals eat plants to obtain energy
carnivores eat other animals to obtain energy
fossil fuels are formed from the remains of dead plants and animals

Nuclear energy

Some substances such as uranium are radioactive. When a reaction takes place in the nucleus of a uranium atom, some radiation is released (see Chapter 21) together with some heat. In a nuclear power station this energy is used to heat water and produce steam. The steam is then used to drive turbines and generate electricity.

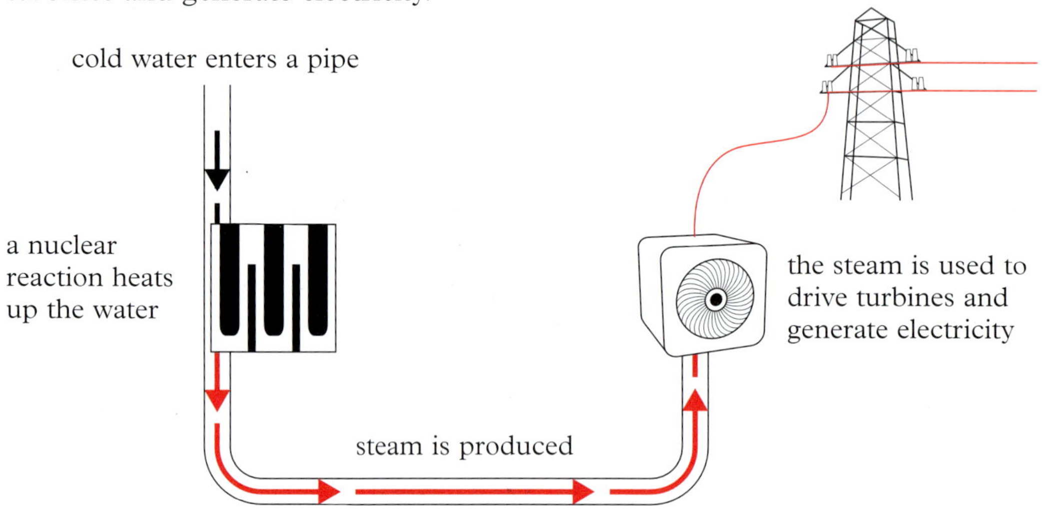
cold water enters a pipe
a nuclear reaction heats up the water
steam is produced
the steam is used to drive turbines and generate electricity

One advantage of nuclear energy is that only a small amount of uranium is needed to produce large amounts of energy. But the radiation emitted by radioactive materials is extremely dangerous to all living things and the waste products from a nuclear power station will remain dangerously radioactive for thousands of years.

Summary

Most of the energy needs of the industrialised world are met by burning fossil fuels such as coal, oil and gas. It is not possible to renew these fuels as they take millions of years to form. If we continue to use these fuels at the present rate there will soon be none left. The burning of these fuels in such large amounts is also causing high levels of pollution which is likely to result in long lasting damage to our atmosphere.

To avoid these problems we need to look for alternate sources of energy which are renewable. These include solar energy, wind energy, hydroelectric energy, tidal energy, wave energy, geothermal energy and biomass such as wood.

The Sun is our ultimate source of energy. This energy may be captured by plants and trees. Some animals eat plants and trees to obtain energy. Plants, trees and animals may die and be turned into fossil fuels. Some of the energy we receive from the Sun drives the Earth's weather, causing winds and waves.

Nuclear energy is a new source of energy which has many advantages but also has some very serious disadvantages. It may in the future become our main source of energy or it may be decided that it is too dangerous for us and our environment.

Key words

biomass Material which has grown, like wood and plants.

concentrated energy When there is a lot of energy in a small volume like in a piece of coal.

dilute energy Energy which is spread out thinly and is of little use, like the heat and light produced by burning a piece of coal yesterday.

fossil fuel A fuel which is the fossilised remains of living things, like coal, oil and gas.

fuel A substance which will release energy when it is burned.

non-renewable energy Energy which cannot be replenished as quickly as it is being used and will eventually run out.

photosynthesis A chemical reaction which captures solar energy and stores it as chemical energy. This reaction takes place in all plants and trees.

renewable energy Energy which can be replenished as quickly as it is being used.

End of Chapter 3 Questions

1 a) What is a fuel?
b) Name two fossil fuels.
c) Name one fuel which is renewable.
d) Name one renewable source of energy which can be used in any weather.

2 The diagram below shows a pumped storage power station. During the day time water is released from the upper lake and is used to drive the turbines to generate electricity. During the night electrical energy from another power station is used to pump some of the water back up to the upper lake.
a) What type of energy does the water in the upper lake have before it is released?
b) What type of energy does the water have as it enters the turbine?
c) Why is it important that the pumped storage power station generates electricity during the day?
d) Give one advantage and one disadvantage of this energy source.

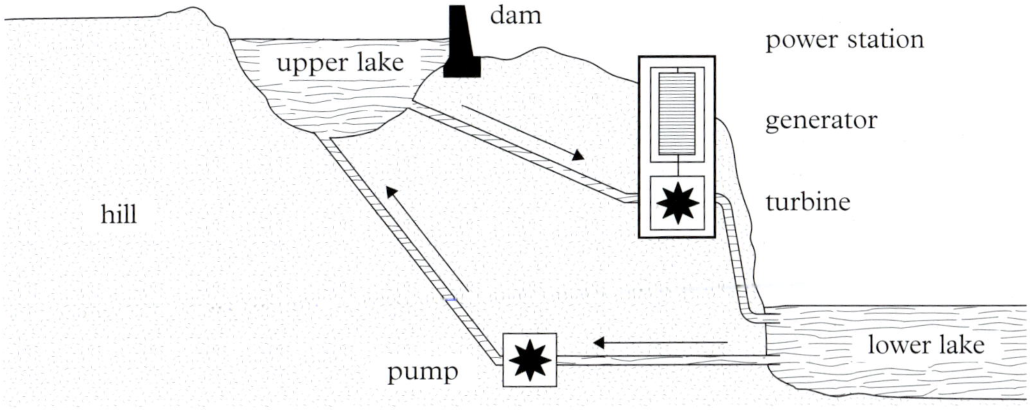

3 a) Describe how the energy which a lion receives from eating a deer originally came from the Sun.
b) The diagram below shows a machine which is used to generate electrical energy using tidal energy. Explain how this machine works.
c) Give one advantage and one disadvantage of this energy source.

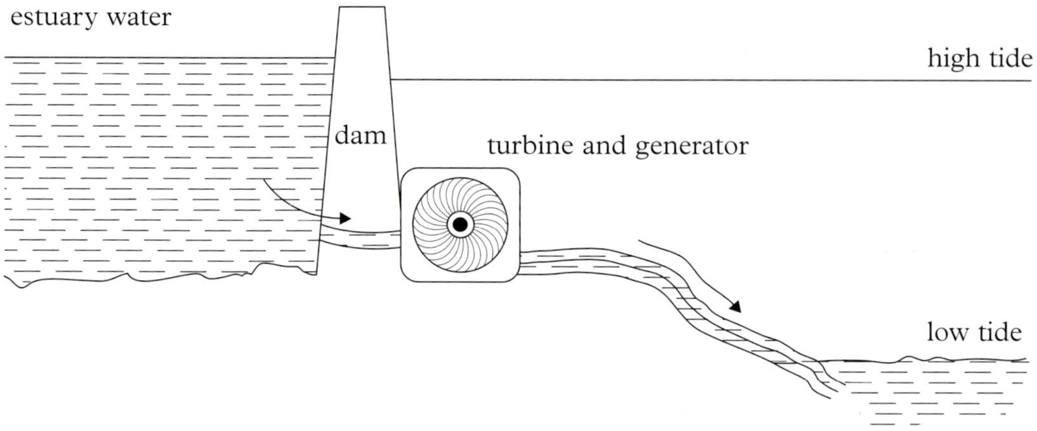

4 The diagram below shows an island community which has to make use of its own energy resources.
a) Name three possible sources of energy the islanders could use.
b) Write down one advantage and one disadvantage for each of the energy sources you have chosen.

5 a) Name three fossil fuels.
b) Where did the energy stored in these fuels originally come from?
c) Describe how a fossil fuel is formed.
d) Explain why fossil fuels are called 'non-renewable' sources of energy.
e) How is the energy released from a fossil fuel?
f) What effect might the use of fossil fuels have on the environment?

4 Thermal Energy on the Move

Conduction

If one end of a metal rod is heated its temperature will increase. After several minutes parts of the rod which are not being heated directly by the flame will also become noticeably warmer. This change in temperature is due to thermal energy travelling along the rod. We say that heat is being **conducted** along the rod.

How things conduct

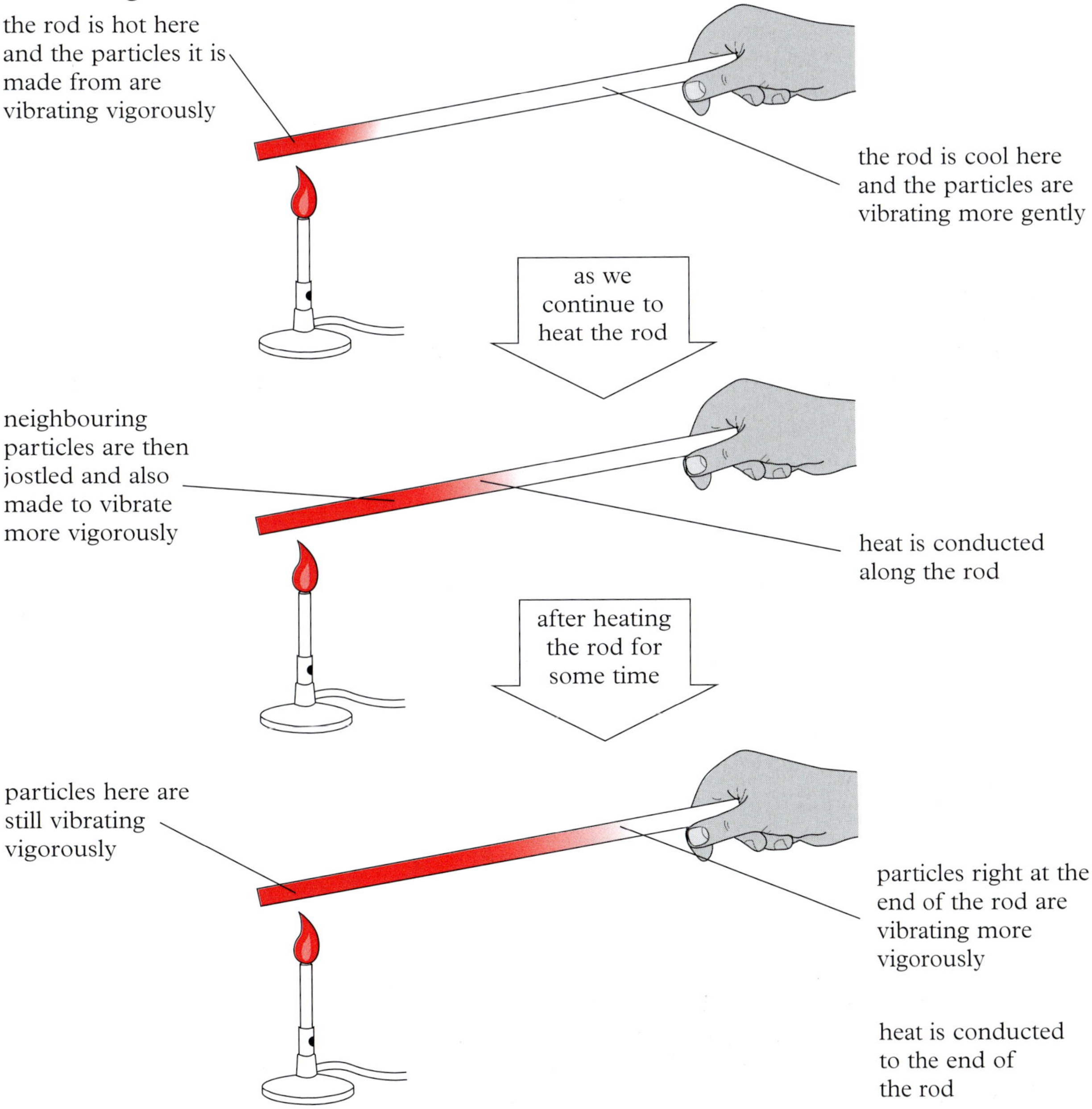

the rod is hot here and the particles it is made from are vibrating vigorously
the rod is cool here and the particles are vibrating more gently
as we continue to heat the rod
neighbouring particles are then jostled and also made to vibrate more vigorously
heat is conducted along the rod
after heating the rod for some time
particles here are still vibrating vigorously
particles right at the end of the rod are vibrating more vigorously
heat is conducted to the end of the rod

Eventually even the particles at the end of the rod furthest from the flame will be vibrating more vigorously. Heat has travelled the full length of the rod.

This movement of heat through vibrations is called **conduction**.

All metals allow heat to travel easily through them in this way. They are **good conductors** of heat. We often use metals in situations where we want good transfer of heat. We make saucepans from metals such as copper and steel because they conduct the heat from the flame to the food quickly.

Insulators

In situations where we wish to prevent the movement of heat, like for example, through the handle of a saucepan, we use materials such as plastic or wood. Materials like these which do not allow heat to travel through them easily are called **insulators**.

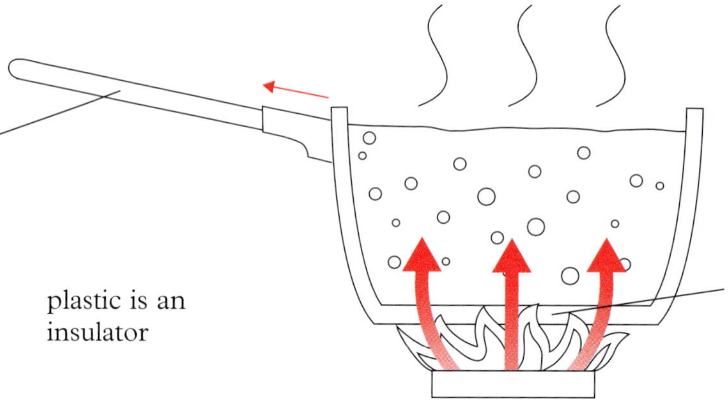
the plastic handle remains cool because it is a poor conductor
plastic is an insulator
the metal saucepan conducts heat quickly from the flame to the food

Radiators
A radiator has hot water flowing through its metal pipes.

In order to warm a room, heat from the hot water must travel through the walls of the radiator. To ensure that this transfer can take place easily and quickly the radiator is made of metal.

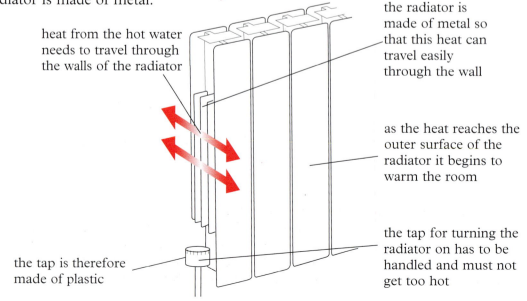

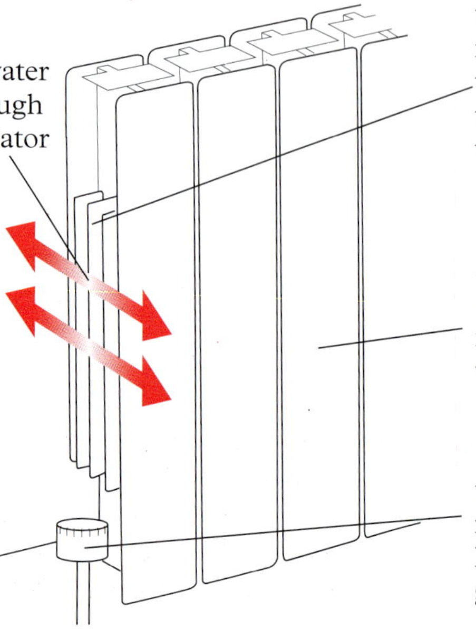
heat from the hot water needs to travel through the walls of the radiator
the radiator is made of metal so that this heat can travel easily through the wall
as the heat reaches the outer surface of the radiator it begins to warm the room
the tap for turning the radiator on has to be handled and must not get too hot
the tap is therefore made of plastic

Heat moves from places where the temperature is higher to places where the temperature is lower. Your body temperature is normally higher than the temperature of your surroundings. Heat is therefore continually being lost from your body. Metals are cold to touch because they conduct heat away from your body quickly. Objects which are made from insulators do not allow heat to escape from your body quickly and so do not feel cold when we touch them.

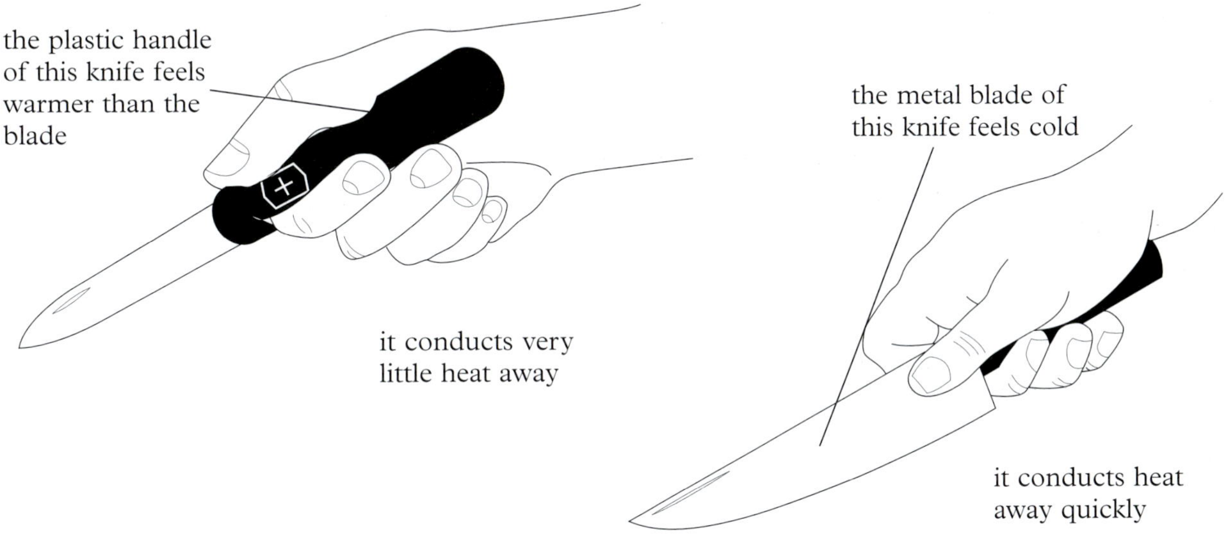
the plastic handle of this knife feels warmer than the blade
it conducts very little heat away
the metal blade of this knife feels cold
it conducts heat away quickly

Conduction of heat through different metals

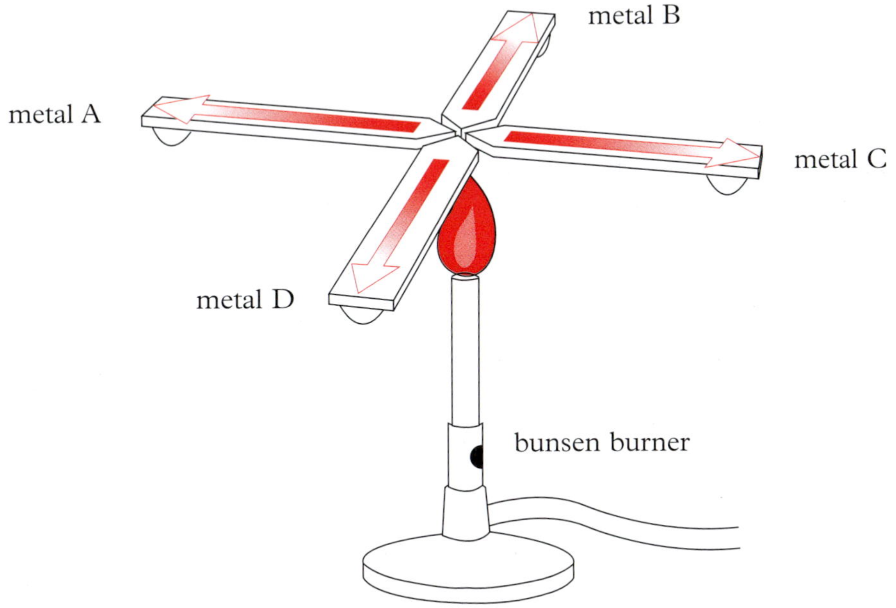
metal B
metal A
metal C
metal D
bunsen burner

The outer ends of each metal bar have a marble attached to them using a piece of candle wax. The inner ends of each of the four metal bars are heated equally. Heat is conducted along the metal bars. Eventually the wax becomes so warm it melts and the marbles fall. The best conductor will transfer heat most quickly and is identified as the bar whose marble falls first.

Questions

1 Explain what happens if one end of a metal rod is placed in a fire and left there for 30 minutes.

2 a) What is a good conductor?
b) Name three materials that are good conductors.
c) Describe one practical use of a good conductor.

3 a) What is an insulator?
b) Name three materials that are good insulators.
c) Describe one practical use of a good insulator.

4 Explain why a metal window frame feels cold when you touch it but a plastic window frame does not feel so cold.

Conduction in liquids

Most liquids are poor conductors of heat. The experiment below shows that water is a very poor conductor. The ice at the bottom of the boiling tube melts very very slowly even though there is boiling water just a few centimetres above it. The water between the top and bottom is a very poor conductor and very little heat is able to move between the two.

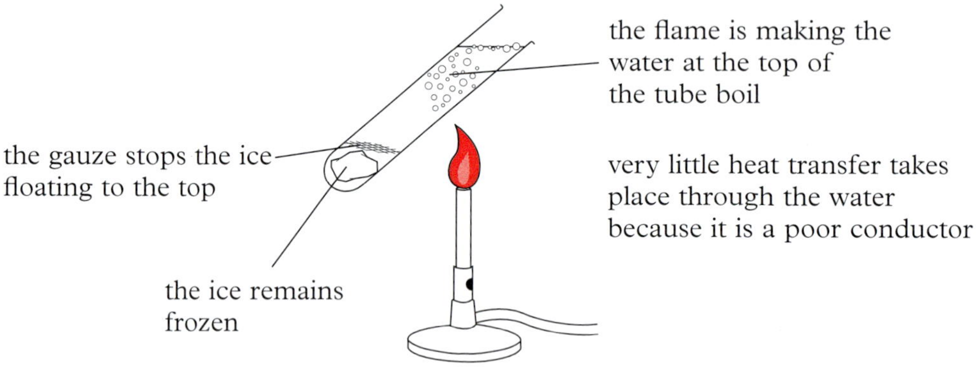
the flame is making the water at the top of the tube boil
the gauze stops the ice floating to the top
very little heat transfer takes place through the water because it is a poor conductor
the ice remains frozen

Conduction in gases

Gases are very poor conductors of heat. In fact they are excellent insulators and are often used to try to prevent heat transfer.

A good example of this is 'double glazing'.

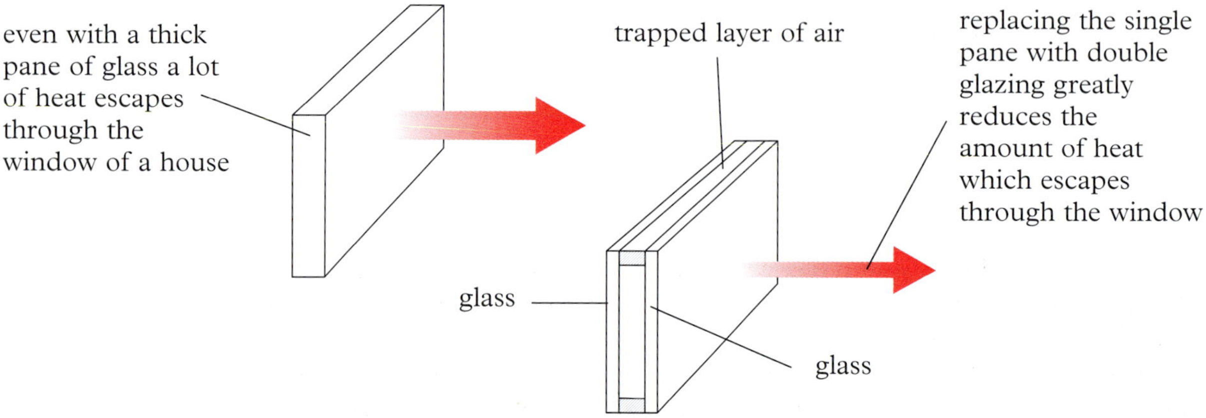
even with a thick pane of glass a lot of heat escapes through the window of a house
trapped layer of air
replacing the single pane with double glazing greatly reduces the amount of heat which escapes through the window
glass
glass

It is the insulating properties of the layer of trapped air between the two panes of glass which reduces heat loss through the window.

Heat loss from a house through the roof can be greatly reduced using fibreglass insulation. The fibreglass has large amounts of air trapped inside it. It is the insulating properties of the air which prevents heat loss by conduction.

Questions

1 Explain why two sheets of glass separated by a thin layer of air (double glazing) will provide better insulation for your house than a single piece of glass which has twice the thickness.

2 Why in the winter do birds 'fluff up' their feathers to help them keep warm?

3 Most of the clothes you wear are made from fibres such as wool, cotton or manmade fibres like nylon. Explain why these clothes help to keep you warm.

Convection

The movement of heat through liquids and gases by conduction is very difficult. There is however another way in which liquids and gases can transfer energy. The heater in the diagram below contains no fan but is still able to warm the whole of the room. Energy transfer takes place by **convection**.

Convection in gases

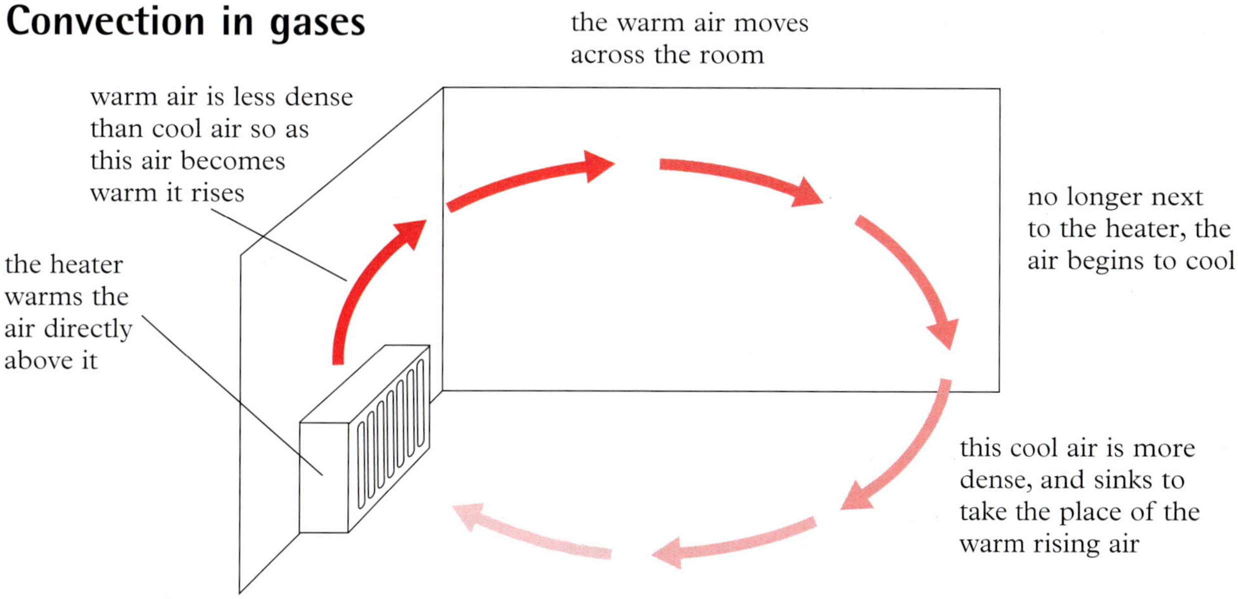
the warm air moves across the room
warm air is less dense than cool air so as this air becomes warm it rises
the heater warms the air directly above it
no longer next to the heater, the air begins to cool
this cool air is more dense, and sinks to take the place of the warm rising air

The air immediately above the heater becomes warm and expands. It is now a little less dense than the air around it so it rises and moves away from the heater. Cooler, more dense air then moves in to take the place of the rising air. It too becomes warm and the process continues carrying heat to all parts of the room. This movement of air is called a **convection current**.

Sea breezes

Energy from the Sun can cause convection currents. These are particularly noticeable at the seaside.

Daytime

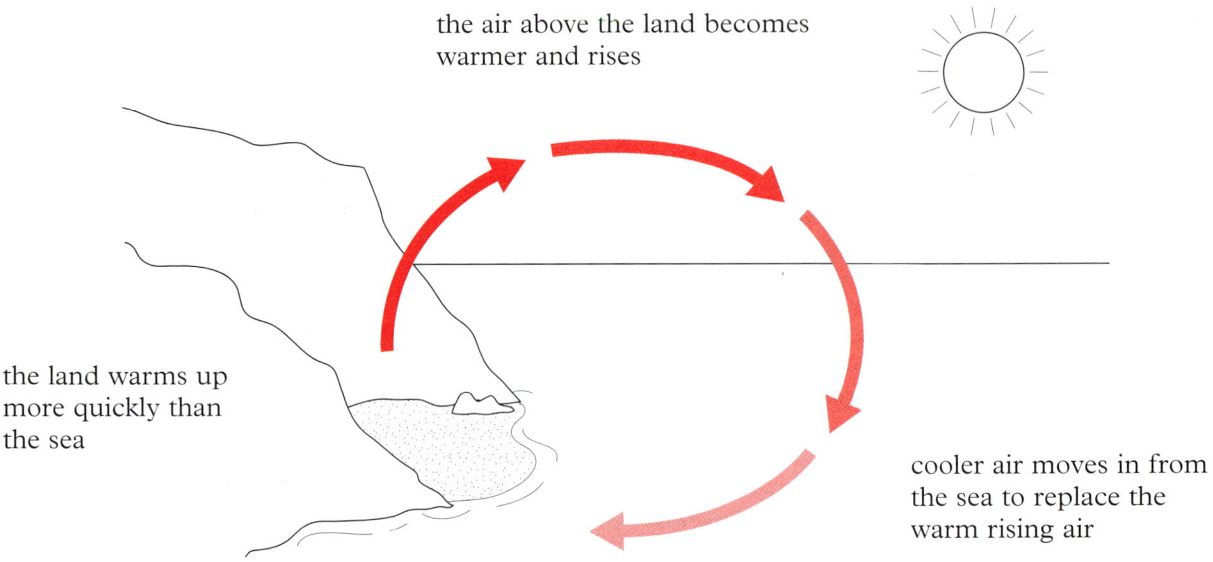
the air above the land becomes warmer and rises
the land warms up more quickly than the sea
cooler air moves in from the sea to replace the warm rising air

Bathers on the beach will feel this convection current as an onshore breeze.

Nighttime

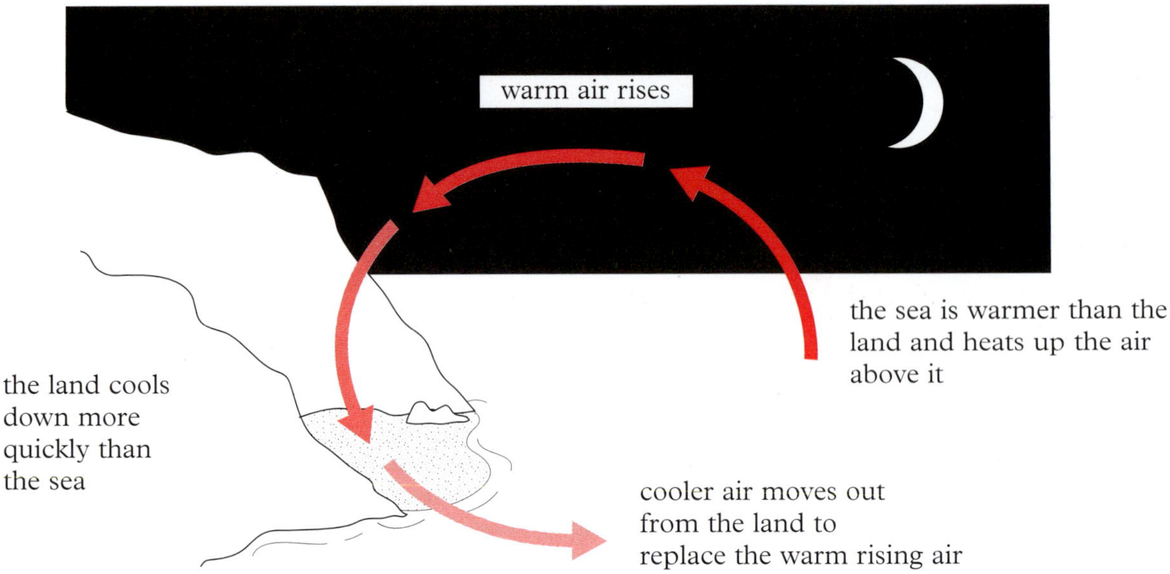
warm air rises
the sea is warmer than the land and heats up the air above it
the land cools down more quickly than the sea
cooler air moves out from the land to replace the warm rising air

This convection current is now felt as an offshore breeze.

Ovens and fridges

The heater in an oven is placed at the bottom. The convection current it then creates transfers heat to all parts.

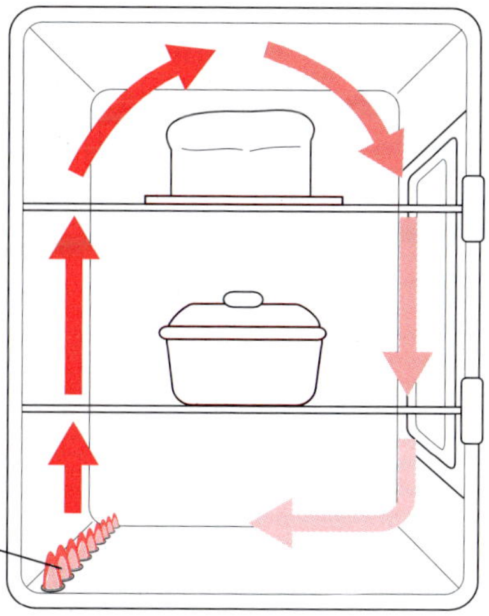
warm air rises
the cooler air falls
the oven is heated in one corner
the cooler air becomes reheated

The freezer compartment in a fridge is placed at the top. The convection current it creates cools all parts of the fridge.

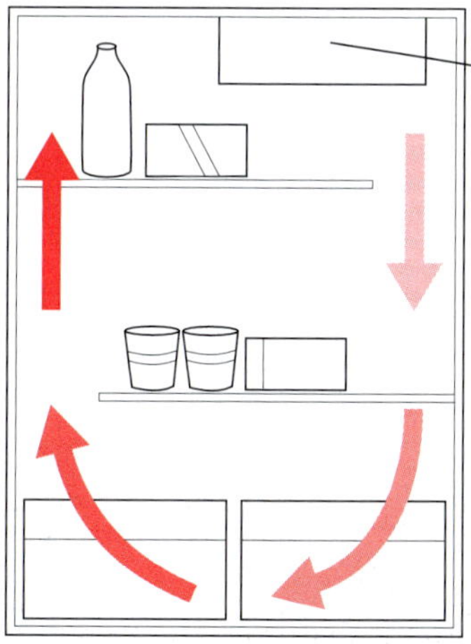
freezer compartment cools the air
the warmer air rises, and is then cooled by the freezer again
cool air sinks
further away from the freezer, the air begins to warm slightly

Convection in liquids

The experiment below shows how heat can also be transferred through a liquid by convection.

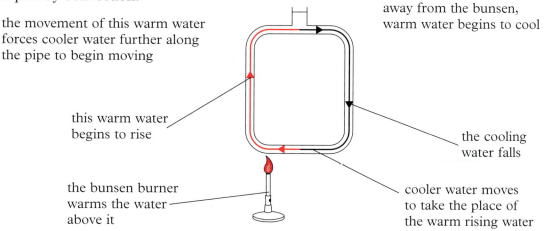

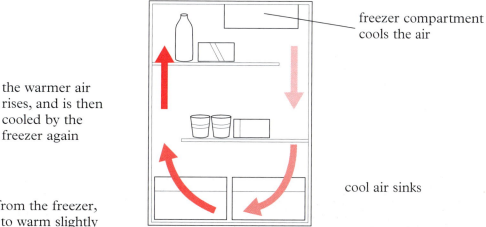

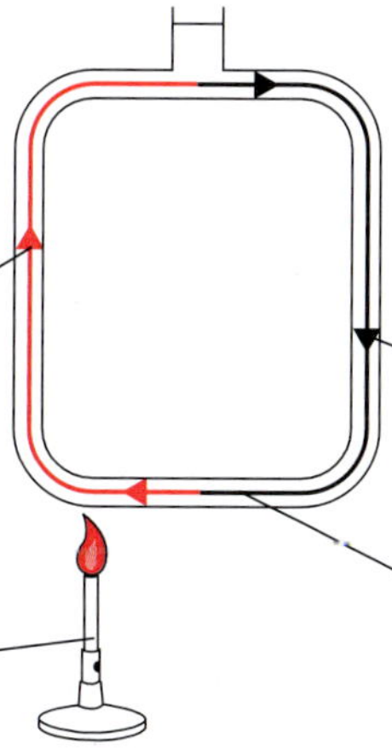
the movement of this warm water forces cooler water further along the pipe to begin moving
away from the bunsen, warm water begins to cool
this warm water begins to rise
the cooling water falls
the bunsen burner warms the water above it
cooler water moves to take the place of the warm rising water

Domestic hot water system

Although most houses now have hot water systems which are driven by pumps, some still rely on natural convection currents to move the heat around. Whichever system is used it is important that the pipes are connected correctly.

Using convection currents

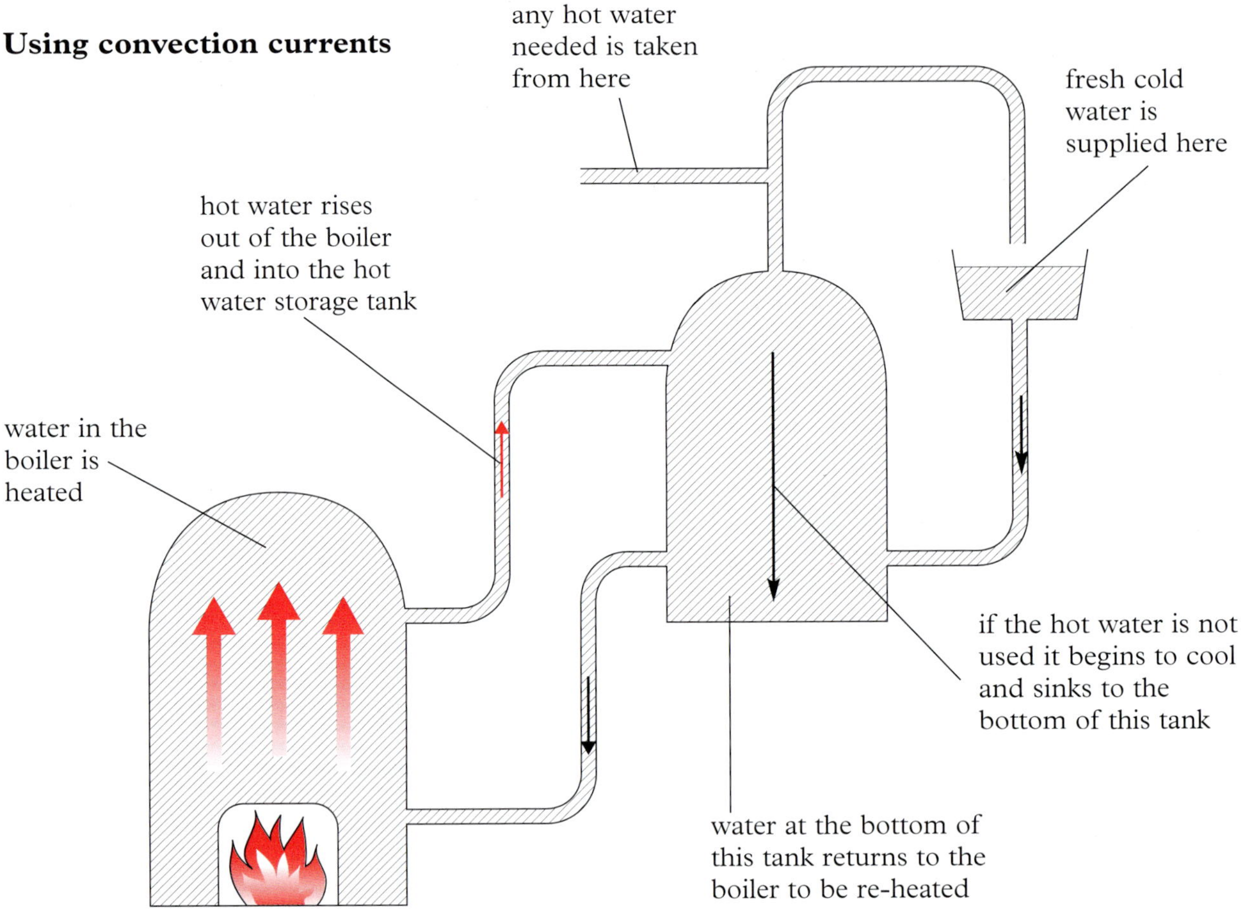
any hot water
needed is taken
from here
fresh cold
water is
supplied here
hot water rises
out of the boiler
and into the hot
water storage tank
water in the
boiler is
heated
if the hot water is not
used it begins to cool
and sinks to the
bottom of this tank
water at the bottom of
this tank returns to the
boiler to be re-heated

Convection in solids

The particles of a solid are unable to move around like those of a liquid or a gas. This means it is impossible for a solid to transfer heat by convection.

Questions

1 Hot water is poured from a kettle into a tall beaker. After five minutes the temperature of the water at the top, in the middle and at the bottom of the beaker is measured. Would you expect all these readings to be the same? Explain your answer.

2 Explain why this spiral rotates when the candle is lit.

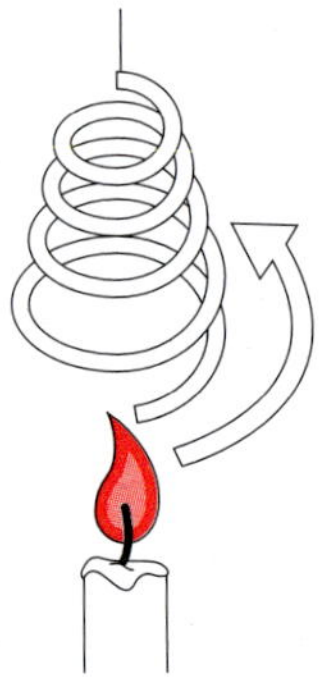

Questions (continued)

3 What would happen if the hot water from the boiler in a domestic hot water system was fed into the bottom of the hot water storage tank?

4 Why is the cold water fed through the bottom and not the top of the storage tank?

5 What is the main advantage of pumping the water between the boiler and the storage tank rather than relying on convection currents?

6 Explain why during the daytime there is often an onshore breeze at the coast and yet during the night there is often an offshore breeze.

7 a) Explain what would happen if the heater was placed at the top of an oven.
b) Explain what would happen if the freezing compartment was placed at the bottom of a fridge.

Radiation

Most of the thermal energy we receive here on the Earth comes from the Sun. It has travelled through 150 million **kilometres** of empty space. The energy travels through space as **infra red radiation**. Radiation takes the form of rays or waves. It is the only way in which heat can travel through a vacuum.

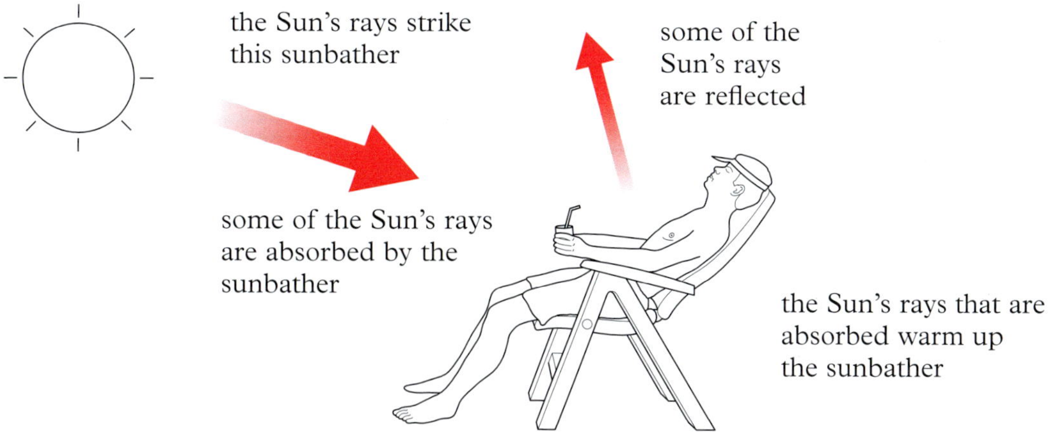

the Sun's rays strike this sunbather
some of the Sun's rays are reflected
some of the Sun's rays are absorbed by the sunbather
the Sun's rays that are absorbed warm up the sunbather

How much energy an object absorbs depends up on the nature of its surface.

Dark rough surfaces **absorb most of the radiation** and **reflect very little.**

Light shiny smooth surfaces **absorb only a small amount of radiation** that hits it**. They reflect most of it.**

The experiment on the next page demonstrates this. A heater is placed directly between two metal plates. Each plate receives the same amount of energy from the heater. Each plate has a marble stuck to it with a piece of wax.

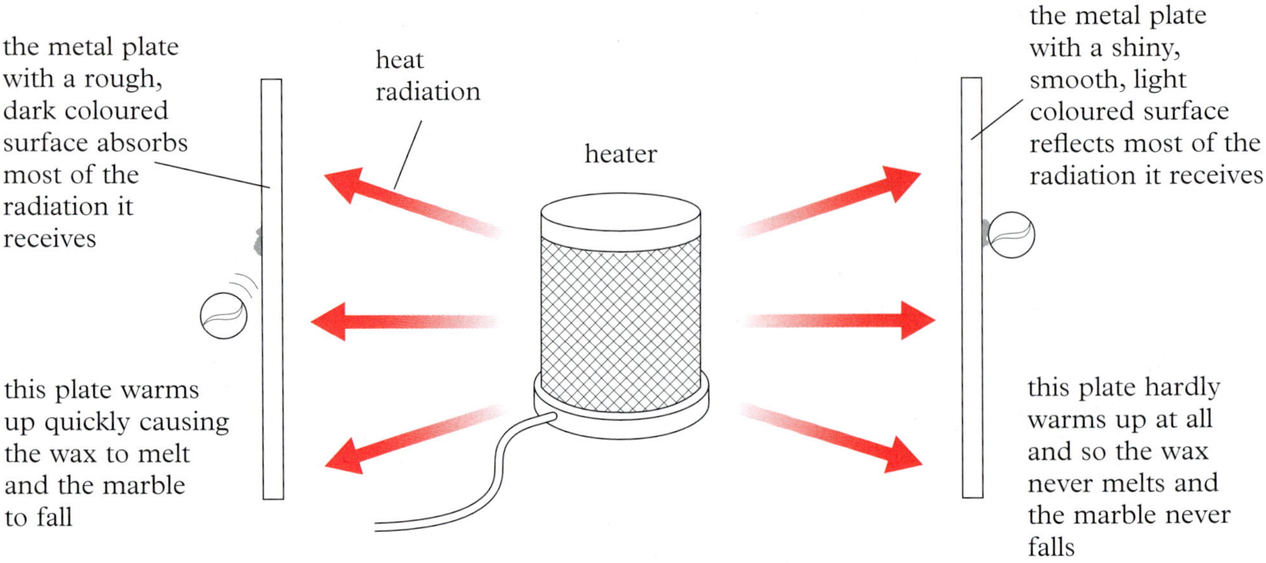
the metal plate with a rough, dark coloured surface absorbs most of the radiation it receives
heat radiation
heater
the metal plate with a shiny, smooth, light coloured surface reflects most of the radiation it receives
this plate warms up quickly causing the wax to melt and the marble to fall
this plate hardly warms up at all and so the wax never melts and the marble never falls

Shiny, light surfaces are used with electric fires and radiators to stop heat energy being wasted.

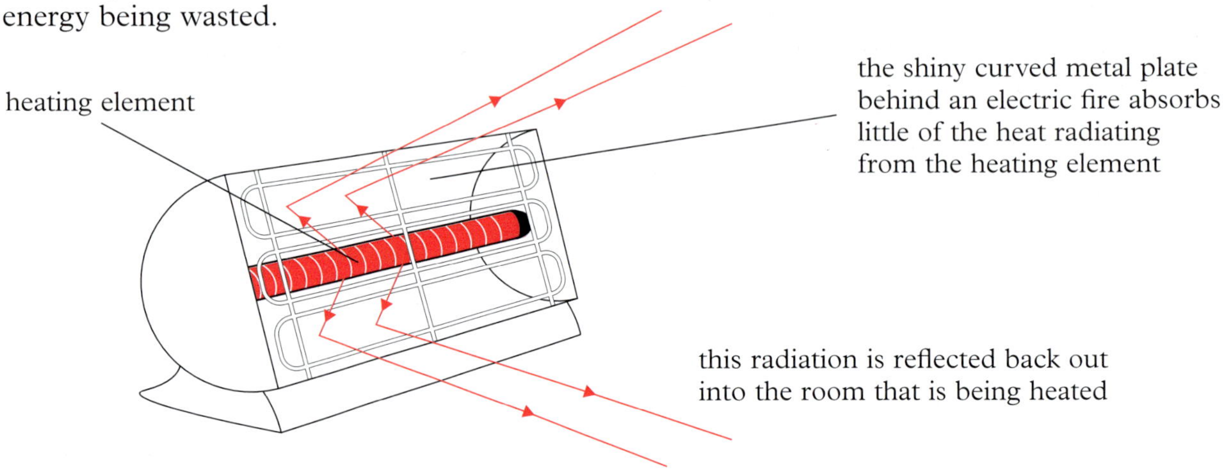
heating element
the shiny curved metal plate behind an electric fire absorbs little of the heat radiating from the heating element
this radiation is reflected back out into the room that is being heated

Placing a shiny piece of foil behind a radiator prevents heat being absorbed by the wall. Instead the heat is reflected back into the room.

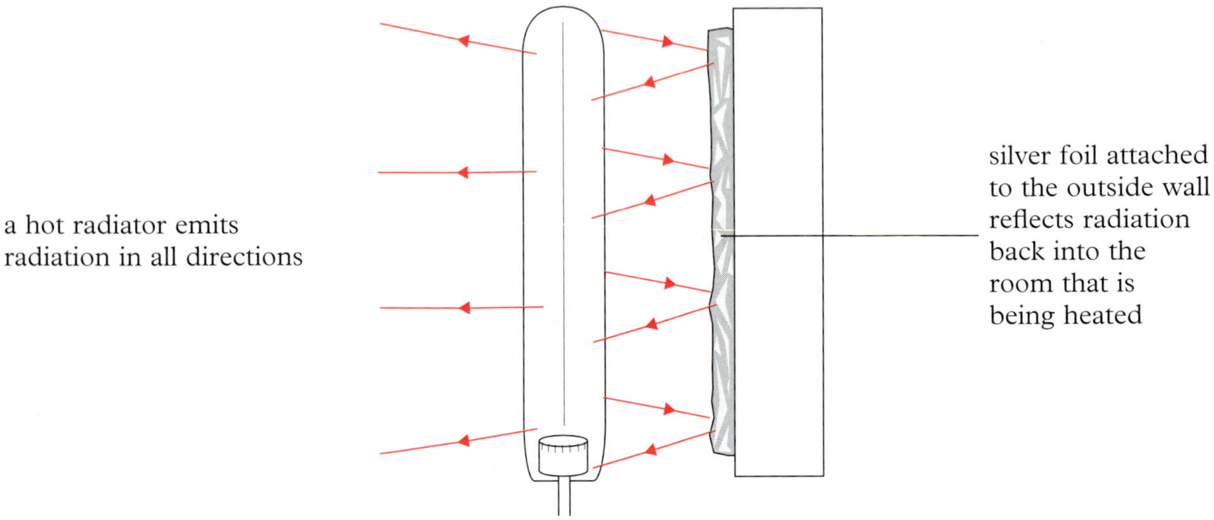
a hot radiator emits radiation in all directions
silver foil attached to the outside wall reflects radiation back into the room that is being heated

In very hot countries houses are often painted white. This light colour helps to reflect radiant heat from the Sun and in this way keep the houses cool.

As well as absorbing radiation most objects emit radiation. At this very moment you are emitting heat radiation into your surroundings. In general the hotter an object the greater the amount of radiation it emits. But the surface of an object also affects the amount of radiation an object emits.

Objects with dark rough surfaces emit lots of radiation.

Objects with light, shiny smooth surfaces emit much less radiation.

The experiment below shows this to be true. A cube is filled with very hot water. The cube has a dark rough side and a light smooth side.

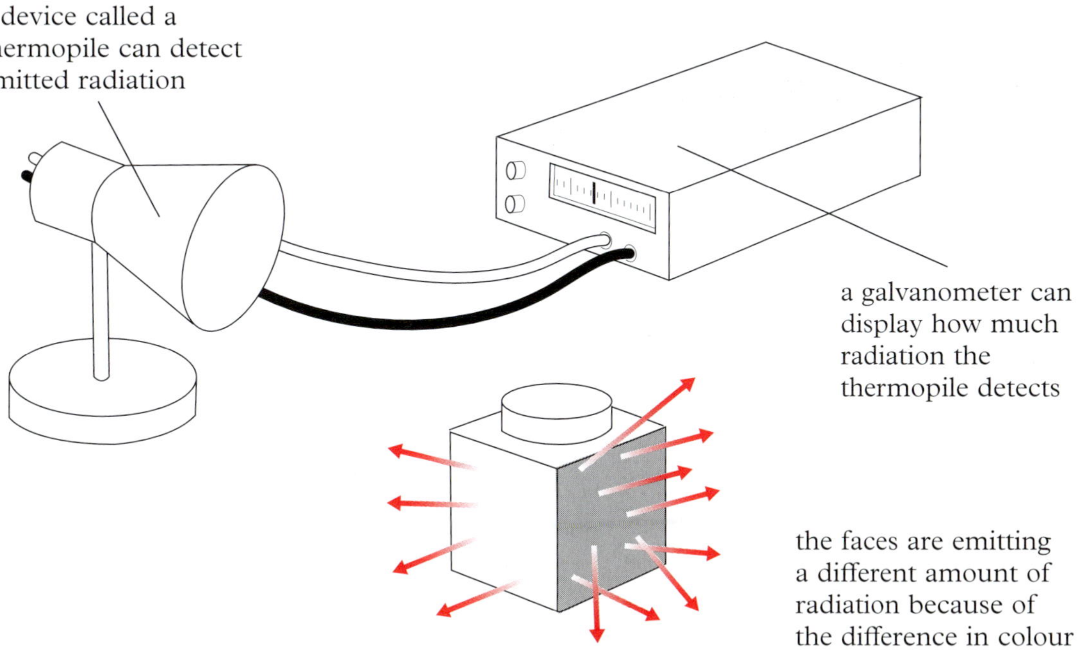
a device called a thermopile can detect emitted radiation
a galvanometer can display how much radiation the thermopile detects
the faces are emitting a different amount of radiation because of the difference in colour

When the dark rough surface is facing the thermopile a high reading is seen on the galvanometer indicating that the surface is emitting a lot of radiation. When the cube is turned so that the light smooth surface is facing the thermopile the galvanometer reading falls. This means that the light smooth surface is emitting less radiation.

This explains why the tea in a silver teapot will stay warmer longer than tea in a dark teapot.

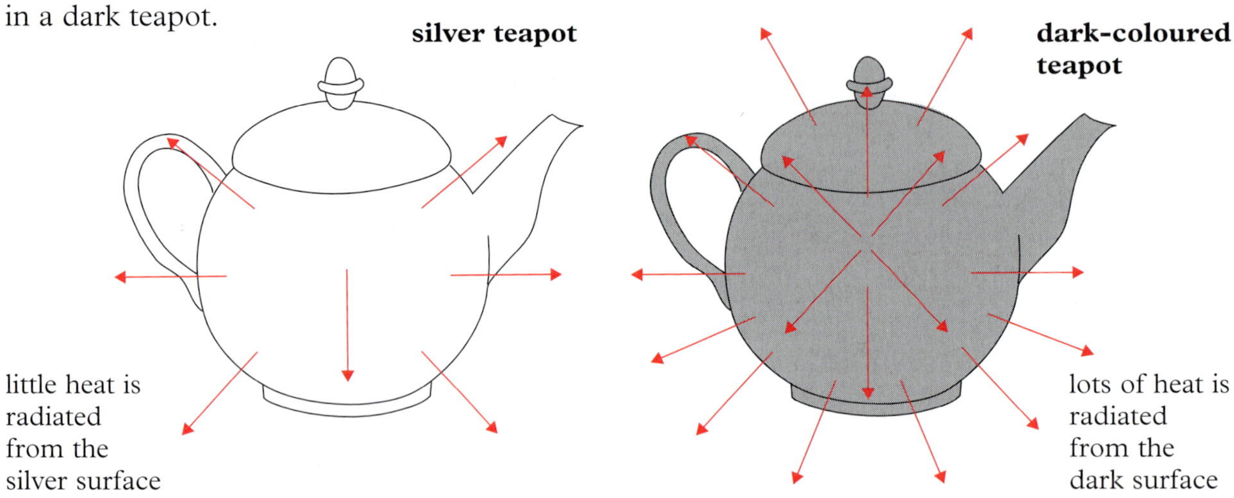

Questions

1 a) Name three methods by which thermal energy can move.
b) Which of these three methods is the only one which will carry thermal energy through a vacuum?

2 Look at the diagram below. Explain what will happen to the thermometers T_1 and T_2 when the heater is turned on.

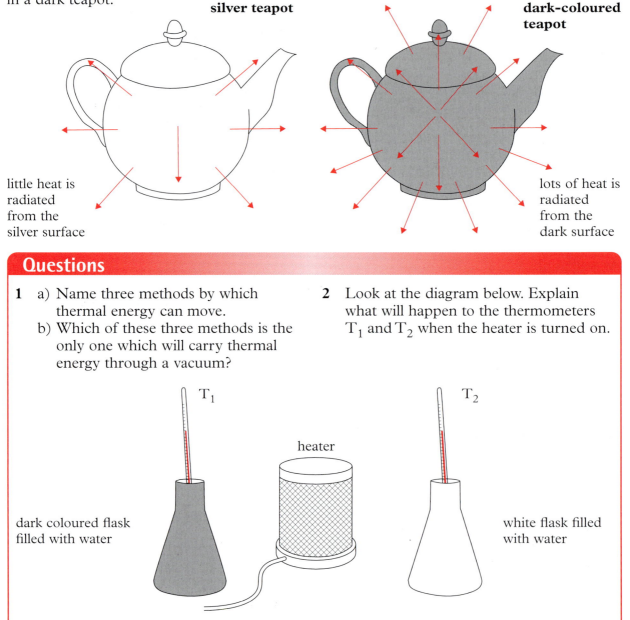

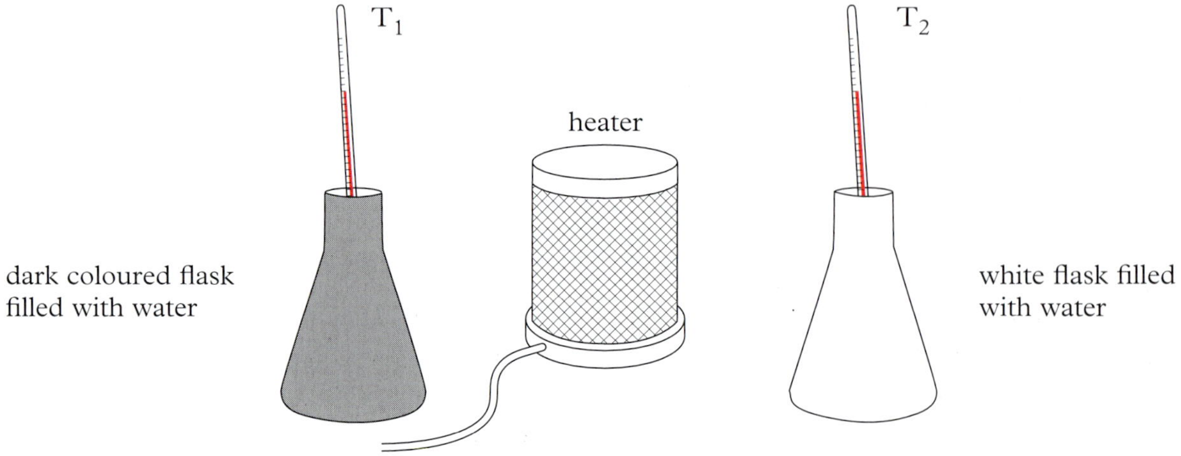

3 Why in summer do we tend to wear light coloured clothes?

4 You are looking to buy a new teapot. You have a choice of three which are identical except for their colour. One is black, one is brown and the third is yellow. Which would you choose and why?

Reducing the heat loss from your home

The diagram below shows how heat might escape from a house which has little or no insulation.

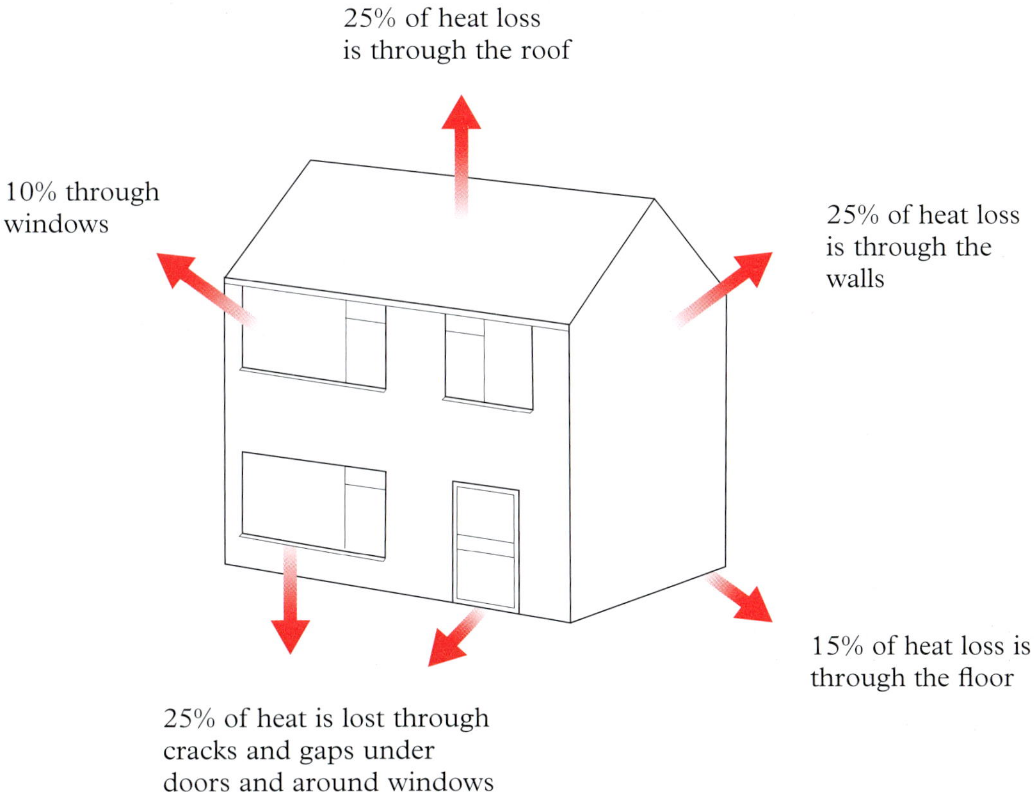
25% of heat loss
is through the roof
10% through
windows
25% of heat loss
is through the
walls
15% of heat loss is
through the floor
25% of heat is lost through
cracks and gaps under
doors and around windows

There are several ways in which we could reduce heat loss and cut our fuel bills.

1. Insulate the loft with fibreglass.
2. Replace single panes of glass with double glazing.
3. Most modern houses have outside cavity walls (i.e. they are built as two layers of brick with an air gap in between). If an insulator such as fibreglass is inserted in the cavity the loss of heat through the walls can be reduced even more.
4. Fit draft excluders and curtains to doors and windows.
5. Fit carpets and underlay to all ground floor rooms.

The vacuum flask (thermos flask)

The vacuum flask is a container which keeps hot things hot and keeps cold things cold. It is able to do this because its construction prevents heat transfer by conduction, by convection and by radiation.

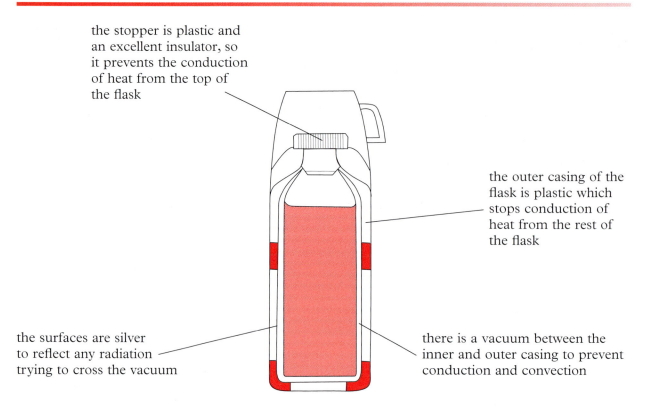

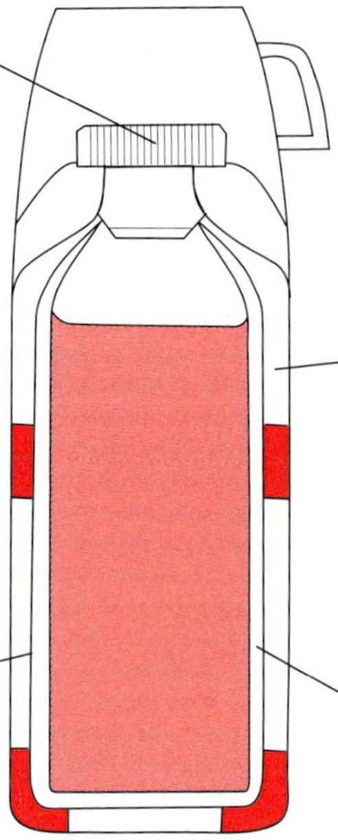
the stopper is plastic and an excellent insulator, so it prevents the conduction of heat from the top of the flask
the outer casing of the flask is plastic which stops conduction of heat from the rest of the flask
the surfaces are silver to reflect any radiation trying to cross the vacuum
there is a vacuum between the inner and outer casing to prevent conduction and convection

Questions

1 a) Suggest five ways in which the new owners of an old house could reduce heat loss during cold weather.
b) For each one explain briefly how these measures work.

2 Draw a labelled diagram of a vacuum flask. Explain how its construction helps to keep an ice cream which is placed inside it cold.

Summary

Thermal energy (heat) moves from regions of high temperature to regions of lower temperature. There are three methods by which it can move. These are conduction, convection and radiation.

By understanding these methods we can encourage or prevent this movement. When we require good transfer of heat by conduction we use metals. When we need to prevent or reduce conduction we use insulators such as plastics. Gases and liquids are poor conductors but if their atoms or molecules are free to move they can transfer heat by convection. Convection cannot take place in a solid.

Warm objects emit a lot of heat in the form of infra red waves. This radiation is able to transfer energy without the use of particles. This means that radiation can transfer heat through a vacuum. Dark, rough surfaces are good absorbers and good emitters of radiation. Shiny smooth surfaces are poor absorbers (they reflect most of the radiation) and poor emitters.

Key words

conduction The transfer of energy through a material but with no visible motion of the material itself.

conductor A substance which allows easy transfer of heat by conduction.

convection The transfer of energy by the movement (flow) of a liquid or gas.

convection current The continuous transfer of energy by the circulation of a liquid or a gas.

insulator A substance which does not allow the easy transfer of heat by conduction.

radiation The transfer of energy by infra red waves.

End of Chapter 4 Questions

1 Which of these statements are true and which are false?

a) Radiators in the home are painted white so that they will radiate heat into a room more quickly.

b) Thermal energy from the Sun travels to the Earth by convection.

c) Heat loss through flat roofs is likely to be greater than heat loss through conventional sloping roofs.

d) Polished silver teapots will keep tea warmer for longer than black ceramic pots.

e) Plastic radiators will soon replace metal ones because they reduce heat loss.

f) Electric fires have silvered reflectors behind the heating elements to improve the transfer of heat into the room by conduction.

g) One reason polar regions are cold is because the landscape is white and therefore it absorbs very little of the Sun's radiation.

h) Houses which have open fires are likely to be more draughty than those that do not have open fires.

i) If silver foil is fixed behind radiators that are fixed to an outside wall this will reduce any heat loss through the walls.

2 Before pumps had been fully developed mines used to be ventilated by digging two shafts and building a large fire at the bottom of one of them. Explain how the fire 'draws' fresh air into the mines.

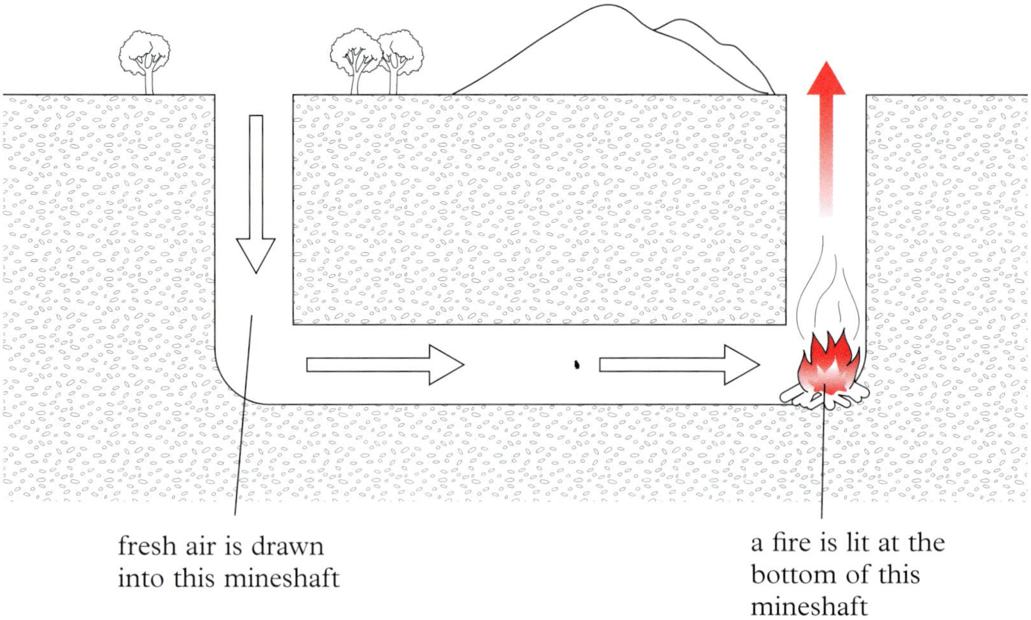
fresh air is drawn into this mineshaft
a fire is lit at the bottom of this mineshaft

3 This diagram shows the basic structure of a simple solar panel. It uses the heat from the Sun to warm the water which flows through the pipes.

a) Why is the pipe fixed to a black absorbing material?
b) Why is the pipe made of copper?
c) Why is the pipe made to zig zag up and down?
d) i) Why is the collector plate fixed to an insulator?
   ii) Suggest a suitable material which could be used as the insulator.
e) Why is the panel front covered with glass?
f) Suggest one advantage and one disadvantage of double glazing the panel.

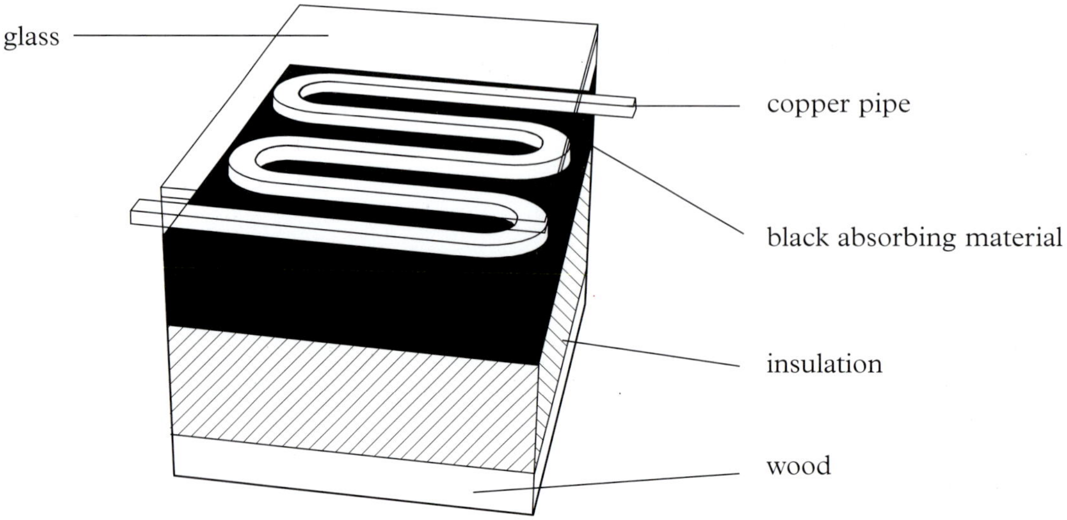
glass
copper pipe
black absorbing material
insulation
wood

4 The plumber who installed this domestic hot water system really didn't know what he was doing. Make a list of all the mistakes he has made and explain why they are wrong. Now redraw the system to show how it should have been installed.

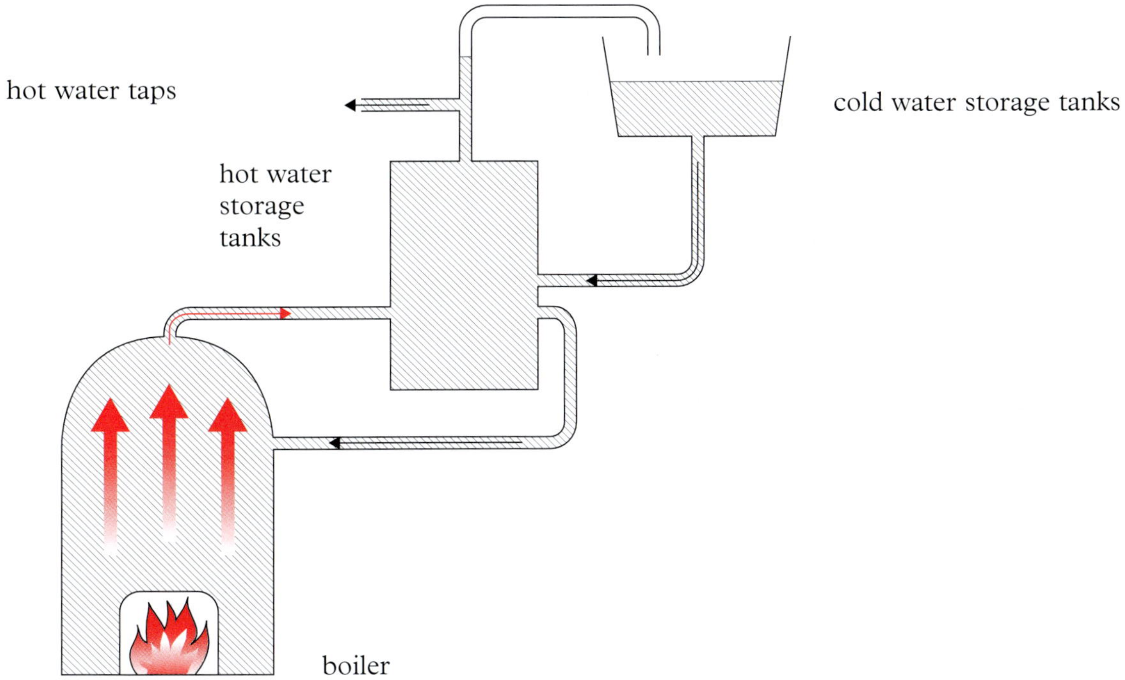

5 Explain why on a bright sunny morning the icicles on the dark sections of the canopy have melted and yet those on the light sections have not.

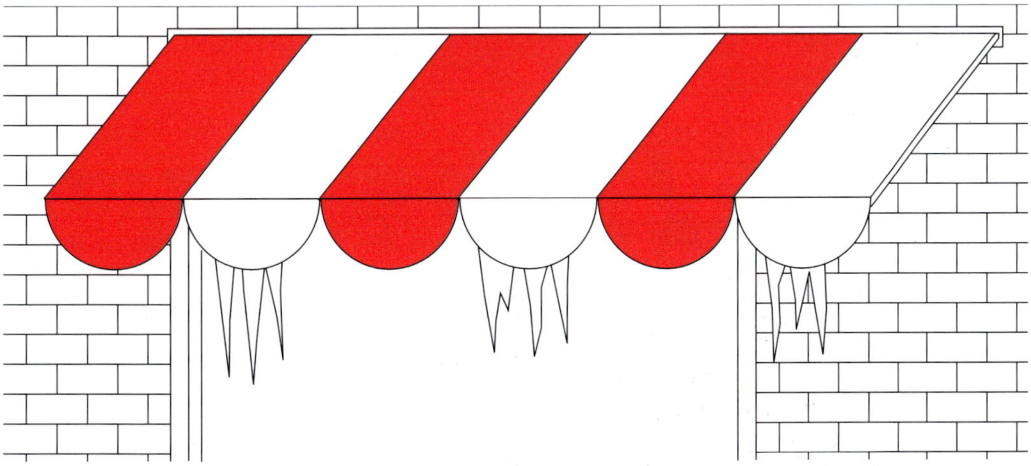

6 There are several ways in which we can reduce the energy loss from our homes. Describe one method for each of the following areas of a house:

a) the roof
b) the walls
c) the windows
d) the floors
e) the doors.

5 Speed, Velocity and Acceleration

Average Speed

this athlete can travel 100 m in 10 s
on average he will travel 10 m each second

In other words his **average speed** is 10 metres per second which is written as 10 m/s.

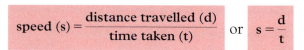

this cyclist can travel around a 400 m track in 20 s
on average he will travel 20 m each second

His average speed is 20 m/s.

To calculate the average speed of an object we need two pieces of information:

1 how far the object has travelled
2 the time it took for the object to travel this distance.

When we have these measurements we use the equation below to calculate the speed of the object:

$$\text{speed (s)} = \frac{\text{distance travelled (d)}}{\text{time taken (t)}} \quad \text{or} \quad s = \frac{d}{t}$$

Examples

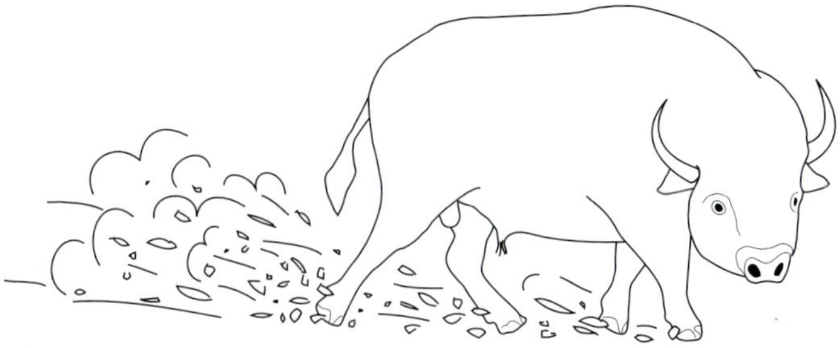

running at its fastest this bull can travel 90 m in just 5 s

Calculate its average speed.

Using s = d / t

s = 90 / 5

s = 18 m/s

The average speed of the bull is 18 m/s.

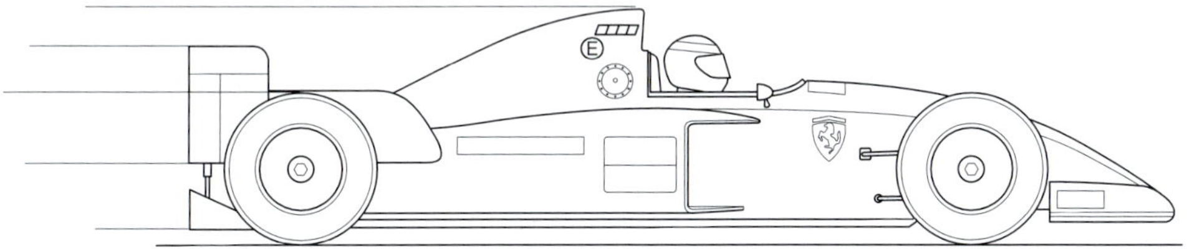

This racing car can complete one lap of a circuit 5.4 km (5400 m) long in 90 s. Calculate its average speed.

Using s = d / t

s = 5400 / 90

s = 60 m/s

The average speed of the racing car is 60 m/s.

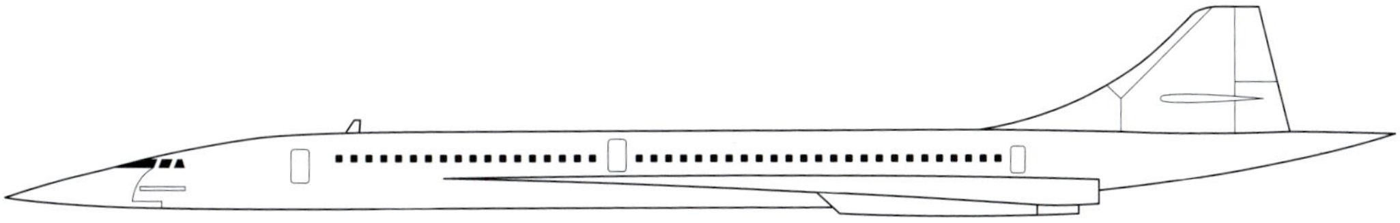

Concorde travels 3000 km in just 1½ hours. Calculate its average speed.

Using s = d / t

s = 3000 / 1.5

s = 2000 km/h

The average speed of Concorde is 2000 km/h.

Questions

1 Calculate the average speed of a hiker who walks 600 m in 200 s.

2 Calculate the average speed of a rocket which travels 1200 m in 30 s.

3 Calculate the average speed of a tortoise which walks 50 cm in 500 s.

4 Calculate the average speed of a bus which takes 30 minutes to complete its route which is 7.2 km long. (Hint:- work out the speed in m/s.)

The table below gives some typical speeds.

Object	Speed
Light	300 000 000 m/s
Rifle bullet	700 m/s
Sound in air	340 m/s
Jumbo jet (maximum speed)	270 m/s
Cheetah	28 m/s
Olympic sprinter	12 m/s
Man walking briskly	2 m/s
Snail	0.001 m/s

The equation s = d / t can be written as a formula triangle. This can then be used to calculate not only the speeds of objects but also the distances travelled and times taken.

the equation is s = d/t so the triangle is written as

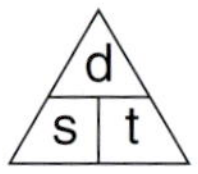

now cover the quantity you want to calculate

the positions of the remaining two letters tell you how the equation should look

to calculate the speed cover the ‘s’

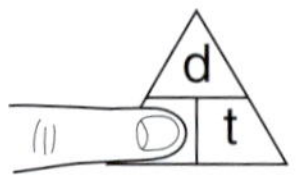

i.e. s = d/t

to calculate the distance cover the ‘d’

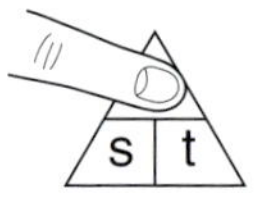

i.e. d = s x t

to calculate the time cover the ‘t’

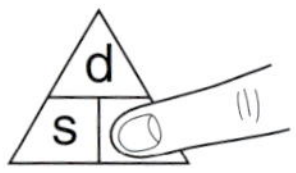

i.e. t = d/s

Examples

1 A hiker walks at an average speed of 2 m/s for 100 s. Calculate the distance she has travelled.

The formula triangle shows that

distance travelled (d) = speed (s) × time taken (t) therefore:

$$d = 2 \times 100$$

distance travelled = 200 m

2 A cyclist travels 1000 m at an average speed of 20 m/s. How long did it take him to complete his journey?

The formula triangle shows that:

time taken (t) = distance travelled (d) / speed (s), therefore:

$$t = 1000 / 20$$

time taken to complete his journey = 50 s

Questions

1 A motorcyclist travels at an average speed of 80 km/h for 5 hours. How far has he travelled in this time?

2 How far will a sound wave travel in 5 s? The speed of sound is 340 m/s.

3 How far will a light wave travel in 5 s? The speed of light is 300 000 000 m/s

4 How long will it take a sound wave to travel 1020 m? The speed of sound is 340 m/s

5 How long will it take a runner to travel 1500 m if her average speed is 7.5 m/s?

6 How long will it take a jumbo jet to travel 250 km if its average speed is 250 m/s?

7 Calculate the missing values in the table below:

a) s = ?	d = 10 m	t = 5 s
b) s = ?	d = 50 m	t = 10 s
c) s = ?	d = 250 m	t = 25 s
d) s = 20 m/s	d = ?	t = 5 s
e) s = 60 m/s	d = ?	t = 10 s
f) s = 15 m/s	d = ?	t = 8 s
g) s = 40 m/s	d = 80 m	t = ?
h) s = 10 m/s	d = 40 m	t = ?
i) s = 15 m/s	d = 60 m	t = ?

Speed and velocity

In everyday life we use the words **speed** and **velocity** as if they have the same meaning. In Science there is a small but important difference between the two words. The speed of an object tells us how quickly it is moving. The velocity of an object tells us how quickly it is moving, and the direction in which it is travelling. Velocity tells us two pieces of information.

The diagram below explains why this difference is important. The two boats are unable to see each other because there is mist on the water but they are able to talk by radio.

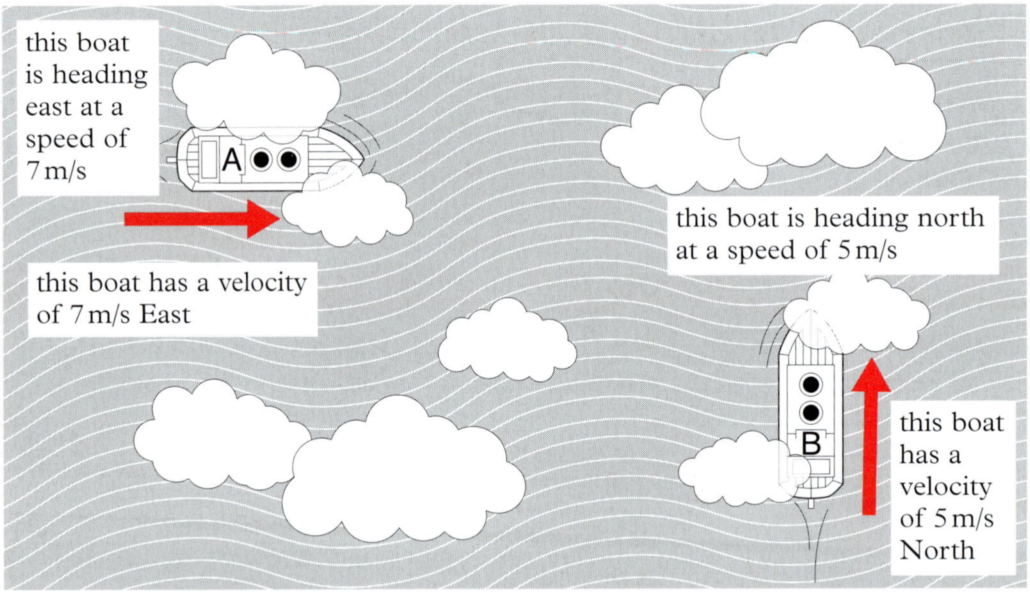
this boat is heading east at a speed of 7 m/s
A
this boat has a velocity of 7 m/s East
this boat is heading north at a speed of 5 m/s
B
this boat has a velocity of 5 m/s North

To ensure that the boats don't crash the captains need to know not only their speeds but also the directions in which they are travelling, i.e. they need to know the velocity of each boat. The speed of boat B is 5 m/s. Its velocity is 5 m/s North. The speed of boat A is 7 m/s. Its velocity is 7 m/s East.

Acceleration

this sledger is going down a steep hill
the sledge is picking up speed all the time

An object which is increasing its speed is **accelerating**.

this dragster has finished its race and needs to slow down before it runs out of road

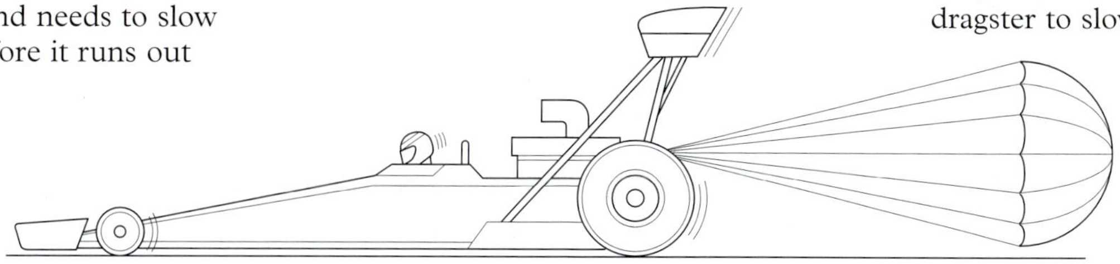

the parachute helps the dragster to slow down

An object which is slowing down is **decelerating**.

If an object increases its speed quickly it has a large acceleration. If an object increases its speed slowly it has a small acceleration.

We can calculate the acceleration of an object using the equation

$$\textbf{acceleration (a)} = \frac{\textbf{change in velocity } (\Delta \textbf{v})}{\textbf{time taken (t)}}$$

The Δ symbol means 'change in'.

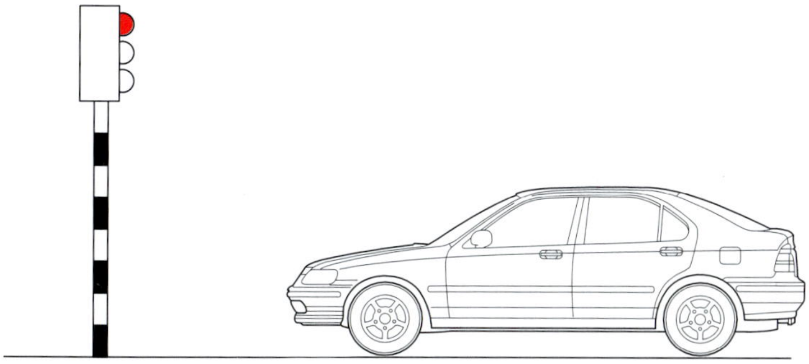

this car can change its velocity from 0 m/s to 100 m/s in 5 s

$$a = \frac{\Delta v}{t}$$

$$a = \frac{100 - 0}{5}$$

$$a = 20\ \text{m/s/s}$$

The acceleration of the car is 20 m/s per second. This means that the car increases its speed by 20 m/s every second. This is normally written as 20 m/s^2 or 20 m/s/s.

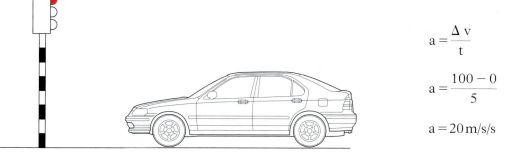

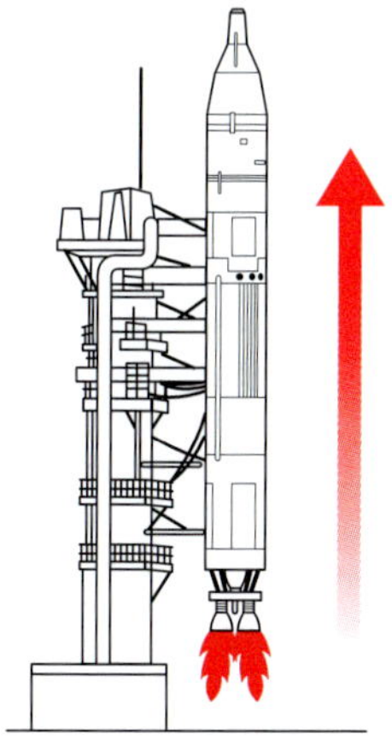

this rocket can change its velocity from 10 m/s to 2010 m/s in 50 s

$$a = \frac{\Delta v}{t}$$

$$a = \frac{2010 - 10}{50}$$

$$a = 40\ \text{m/s/s}$$

The acceleration of the rocket is 40 m/s per second. This means the rocket increases its speed by 40 m/s every second. Again, this can be written as 40 m/s^2.

Questions

1 Calculate the acceleration of a car which increases its velocity from 20 m/s to 100 m/s in 5 s.

2 Calculate the acceleration of a falling stone which increases its velocity from 0 m/s to 50 m/s in 10 s.

3 Calculate the acceleration of a dog which increases its velocity from 3 m/s to 30 m/s in 9 s.

4 Calculate the deceleration of a car which is able to brake from 100 km/h to 10 km/h in 3 s.

5 A racing motorcyclist falls from his bike whilst travelling at 160 km/h. If he slides along the road surface for 4 s before coming to rest unhurt calculate his deceleration.

Using graphs to show motion

It is often useful to show the movement of an object in the form of a graph.

Distance – Time Graphs

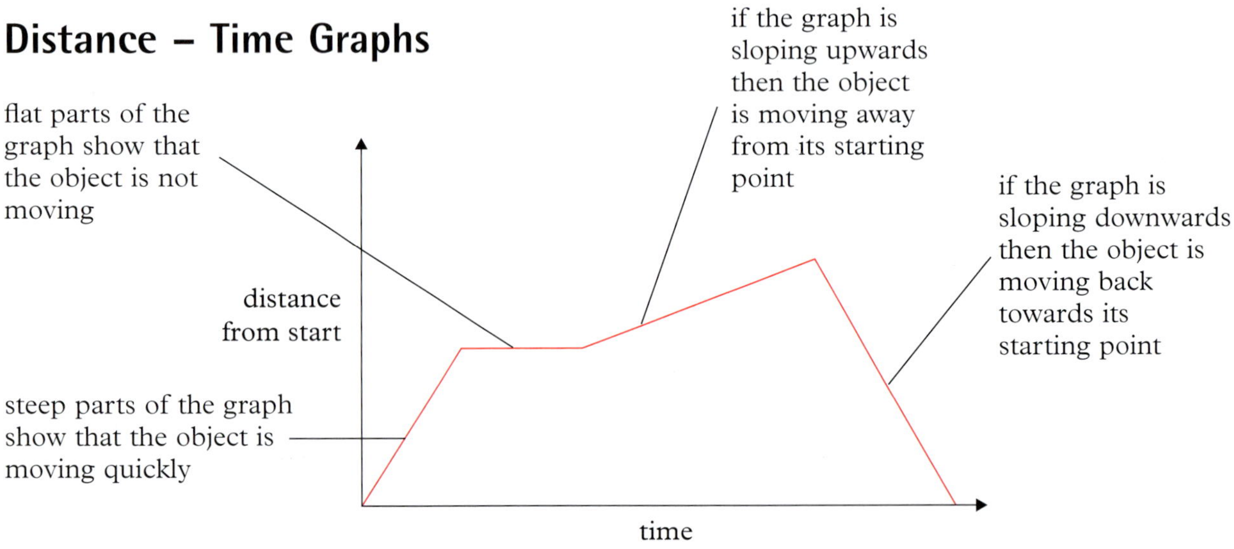
if the graph is sloping upwards then the object is moving away from its starting point
flat parts of the graph show that the object is not moving
if the graph is sloping downwards then the object is moving back towards its starting point
distance from start
steep parts of the graph show that the object is moving quickly
time

We can calculate the speed of an object at different times during its journey using a distance-time graph.

The distance-time graph below shows the journey of a cyclist.

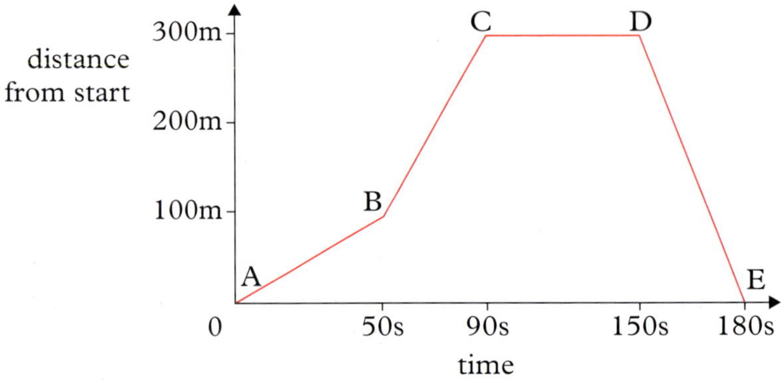
distance from start
300m
200m
100m
C
D
B
A
E
0
50s
90s
150s
180s
time

- Between A and B the cyclist is travelling slowly, away from his starting point. His speed during this part of his journey is distance travelled / time taken = 100 m / 50 s = 2 m/s.
- Between B and C the cyclist is travelling more quickly away from his starting point His speed during this part of his journey is distance travelled / time taken = 200 m / 40 s = 5 m/s.
- Between C and D the cyclist has stopped.
- Between D and E the cyclist is travelling very quickly back towards his starting point. His speed during this part of his journey is distance travelled / time taken = 300 m / 30 s = 10 m/s.

Questions

1 Describe in detail the journey illustrated in the two graphs below.

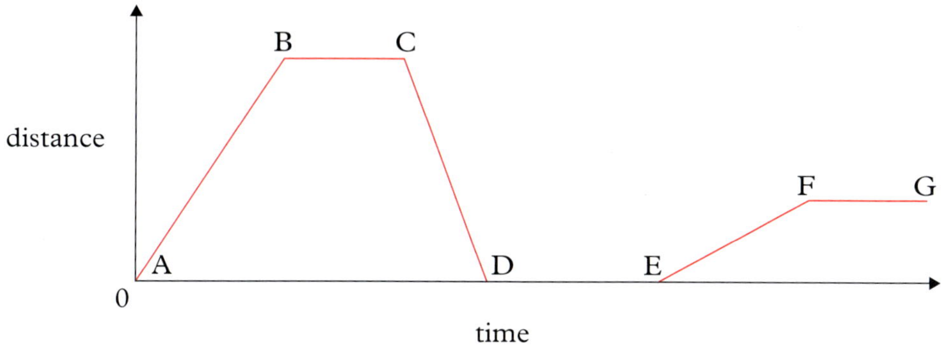

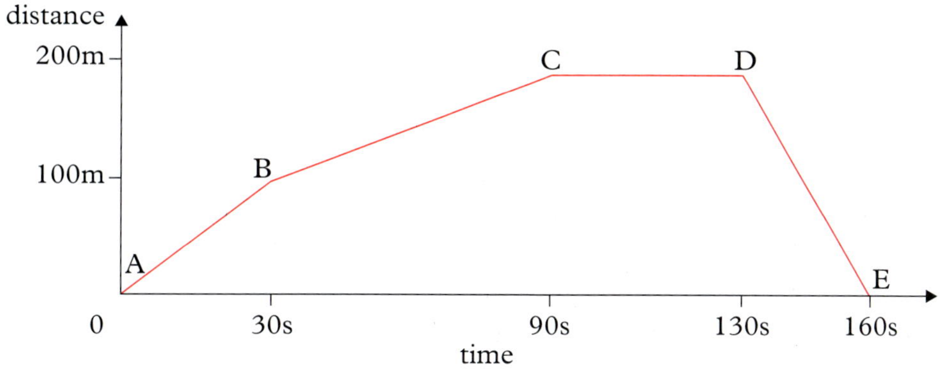

2 Sketch a graph to illustrate the following journey.

a) A car starts from home and travels at a constant speed for 10 minutes.

b) The car stops for 5 minutes.

c) The car continues its journey travelling at a higher constant speed for 15 minutes.

d) The car stops for a further 5 min then returns home travelling at an even higher constant speed.

Velocity – Time Graphs

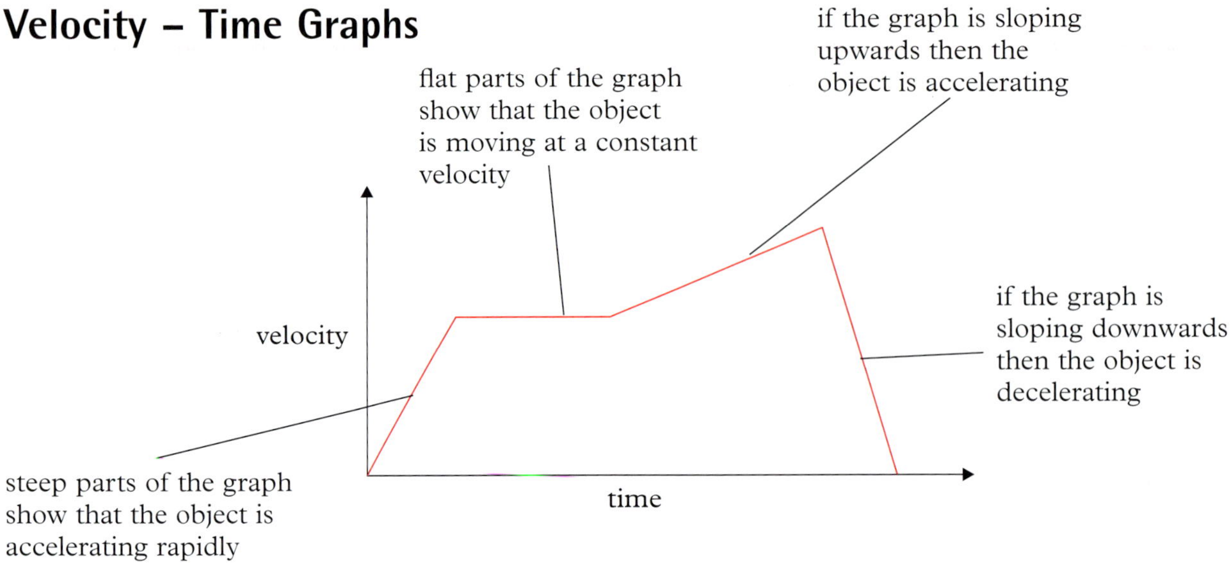
if the graph is sloping upwards then the object is accelerating
flat parts of the graph show that the object is moving at a constant velocity
velocity
if the graph is sloping downwards then the object is decelerating
steep parts of the graph show that the object is accelerating rapidly
time

The velocity–time graph below shows the journey of a runner.

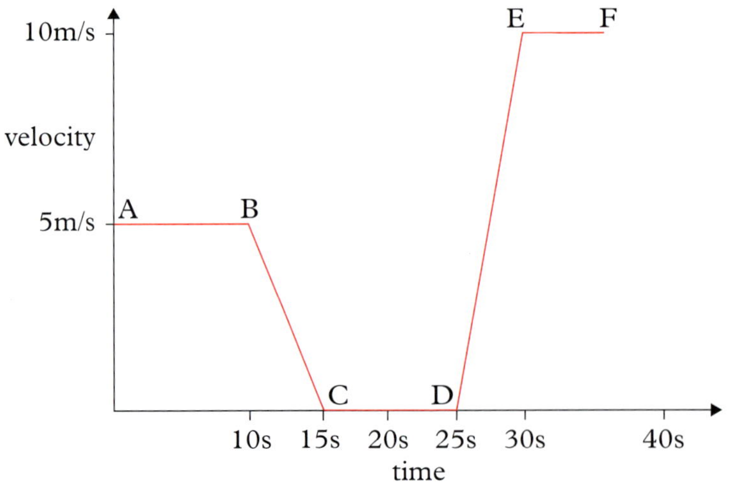
10m/s
velocity
5m/s
A
B
C
D
E
F
10s
15s
20s
25s
30s
40s
time

- Between A and B the runner travels at a constant speed of 5 m/s for 10 s. There is no change in his velocity so his acceleration is zero.
- Between B and C he takes 5 s to slow down and stop. His deceleration during this part of his journey is change in velocity / time taken = 5 / 5 = 1 m/s^2.
- Between C and D he remains stationary for 10 s. His acceleration is again zero.
- Between D and E he increases his speed to 10 m/s in 5 s. His acceleration during this part of his journey is change in velocity / time taken = 10 / 5 = 2 m/s^2.
- Between E and F he travels at constant velocity of 10 m/s for 5 s and his acceleration is zero.

Questions

1 Describe in detail the journey illustrated in the two graphs below.

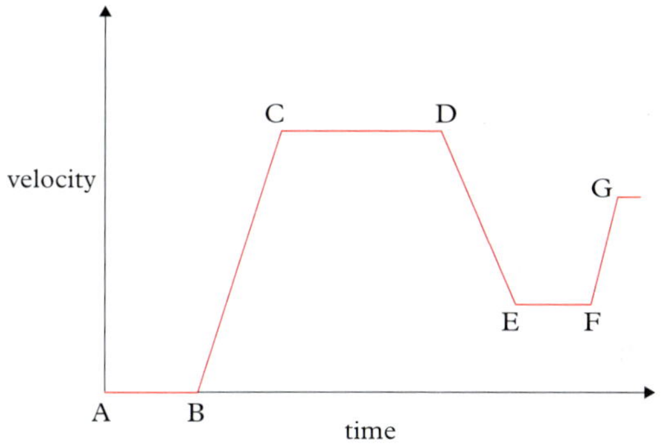

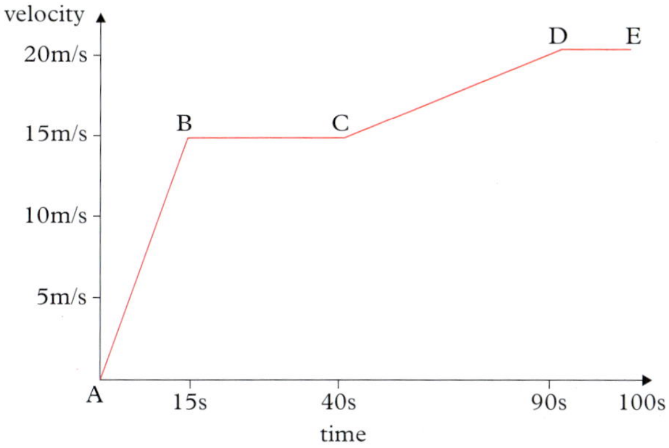

2 Sketch a graph to illustrate the following journey.
   a) A car starts from rest and increases its speed to 30 m/s in 5 s.
   b) The car travels at 30 m/s for 7 s.
   c) The car suddenly brakes coming to a halt in 3 s.
   d) The car remains stationary for 5s.
   e) The car increases its speed to 60 m/s in 10 s.
   f) The car travels at 60 m/s for 20 s.

Velocity – Time Graphs Part 2

We can use Speed or Velocity – Time graphs to work out the distance an object or person has travelled.

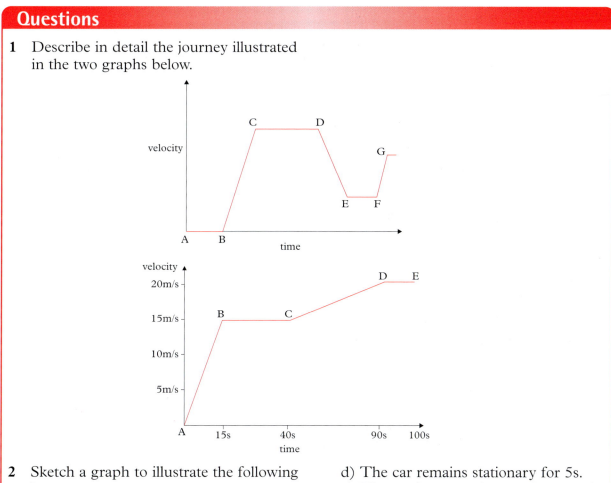

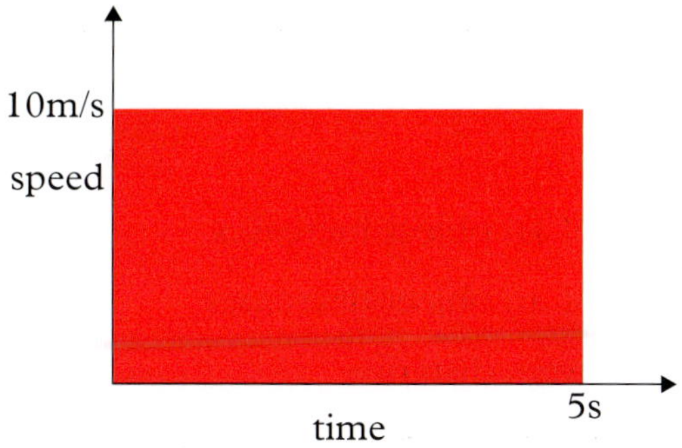

area under graph = 10 m/s × 5 s
= 50 m

This Speed-Time graph is for a sprinter.

This sprinter has run at a constant speed of 10 m/s for 5 s. The total distance he has travelled is 50 m. This is equal to the area under the graph.

This Speed-Time graph is for a car which has a constant acceleration.

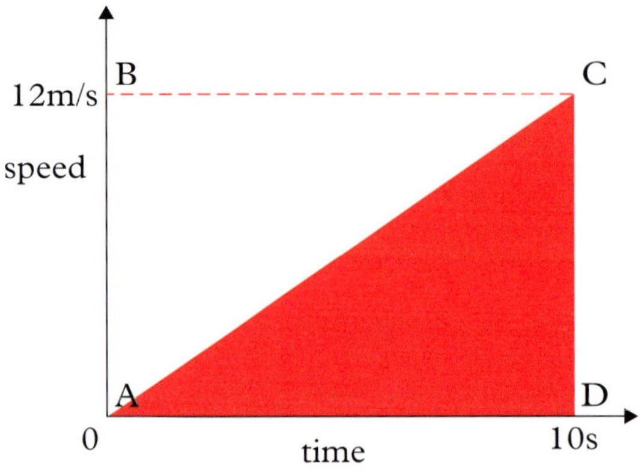

the area under the graph is a triangle

The area under this Speed-Time graph is half the area of the rectangle ABCD.

The area = $\frac{1}{2} \times 12\,\text{m/s} \times 10\,\text{s}$
$= 60\,\text{m}$

In 10 s this car has travelled 60 m.

Questions

1 Using the Speed – Time graph on page 50 for the runner calculate the distances travelled between points:

a) A and B,
b) B and C,
c) C and D,
d) the total distance travelled.

Summary

The average speed of an object can be calculated using the equation **average speed (s) = total distance travelled (d) / time taken (t).**

The acceleration of an object can be calculated using the equation **acceleration (a) = change in velocity (Δv) / time taken (t).**

The motion of an object can be described by distance-time and velocity-time graphs.

Key words

acceleration The rate at which the velocity of an object increases.

deceleration The rate at which the velocity of an object decreases.

speed The rate at which an objects covers distance.

velocity The speed of an object in a certain direction.

End of Chapter 5 Questions

1 The distance-time graph below shows the journey of a long distance runner.

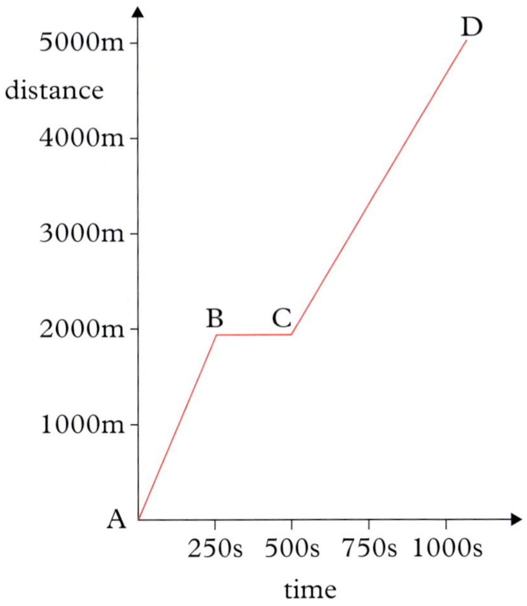

a) Between which two points was the athlete running fastest?
b) Between which two points was the athlete stationary?
c) Write down the equation which links the speed of the runner, the distance he has travelled and the time taken.
d) Calculate the speed of the runner between points A and B.
e) Calculate the speed of the runner between C and D.
f) Calculate the average speed of the runner for the whole of his journey.

2 The journey of a cyclist was recorded in the form of a table.

Time taken in seconds	Distance travelled by cyclist in metres
0	0
100	500
200	1000
300	1000
400	2000
500	3500
600	6000
700	7000

a) Use this information to draw a distance time graph.
b) During which time was the cyclist stationary?
c) Calculate the highest speed the cyclist reached.
d) Calculate the average speed of the cyclist for the whole journey.

3 The diagram below shows a bob sleigh team about to begin their run.

The table below shows how the speed of the sleigh changes with time as it travels down the course.

Time from beginning of run in seconds	Speed of sleigh in metres per second
0	0
5	15
10	25
15	35
20	40
30	40
50	40
55	5
60	0

a) Use this information to draw a speed – time graph.
b) During which time was the sleigh travelling at a constant speed?
c) During which time did the sleigh have the greatest acceleration?
d) During which time did the sleigh decelerate?
e) Calculate the distance travelled by the sleigh in the first 5 s of its journey.

6 Forces

Effects of Forces

There are many different types of force such as pushes, pulls, twists and stretches.

if we apply a force to an object it may make the object start to move

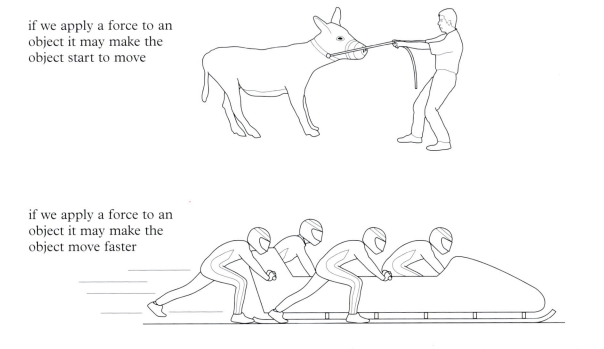

if we apply a force to an object it may make the object move faster

if we apply a force to an object it may make the object move more slowly

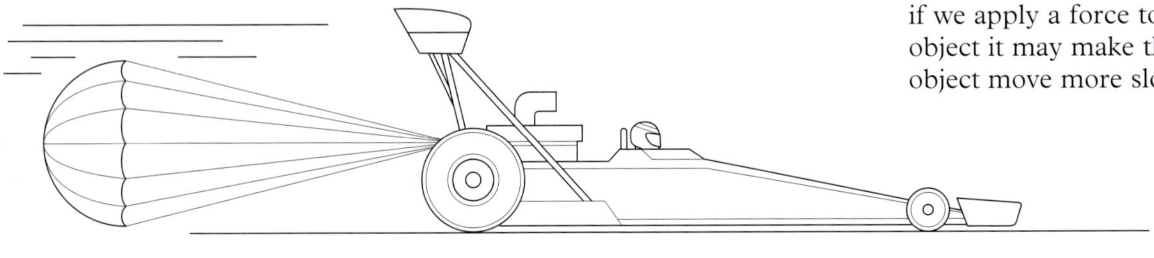

if we apply a force to an object it may make the object stop

if we apply a force to an object it may change the direction in which the object is moving

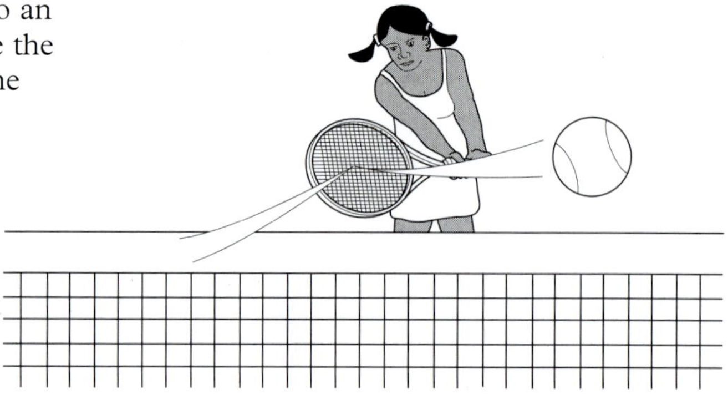

if we apply a force to an object it may change the shape of the object

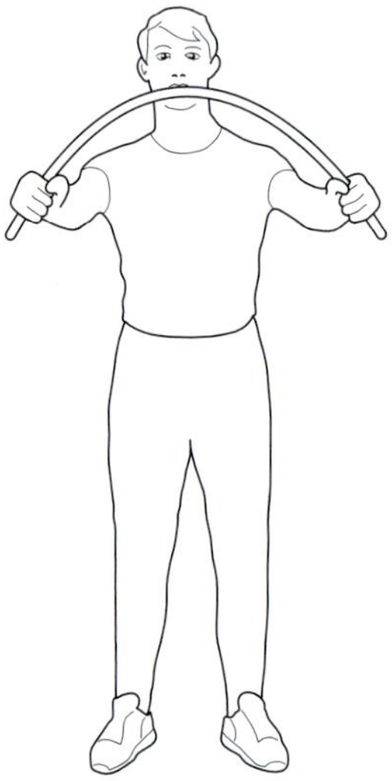

Sometimes it is not necessary to be in contact with an object in order to apply a force to it.

Example 1

these steel screws are being lifted by a magnetic force

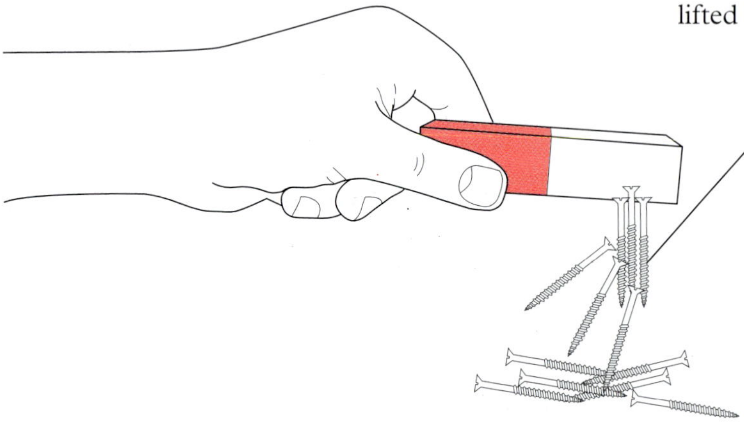

Example 2
The force of gravity pulls objects downwards. e.g. an apple falling from a tree.

Questions

1 Name four different types of forces.

2 Name five things that may happen to an object which has a force applied to it.

3 Give two examples of forces which can be applied to an object without being in contact with the object.

4 Choose five different sports. Describe one situation in each sport where a force is applied and explain what this force does.

Measuring forces

We measure forces in newtons (N). Newtons are named after a famous scientist called Sir Isaac Newton. The examples below will give you an idea of the size of a force.

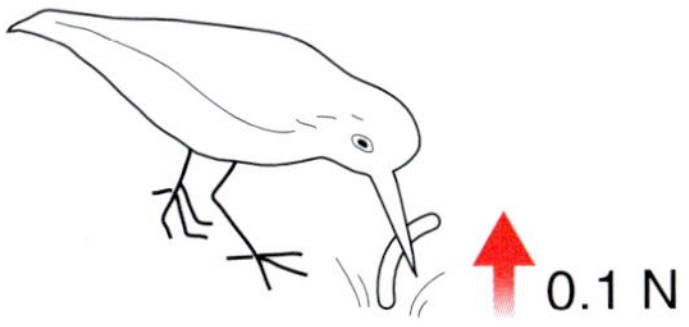

a bird may need to pull with a force of about 0.1 N to remove a worm from the ground

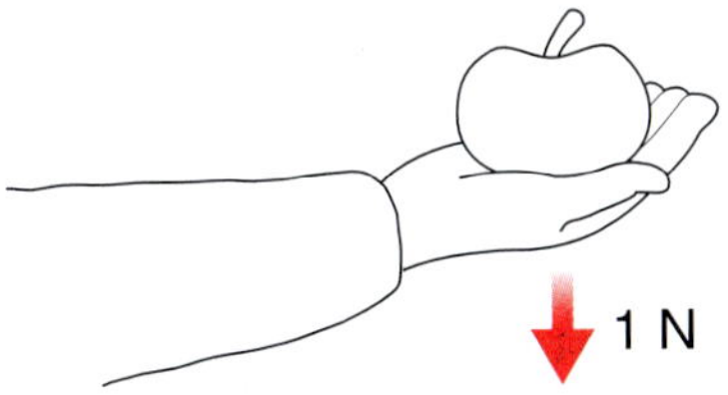

the weight of an average sized apple is 1 N

if you hold an average sized apple in your hand you will feel a downward force of 1 N

you will apply a force of about 10 N when pushing open a door

you will need to pull with a force of about 50 N to stretch the elastic of a strong catapult

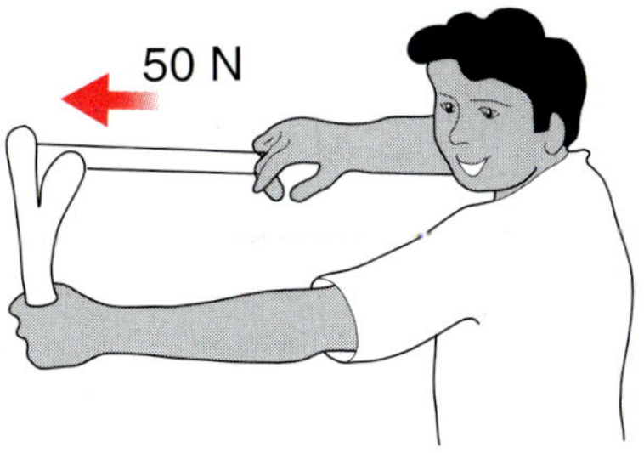

skiing downhill requires little or no effort but on the flat a skier may have to push with a force of about 400 N

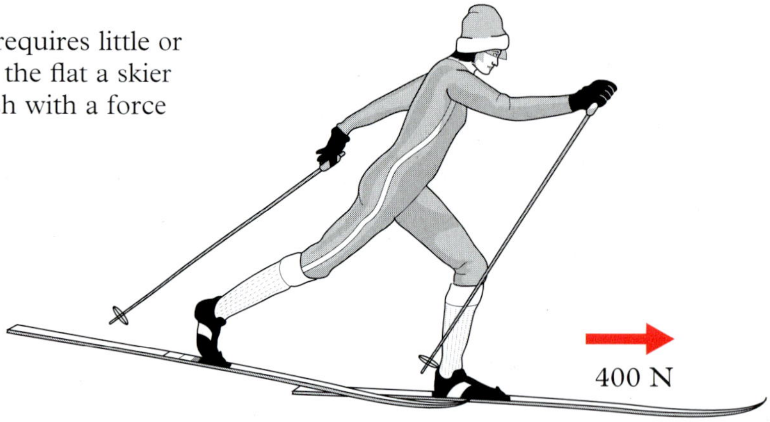

an athlete may push against a force of 1000 N during weight training

to lift his opponent a Sumo wrestler may need to apply a force of 2500 N

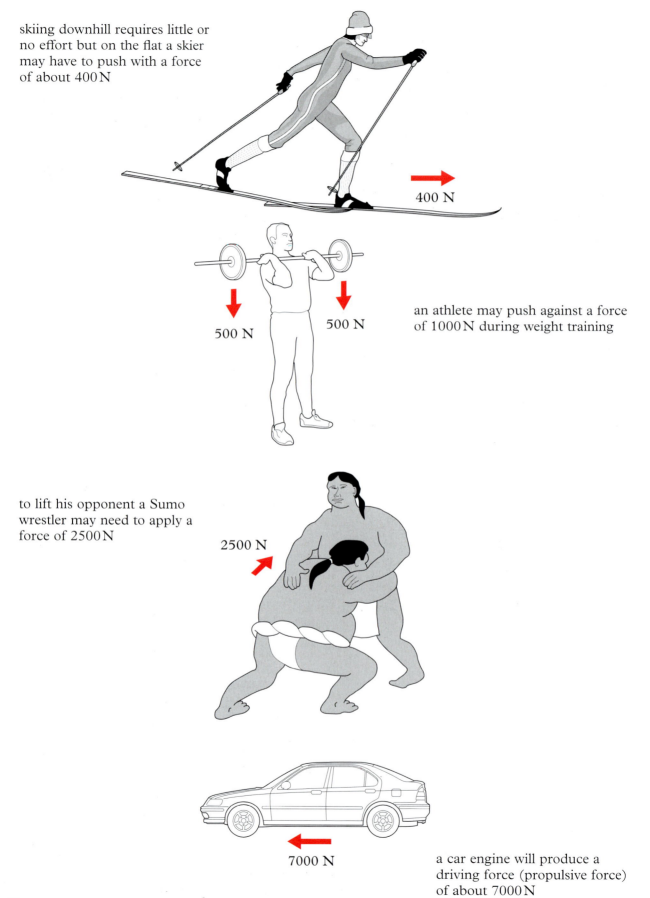

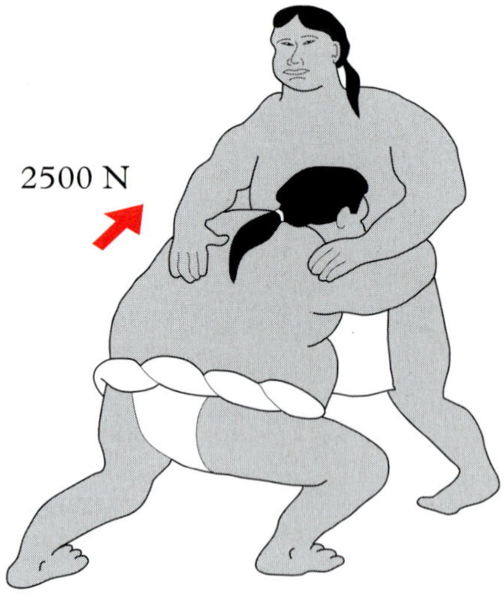

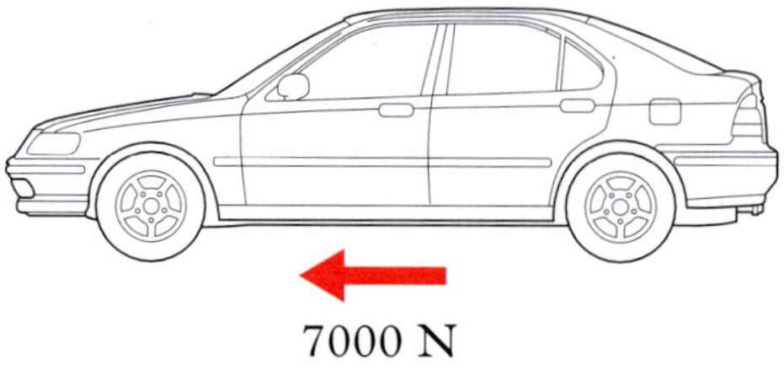

a car engine will produce a driving force (propulsive force) of about 7000 N

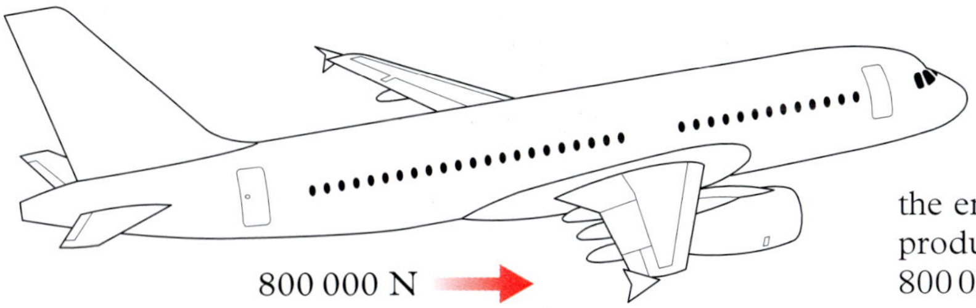
800 000 N
the engines of an airliner may produce a propulsive force of 800 000 N

Questions

1 Estimate the size of the force in newtons needed to:
a) lift a newspaper
b) open a car door
c) turn on a water tap
d) squeeze toothpaste from a tube
e) push a lawnmower
f) drag a heavy sledge uphill.

Stretching springs and wires

If we apply a force to a spring (or a length of wire) it will change shape. It will stretch. If we apply a bigger force it will stretch even more. A scientist called Robert Hooke discovered that the amount a spring stretches is proportional to the size of the force applied to it. If the applied force is doubled the amount by which the spring stretches also doubles.

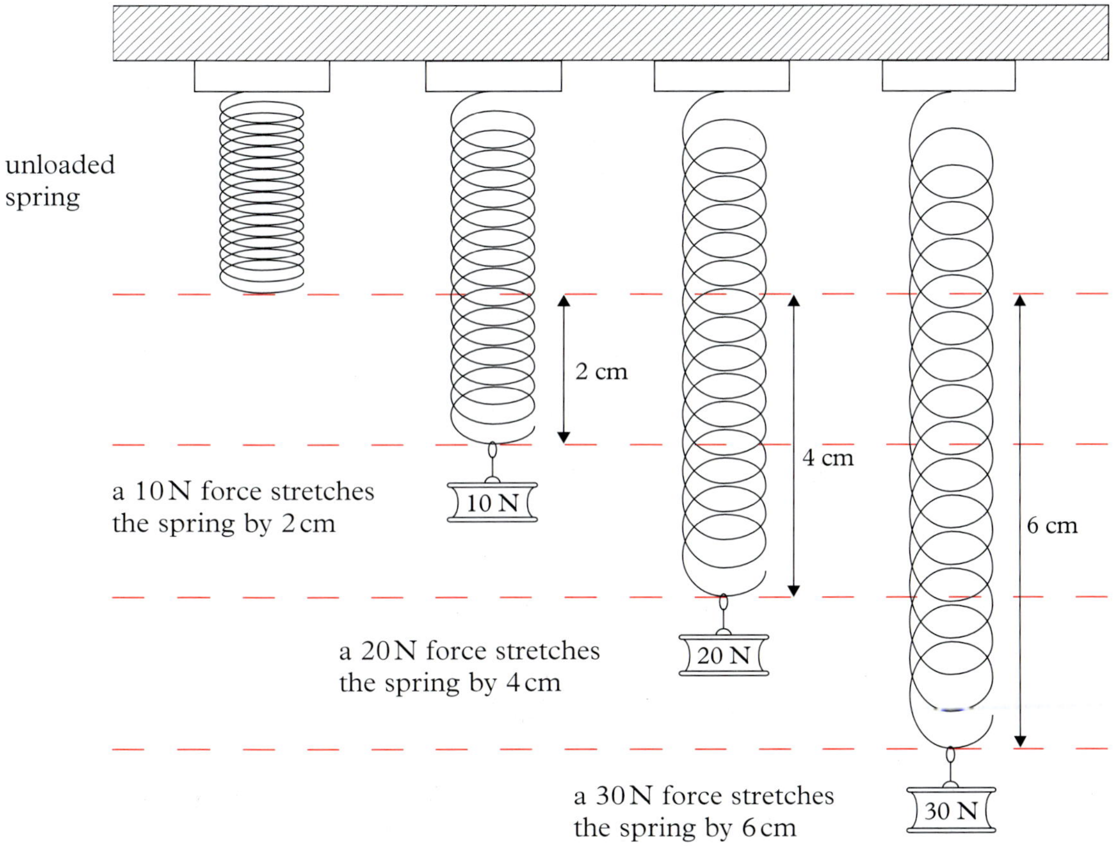
unloaded spring
2 cm
4 cm
6 cm
10 N
20 N
30 N
a 10 N force stretches the spring by 2 cm
a 20 N force stretches the spring by 4 cm
a 30 N force stretches the spring by 6 cm

But if too much force is applied the spring no longer obeys Hooke's Law and becomes permanently deformed. It does not return to its original shape when the forces are removed.

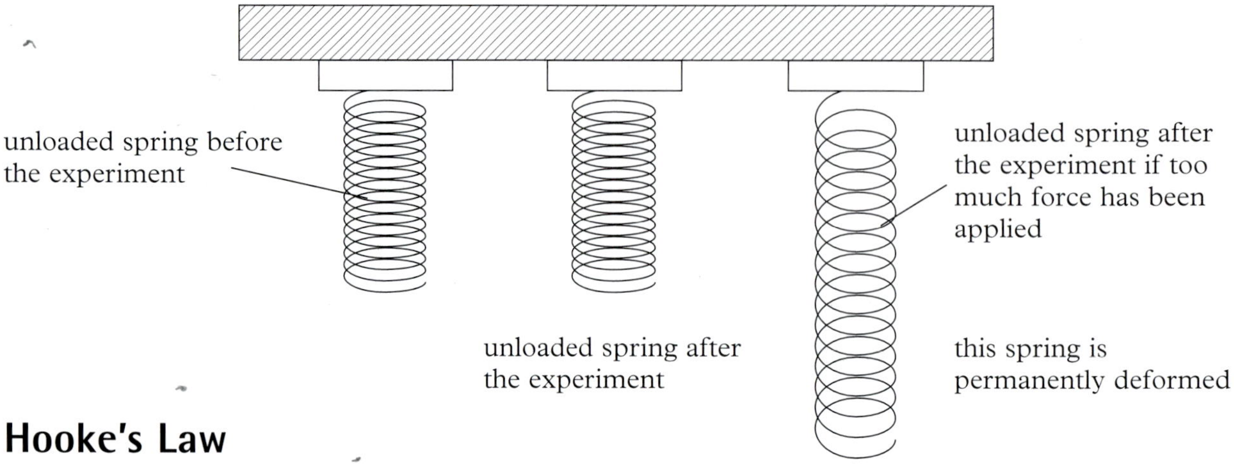
unloaded spring before the experiment
unloaded spring after the experiment
unloaded spring after the experiment if too much force has been applied
this spring is permanently deformed

Hooke's Law

The extension of a spring is proportional to the applied force providing the force is not too large

The maximum force that should be applied to the spring is called the **limit of proportionality**. Increasing the force further would mean the spring no longer obeys Hooke's Law, and wouldn't return to its original shape when all the applied forces are removed.

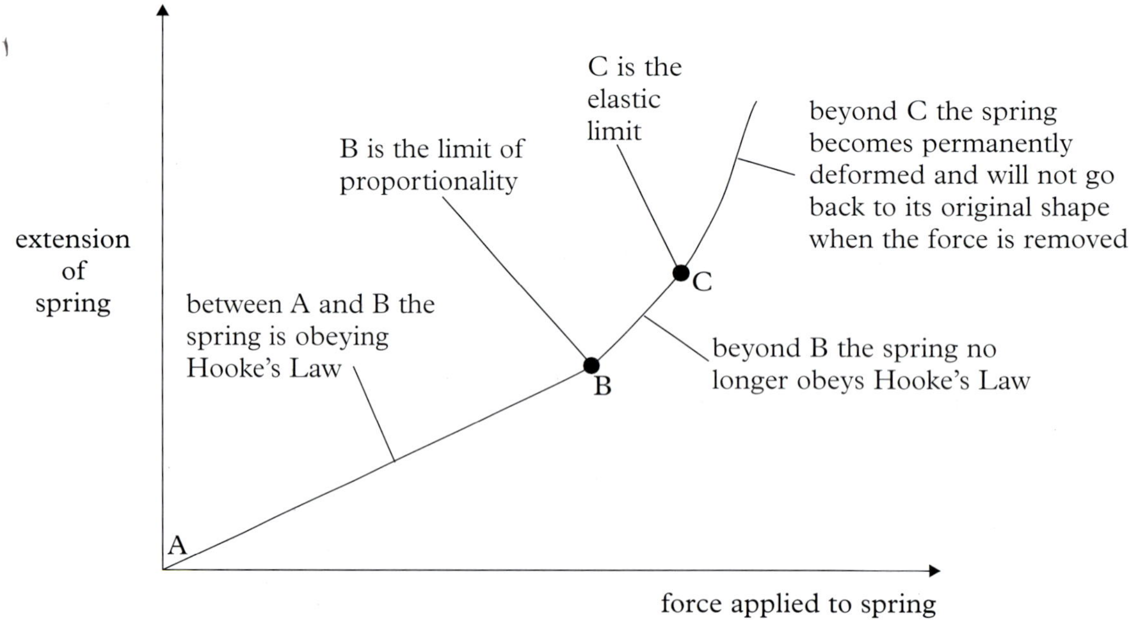
extension of spring
force applied to spring
A
B
C
C is the elastic limit
B is the limit of proportionality
beyond C the spring becomes permanently deformed and will not go back to its original shape when the force is removed
between A and B the spring is obeying Hooke's Law
beyond B the spring no longer obeys Hooke's Law

The force that is applied to the spring is called the **load**.

The newtonmeter

We can use the stretching of a spring to measure the size of a force. The larger the force, the more the spring stretches. A spring used in this way is called a **newtonmeter**.

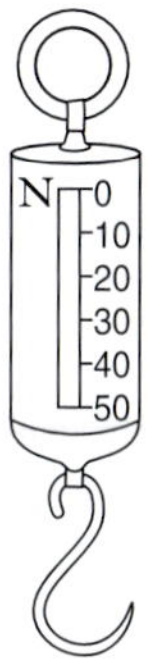
N
0
10
20
30
40
50

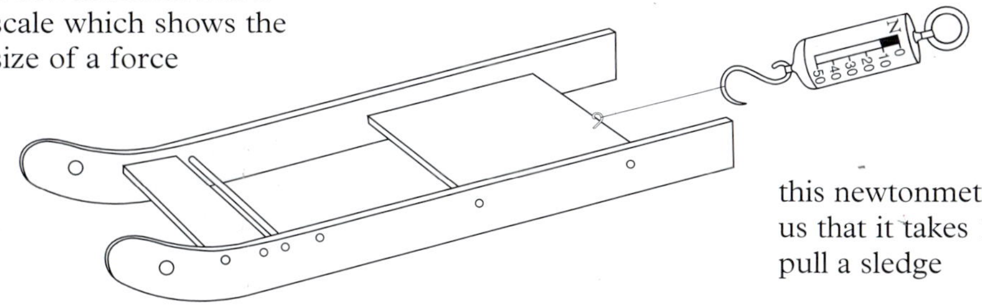
a newtonmeter has a scale which shows the size of a force
this newtonmeter shows us that it takes 10N to pull a sledge

Questions

1 The table below shows how a spring extended when tensile forces (stretching forces) were applied to it.

Tensile force in newtons	Extension in centimetres
0	0
10	2.0
20	4.0
30	6.0
40	9.0
50	13.0
60	18.0

a) Draw an extension-force graph using the results in the table.
b) Describe the shape of the graph.
c) Which part of the graph shows that the spring is obeying Hooke's Law?
d) Mark on your graph the limit of proportionality.
e) What will happen if all the forces applied to the spring are removed at the end of the experiment?

Balanced and Unbalanced Forces

In everyday life it is rare for an object to be acted upon by no forces or just one force. It is much more likely it will be experiencing several forces. These forces may be **balanced** or **unbalanced**. We can see more clearly the effects of several forces on an object by drawing a force diagram.

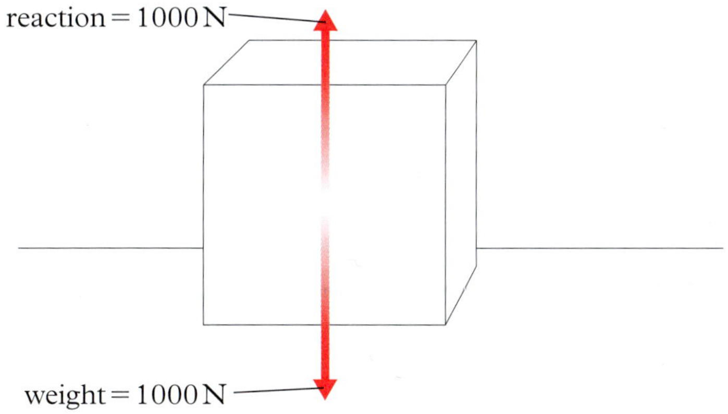
reaction = 1000 N
weight = 1000 N

The crate on the previous page is pushing down on the floor with a force of 1000 N. (This is the weight of the crate). This causes an upward reaction force of 1000 N from the floor. The two forces are balanced and the crate remains stationary.

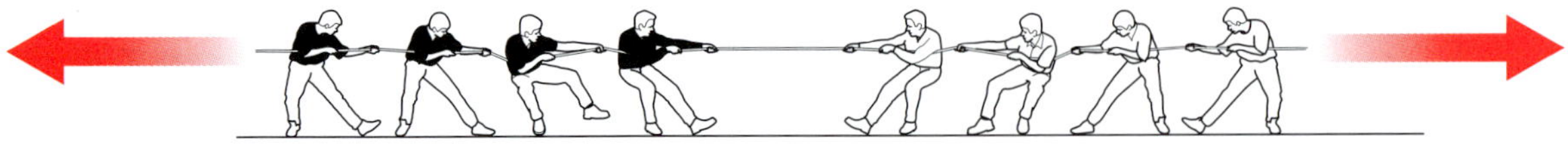

if these two tug of war teams pull with the same force, the forces will be balanced and there will be no movement

If one person leaves the team in black, the white team begins to win the tug of war.

now the team in white are pulling harder than the team in black

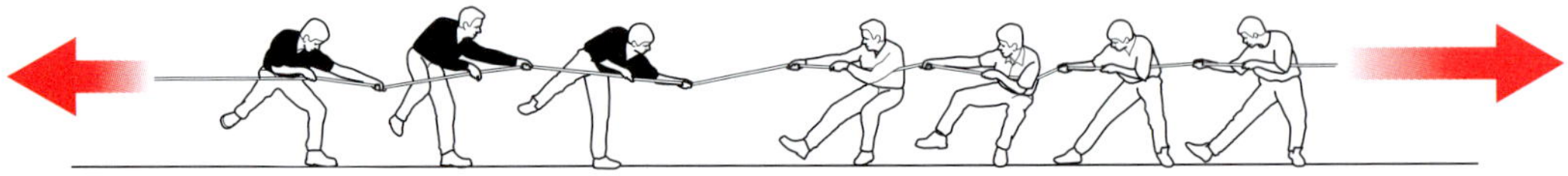

the forces on the rope are unbalanced which causes movement

Unbalanced forces will cause a stationary object to move

Questions

1 Which of the following are examples of balanced forces?
a) a rocket taking off
b) someone lying down sunbathing
c) a car braking
d) an arrow as it is released from a bow
e) a weight hanging from a spring.

2 Draw a force diagram for each of the above examples. Include in your diagram all the forces that are acting.

Accelerating and decelerating forces

Look at the rocketman in the diagrams on the next two pages. His motion depends on the forces produced by his rocket. In each case we have assumed that the only forces present are those marked on each diagram, apart from gravity which is keeping him on the ground.

the rocket is not turned on

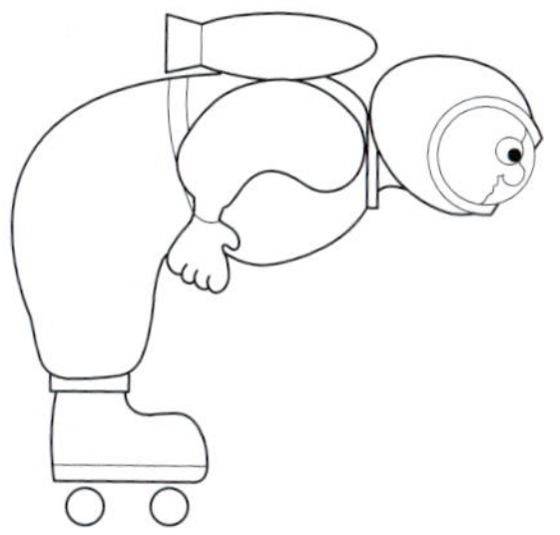

no forces are acting on the rocketman

the rocketman doesn't move

the rocket has been turned on

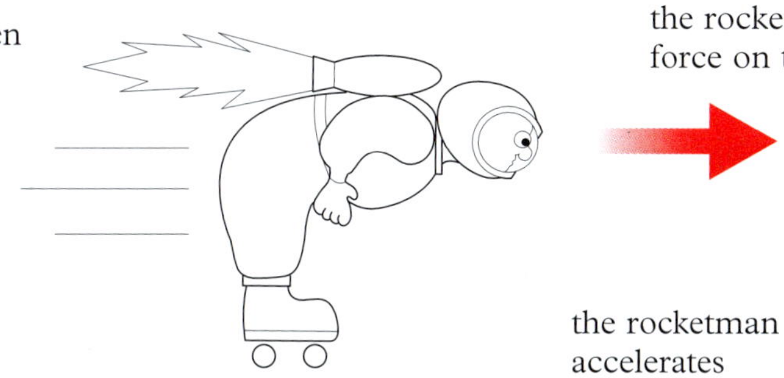

the rocket produces a forward force on the rocketman

the rocketman accelerates

another identical rocket has been added to the rocketman's back

the two rockets produce a forward force on the rocketman

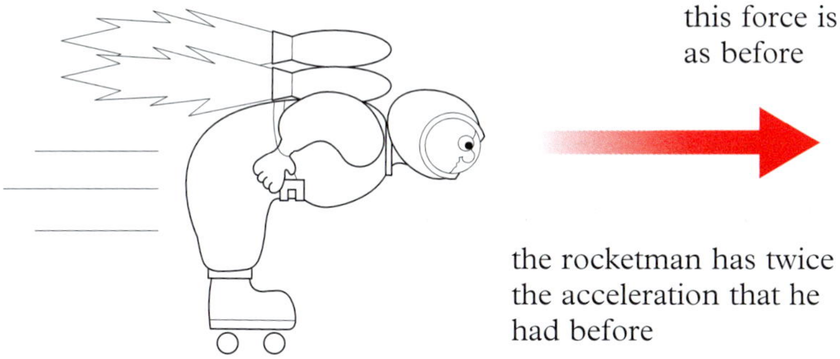

this force is twice as big as before

the rocketman has twice the acceleration that he had before

both rockets have been turned off

no forces are acting on the rocketman

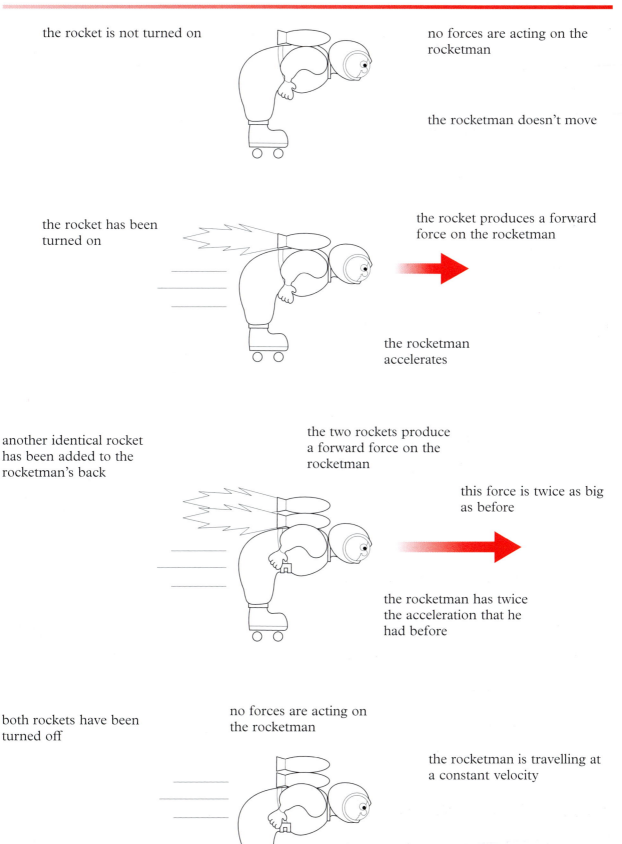

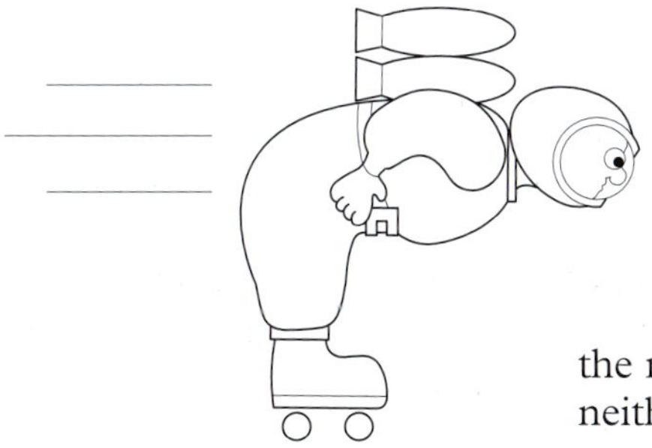

the rocketman is travelling at a constant velocity

the rocketman is freewheeling, he is neither accelerating nor decelerating

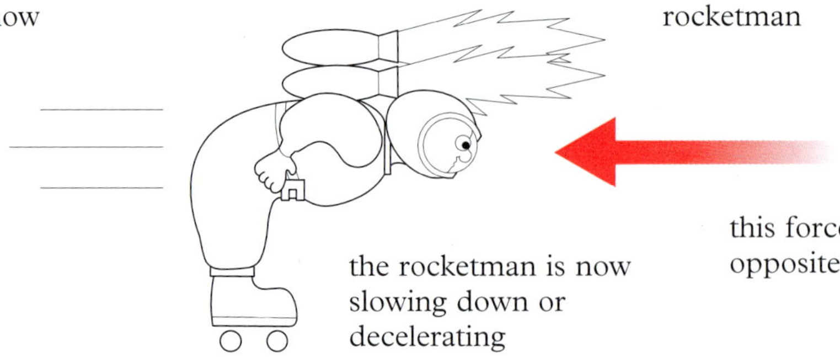

both rockets have now been reversed
the two rockets produce a backward force on the rocketman
this force is acting in the opposite direction to before
the rocketman is now slowing down or decelerating

We can summarise this in two statements:

- The velocity of an object **will not** change if acted upon by balanced forces.
- The velocity of an object **will** change if acted upon by unbalanced forces.

The size of the acceleration or deceleration depends upon the size of the unbalanced forces.

Questions

1 Assuming that there are no other forces present describe the motion of this garden roller if the man

a) pushes it with a force of 100 N
b) pushes it with a force of 300 N
c) pushes it with a force of 100 N and then releases it (applies no force)
d) pushes it with a force of 300 N and then after several seconds pulls with a force of 200 N.

Effects of Friction

Friction is a force. It is present whenever an object moves or tries to move. Friction always acts in a direction which opposes the movement of an object. As an object moves more quickly, the friction forces opposing it increase. Friction forces depend on the speed of the object.

this man is pulling this crate to the left

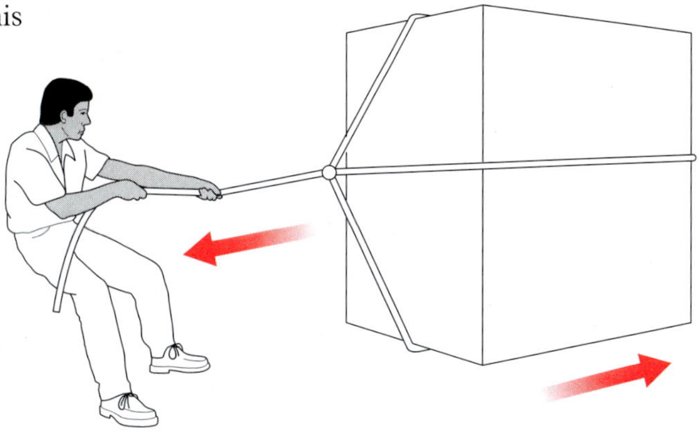

frictional forces between the crate and the floor make it harder for the man to move the crate

frictional forces act in the opposite direction to the motion

Friction can be helpful

On some occasions friction can be very useful. To increase the friction between two surfaces we should make the surfaces rough and dry so that they can 'grip'.

this climber needs a lot of friction between his hands and the rope if he is not to slip

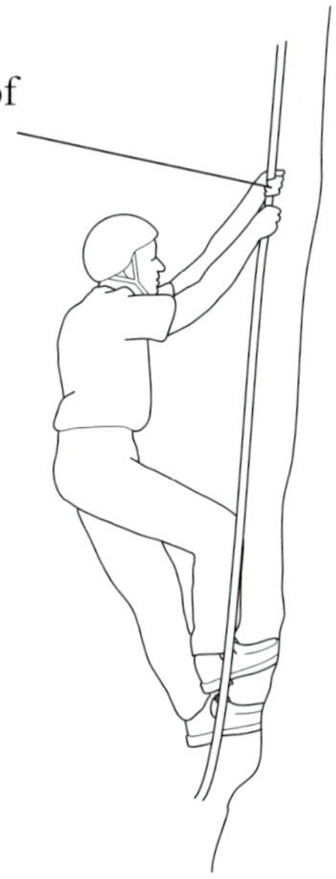

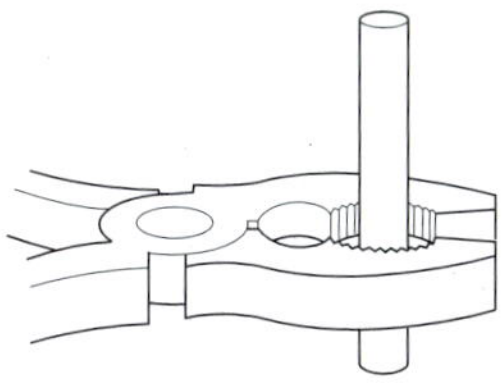

the friction between the jaws of these pliers and the rod allows the rod to be held without slipping

Friction can be unhelpful

On some occasions we may want friction to be as small as possible.

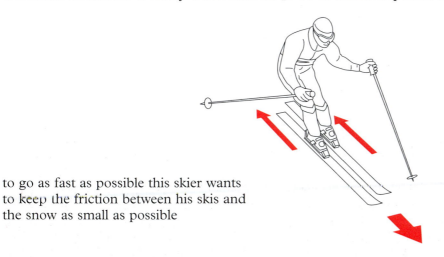

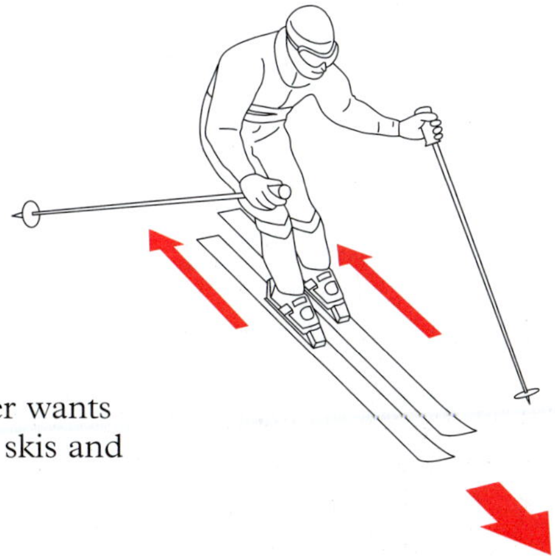

to go as fast as possible this skier wants to keep the friction between his skis and the snow as small as possible

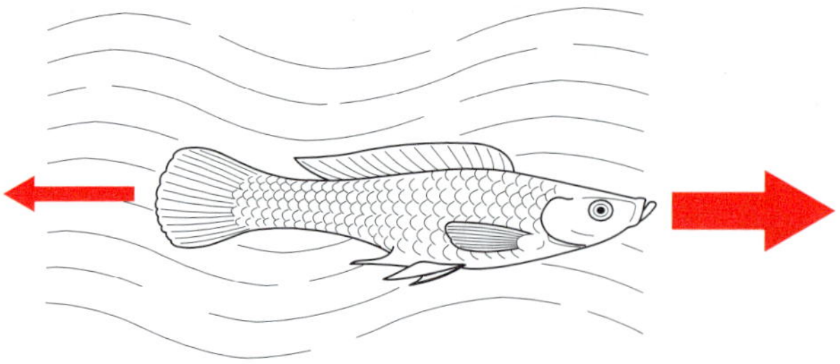
as fish swim there is friction between them and the water
to keep friction small, fish have a smooth streamlined shape

To decrease the friction between two surfaces we should make the surfaces as smooth as possible and add a **lubricant** such as oil. Friction between two surfaces will produce heat and may cause one or both surfaces to wear away. A lubricant helps to reduce this heat and wear. This is why it is important to lubricate machinery.

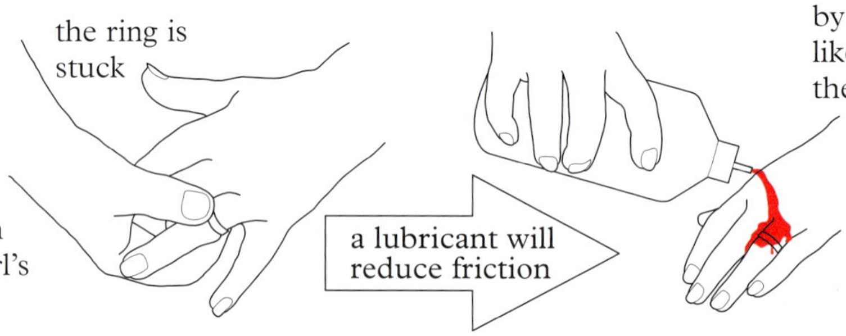
the ring is stuck
by adding a lubricant like washing-up liquid the ring will slip off
the friction between the ring and this girl's finger is too great
a lubricant will reduce friction

Questions

1 What is friction and what does it do?

2 Describe two situations where it is important to have friction.

3 Describe two situations where it is important to keep frictional forces as small as possible.

4 Describe two ways of increasing the friction between two surfaces and two ways of decreasing the friction between two surfaces.

Friction and the motor car

In order to control a car, that is change its speed or direction, there must be friction between its tyres and the road surface. A new tyre with lots of tread has a rough surface and will provide lots of grip even on a wet surface (photograph on the left). A smooth, worn tyre will provide less grip and the car will be much more difficult to control (photograph on the right).

Having tyres which are in good condition is important if you need to stop quickly but there are several other important factors which will also affect how quickly a driver can stop. These other factors are:

1. the reactions of the driver
2. the force applied by the braking system
3. the speed of the car
4. the weather/road conditions

The diagrams below show how the shortest possible stopping distances under ideal conditions increase dramatically with speed.

at 13 m/s (30 mph)

total stopping distance 23m

thinking distance 9m braking distance 14m

at 22 m/s (50 mph)

total stopping distance 53m

thinking distance 15m braking distance 38m

at 30 m/s (70 mph)

thinking distance 21m braking distance 75m total stopping distance 96m

Questions

1. Name four things which affect the stopping distance of a car.
2. How might the weather conditions affect the stopping distance?
3. If a driver is tired or has been drinking alcohol how might this affect his stopping distance?
4. Why is it important to have speed limits in towns and villages?

Terminal velocity

When the driver of this stationary car presses the accelerator pedal, the car accelerates.

the frictional forces are small

the force driving the car forwards is large

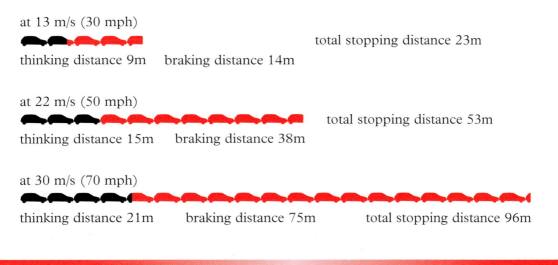

there is a large difference between the forces and so there is a large acceleration

As the car's speed increases so do the frictional forces.

the frictional forces due to **air resistance** are increasing

the same force is driving the car forwards

the difference in the forces is not so large so the acceleration is smaller

Eventually the frictional forces become as large as the propulsive forces from the engine.

the frictional forces are equal to the driving force

the same force is driving the car forwards

the forces are the same and so the car is no longer accelerating

This maximum constant velocity is called the **terminal velocity** of the car.

To travel faster than this speed we could replace the engine with one which produces a larger propulsive force or we could make the car more **streamlined**. A more streamlined car lets the air pass over it more easily, which decreases air resistance.

frictional forces are reduced because the car is more streamlined

a more streamlined car can accelerate to higher speeds before the friction forces equal the driving force

the car accelerates to a higher terminal velocity

This velocity-time graph shows a car accelerating from rest. As its velocity increases the graph gets less steep. This shows that the car's acceleration is decreasing.

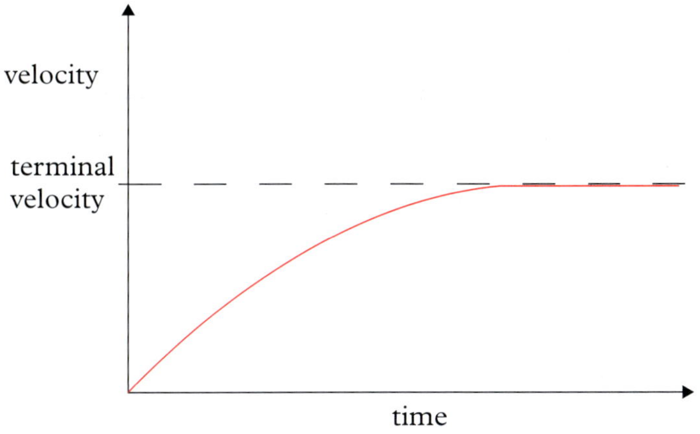
velocity
terminal velocity
time

Acceleration due to gravity

If we drop an object such as an apple from the top of a tall tower, its velocity increases as it falls. This acceleration is caused by gravitational forces which are pulling the apple downwards. Assuming there are no other forces present, the acceleration due to gravity on the surface of the Earth is the same for all objects and has the value $9.81\,m/s^2$. In real life, as an object falls it will experience frictional forces due to air resistance.

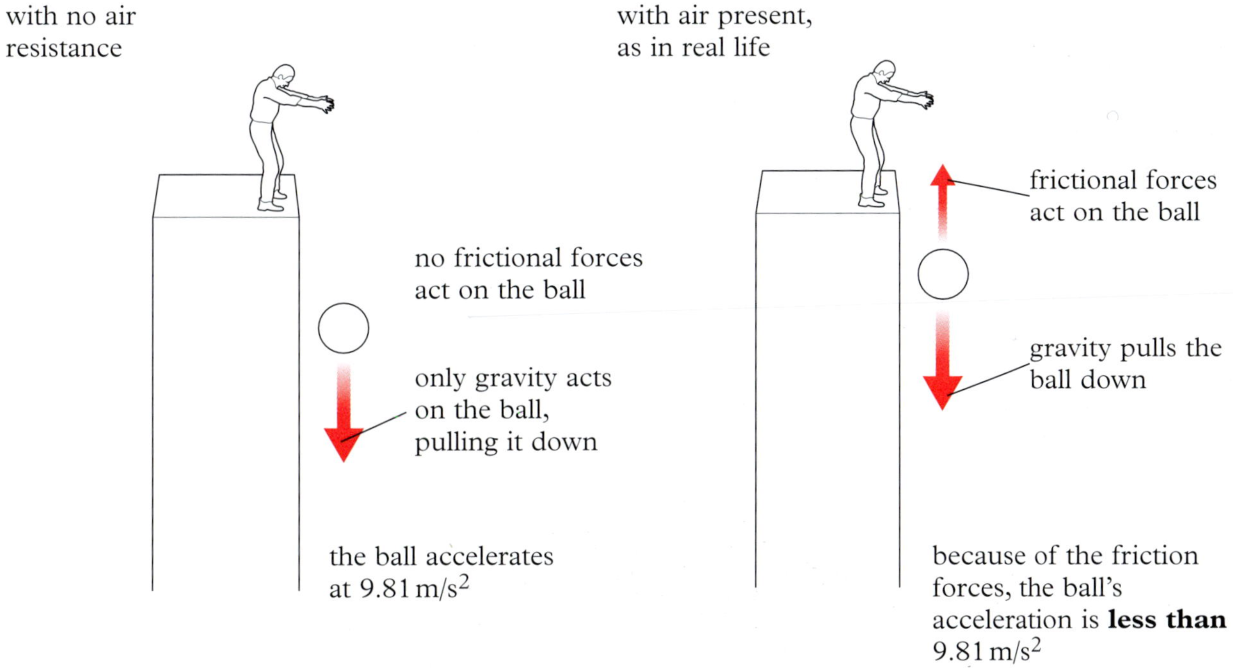
with no air resistance
no frictional forces act on the ball
only gravity acts on the ball, pulling it down
the ball accelerates at 9.81 m/s²
with air present, as in real life
frictional forces act on the ball
gravity pulls the ball down
because of the friction forces, the ball's acceleration is **less than** 9.81 m/s²

Friction forces will reduce this ball's acceleration. The faster the ball falls the larger these frictional forces become until eventually the ball no longer accelerates. It has reached its terminal velocity.

Parachutists use air resistance to prevent them from falling too fast.

when a parachute is first opened, the air resistance is larger than the pull due to gravity
the parachutist decelerates

As his speed decreases, the air resistance lessens.

once the parachute is open the parachutist decelerates
after a few seconds air resistance balances the pull of gravity
the downward force (gravity) equals the upwards force (air resistance)
the parachutist has reached his terminal velocity

When the downward force and the frictional forces are balanced, his speed remains constant. A parachute is designed so that its terminal velocity is low enough to allow the parachutist to land safely.

Questions

1 What is air resistance?

2 Give one situation when it is useful to have air resistance and one situation when it is a disadvantage to have air resistance.

3 How could you decrease the air resistance of an object?

4 Sketch a Velocity-Time graph for a cushion dropped from a high tower.

5 What is meant by the following sentence? *The car has a terminal velocity of 150 km/h.*

6 Suggest two ways in which the terminal velocity of this car could be increased.

Summary

If the forces acting upon a body do not cancel each other out there will be a resultant force. This force will cause the object to accelerate. The larger the resultant force the greater the acceleration of the object.

Friction is a force which acts when two surfaces slide over each other or an object moves through air, or a liquid such as water. To increase the frictional forces between objects their surfaces should be dry and rough. To decrease the frictional forces their surfaces should be smooth and lubricated.

As an object accelerates through air or water the frictional forces increase until they are equal to the accelerating forces. Because the forces are balanced the object will then travel at a constant velocity called the terminal velocity.

Key words

air resistance Frictional forces experienced by an object as it travels through air.

balanced forces Forces which, when combined, cancel each other out producing no resultant force and therefore no acceleration.

friction A force which opposes motion.

lubricant A substance such as oil or soap which is used to reduce friction between two surfaces.

streamlined When an object is shaped so that it can move through air or a liquid with little friction.

unbalanced forces Forces which, when combined, do not cancel each other out. They produce a resultant force which will cause an object to accelerate.

End of Chapter 6 Questions

1 The sky-diver shown below jumps from a stationary hot-air balloon. The arrows X and Y show the forces acting on the sky-diver as she falls.

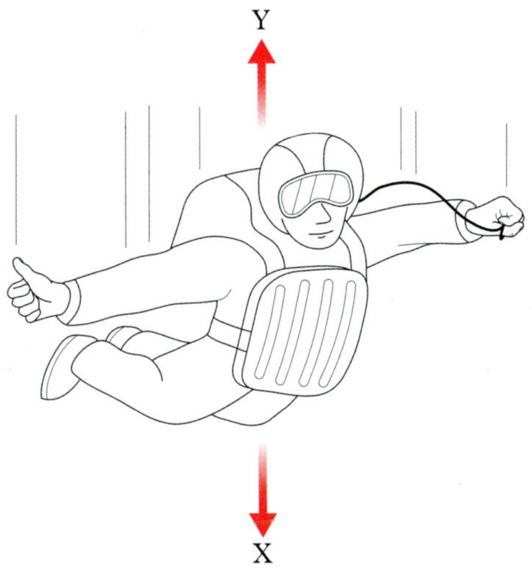

a) Name the forces X and Y.
b) i) Explain why the sky-diver accelerates immediately after leaving the balloon.
   ii) Explain why the sky-diver travels at a constant speed some time later.
   iii) What name is given to this constant speed?
c) How could she increase her speed?
d) What happens to each of the forces X and Y and the motion of the sky-diver when she opens her parachute?

2 The table below shows how the speed of a free-falling sky-diver changed with time.

Time in s	Speed in m/s	Time in s	Speed in m/s
0	0	6	50
1	10	7	52
2	20	8	52
3	30	9	52
4	38	10	52
5	44		

a) Plot a graph of speed against time
b) What is happening to the sky-diver during the first 7 s of his fall?
c) What is happening to the sky-diver 8 s after he has started his jump?
d) What will happen to his speed when he opens his parachute?

3 It is important that a driver knows the stopping distance of his car. The Highway Code tells us that the total stopping distance has two parts. The thinking distance and the braking distance.

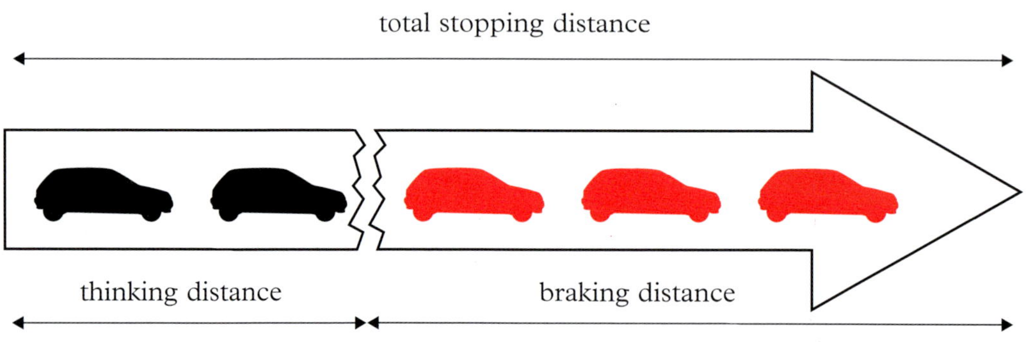

thinking distance + braking distance = total stopping distance

a) Name two things which could affect the thinking distance.
b) Name three things which could affect the braking distance.
c) A driver's reaction time is 0.6 s. During this time the car has travelled 24 m. Calculate the speed of the car.
d) The driver brakes from this speed and brings the car to rest after 10 s. Calculate the deceleration of the car.

4 The diagrams below show a steel spring before any force is applied to it and then after a force of 30 N is applied to it.

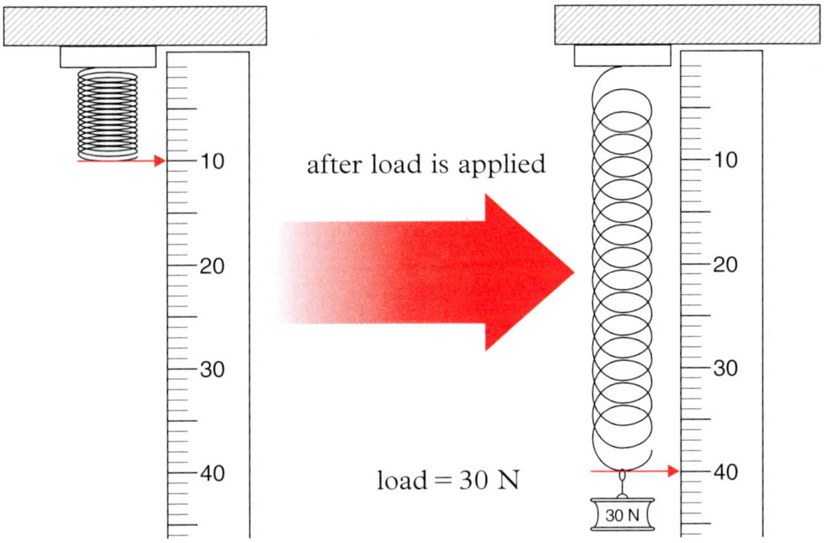

a) What is the extension of the spring caused by the 30 N force?
b) If the spring is obeying Hooke's Law what would be the extension of the spring if a force of 20 N was applied to it?
c) What would be the extension of the spring if a force of 10 N was applied to it?
d) What would happen to the spring if a force larger than the elastic limit is applied to it?
e) Sketch a graph of extension against force for a spring which is extended beyond its elastic limit.

5 The diagram below shows the four forces experienced by an aeroplane in flight.

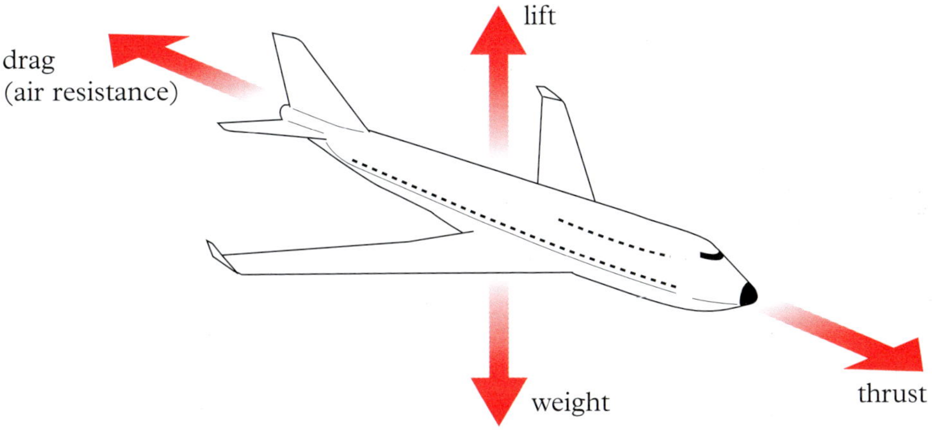

Describe the sizes of these forces in the following situations.

a) i) The aeroplane is accelerating and gaining height.
ii) The aeroplane is in level flight and flying at a constant velocity.
iii) The aeroplane is decelerating and losing height.

b) All modern aeroplanes are streamlined. Explain the meaning of the word **streamlined**.

7 Turning Forces: Moments

Sometimes when we apply a force to an object the object will turn or rotate. This turning effect of a force is called a **moment**. The diagrams below show moments created by forces.

your fingers create moments which turn the tap

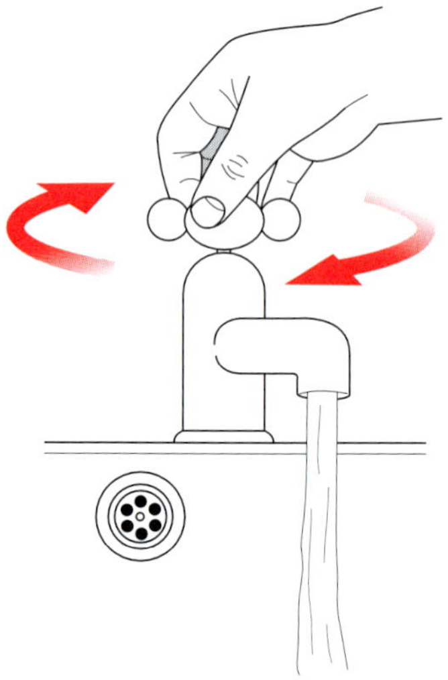

pressing on the pedals creates a moment which turns the cogs and chain

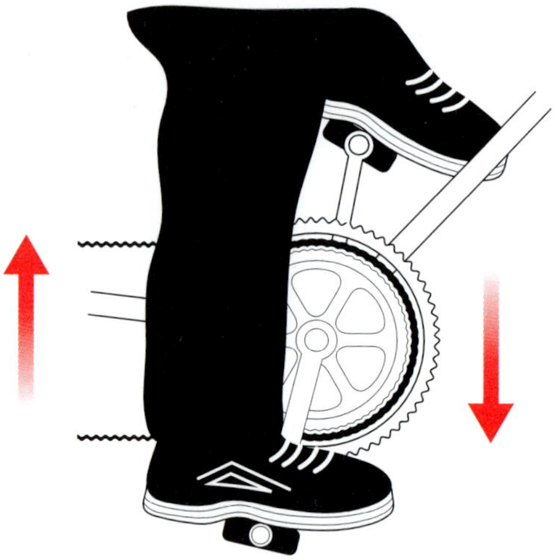

pushing with one hand and pulling with the other creates moments which turn the steering wheel

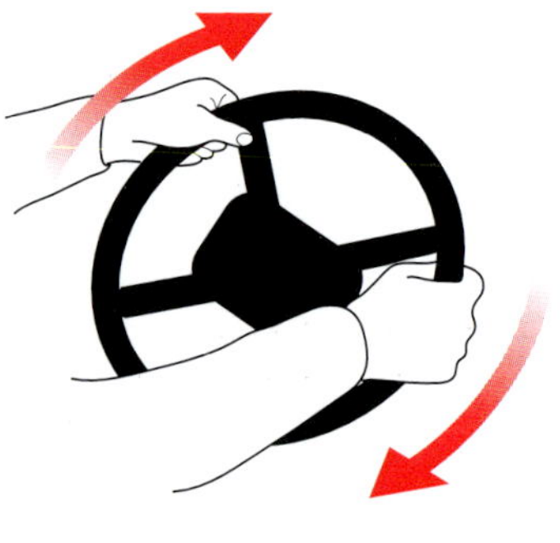

lifting the handles creates a moment which rotates the wheelbarrow around the wheel

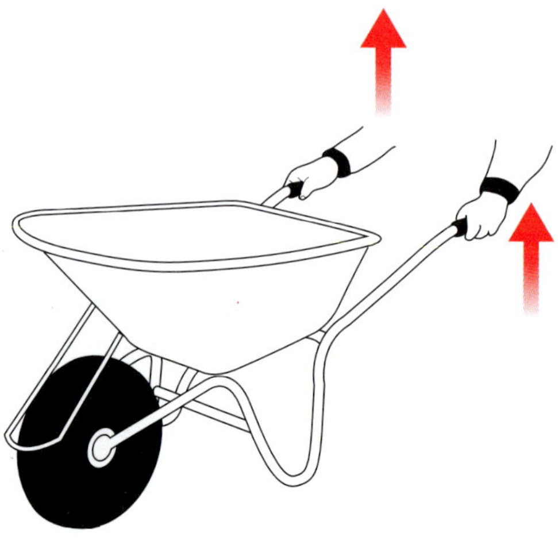

Questions

1 Which of the following activities uses a force to create a moment?
a) pedalling a bike
b) closing a car door
d) making a snowball
e) making a snowman
f) opening a book
g) pushing a sledge
h) winding up a clockwork toy.

The size of a moment depends upon two things:

1 The size of the force. The larger the force the larger the moment.
2 The place where the force is applied.

The direction of the turning effect depends on the direction of the force.

The point around which the force is turning is called the **pivot**.

We can calculate the size of a moment using the equation:

moment of force = force × distance from pivot

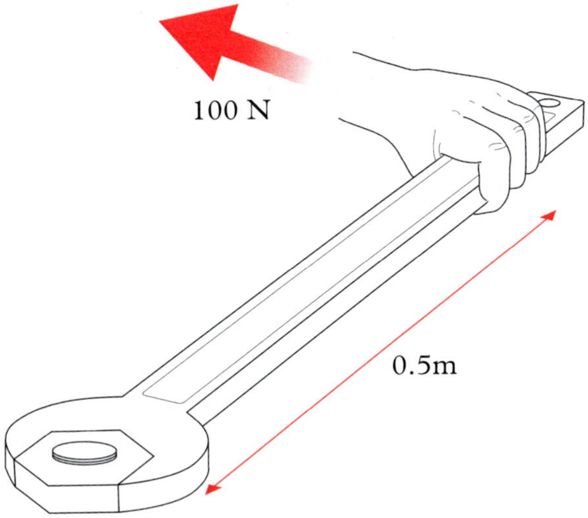

the moment created by the long spanner is $100\,N \times 0.5\,m = 50\,Nm$

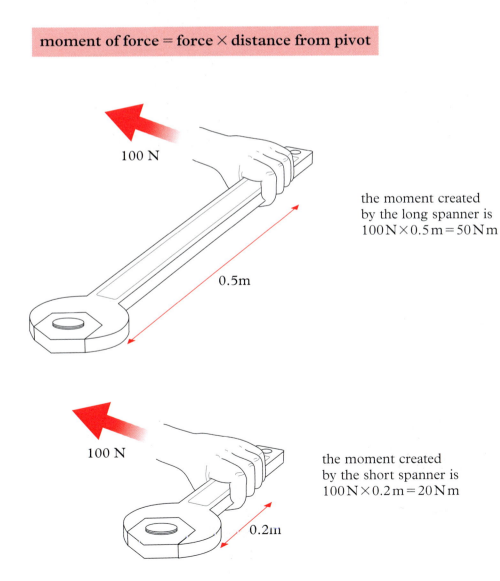

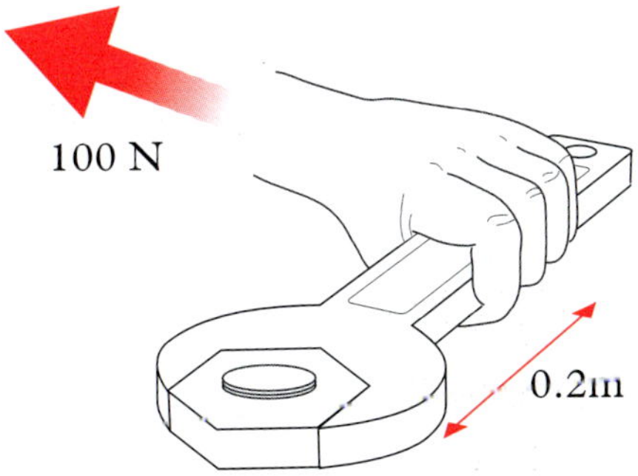

the moment created by the short spanner is $100\,N \times 0.2\,m = 20\,Nm$

We can see from these two examples that if we apply a force a long way from a pivot we will create a large moment. The larger the moment the easier it is to undo a nut. This is why it is much easier to undo a stiff nut using a long spanner rather than a short one.

Questions

1 Calculate the moment created in each of the diagrams shown below.

(a)

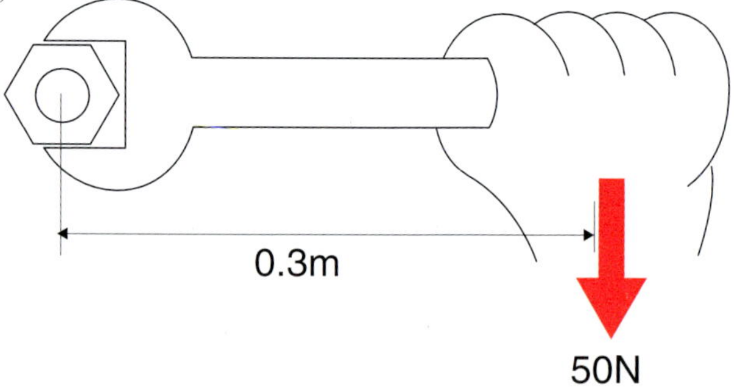

(b)

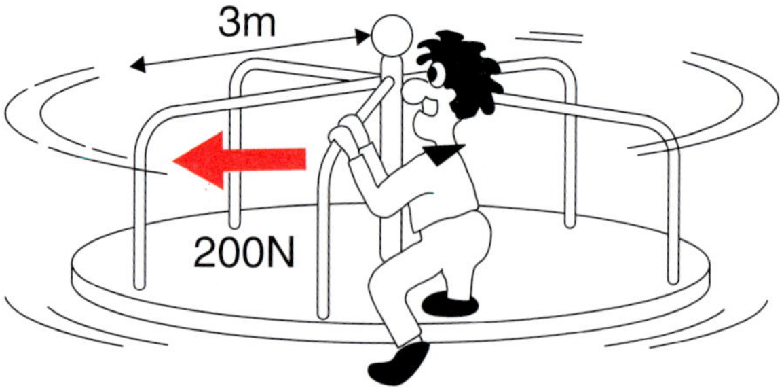

(c)

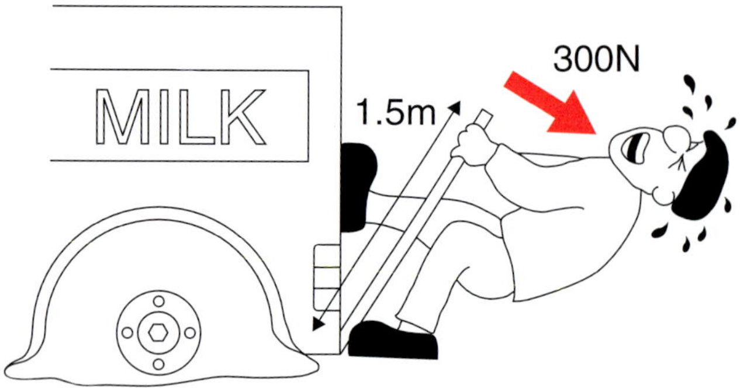

2 Which of these two spanners would you use to undo a stiff nut? Explain your choice.

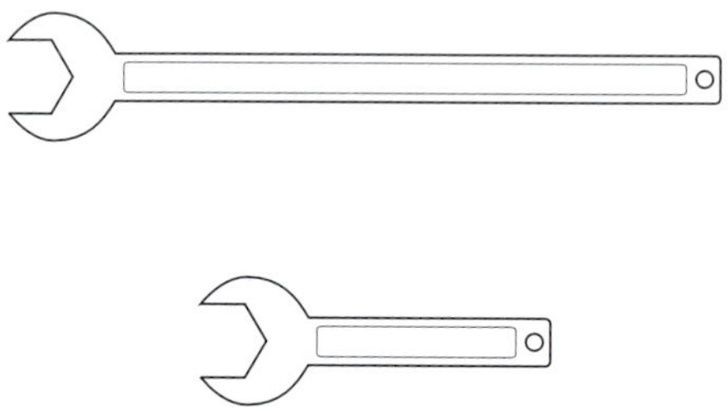

3 Where would you position a handle on this door? Explain your answer.

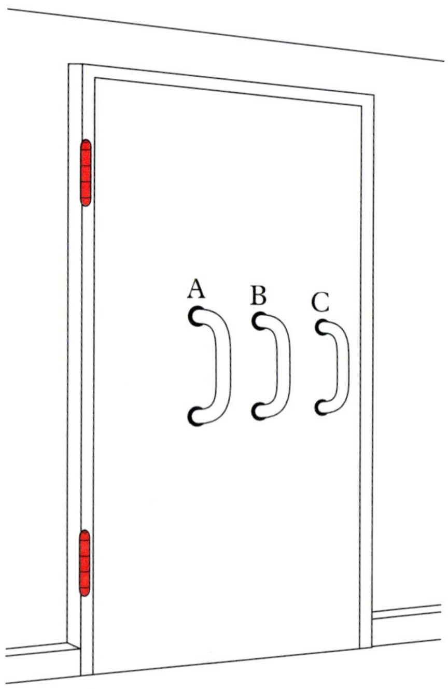

Balancing moments

See-saws show how moments can be balanced.

when this girl sits on the right-hand side of the see-saw her weight creates a moment which tries to turn the see-saw clockwise

this boy's weight on the left-hand side of the see-saw creates a moment which tries to turn the see-saw anticlockwise

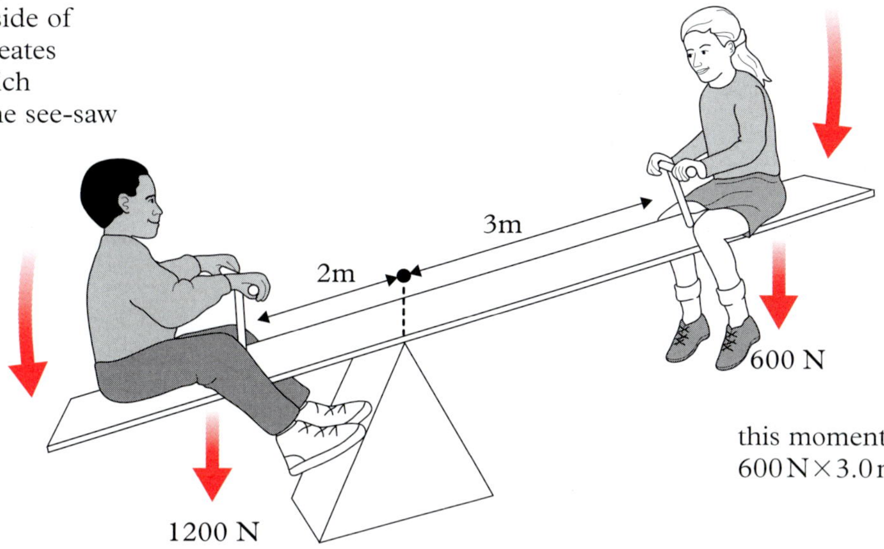

this moment is
600 N × 3.0 m = 1800 N m

this moment is
1200 N × 2.0 m = 2400 N m

The moments are unbalanced. The anticlockwise moment is larger than the clockwise moment.

If the boy moves forward so he is 1.5 m from the centre, the new anticlockwise moment = 1200 N × 1.5 m = 1800 Nm.

if the boy moves forward, the new anticlockwise moment = 1800 N m

the clockwise moment stays the same

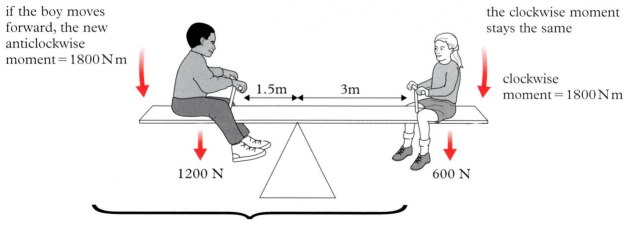

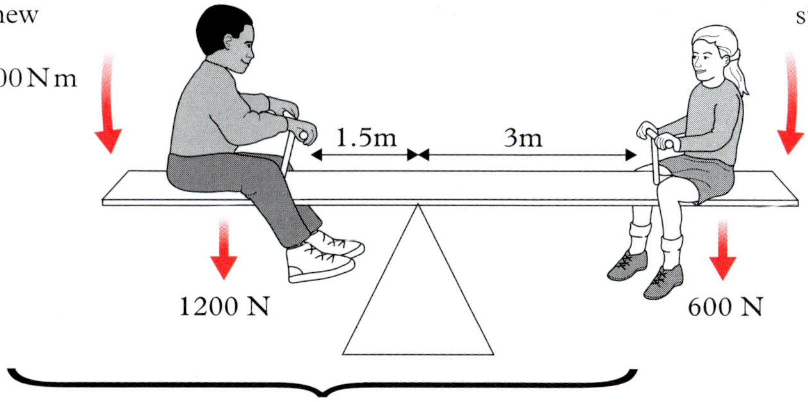

clockwise moment = 1800 N m

anticlockwise moment = clockwise moment

Because the anticlockwise moment equals the clockwise moment, the see-saw is balanced. This is called the **principle of moments**.

End of Chapter 7 Questions

1 Which of the see-saws shown below are balanced?

(a)

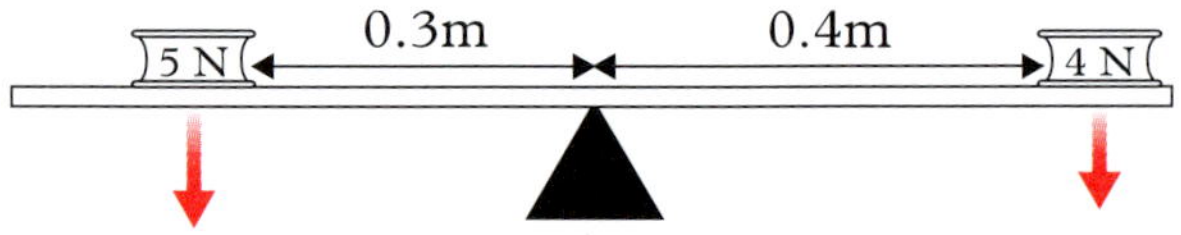

(b)

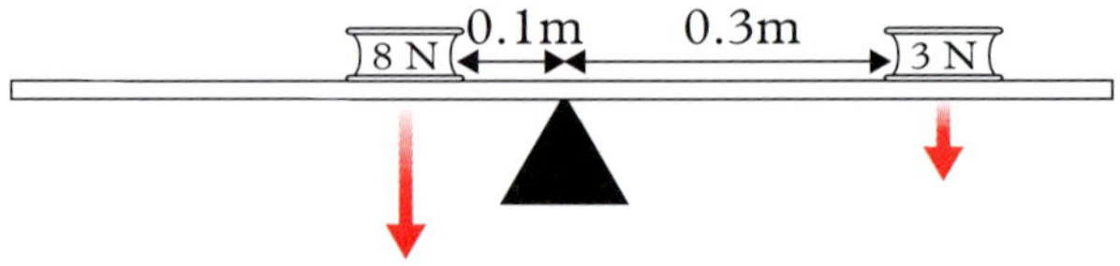

(c)

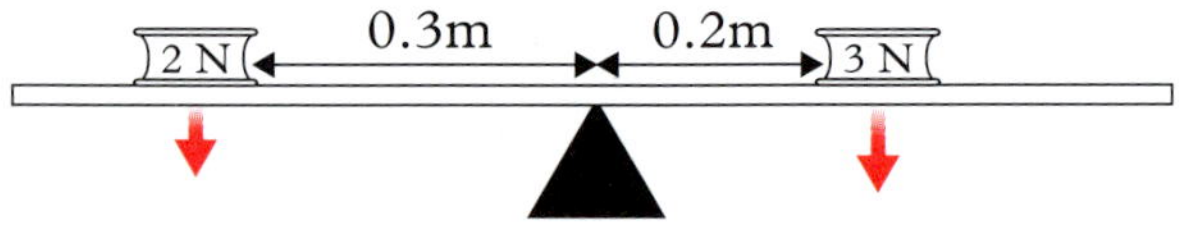

(d)

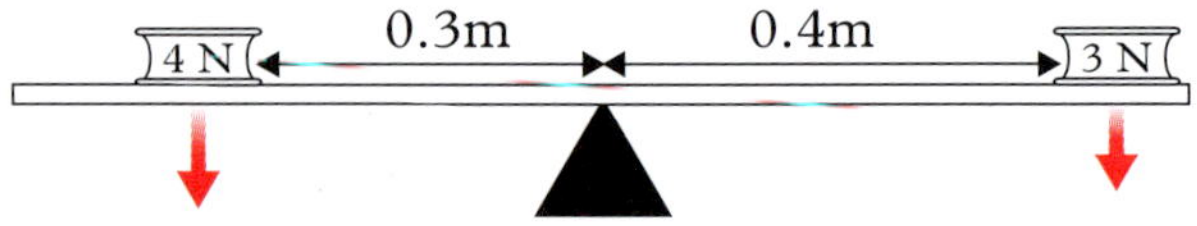

2 If the moments created by these arm-wrestlers are balanced, calculate the value of the force X.

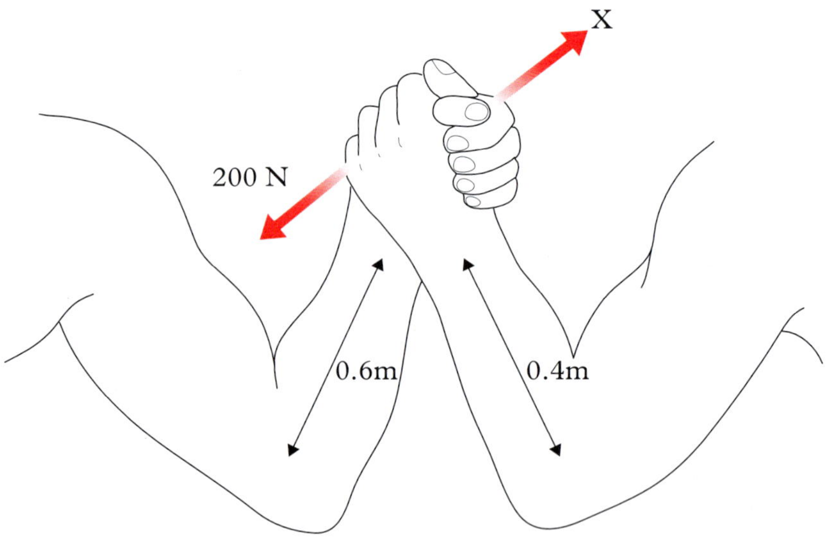

8 Pressure

if you were to stand on someone's foot while wearing stiletto heels, you would cause them considerable pain as all your weight would be concentrated in a small area

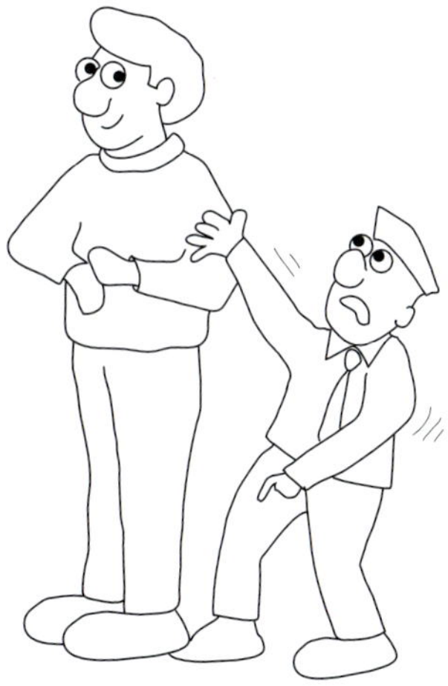

if however you were to stand on someone's foot while wearing hiking boots, you would cause them considerably less pain as your weight would be spread over a large area

if a force is concentrated into a small area it creates a large pressure

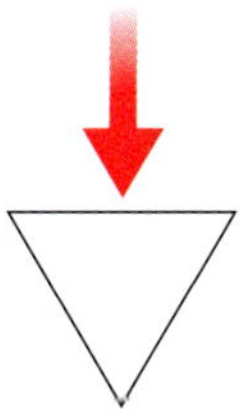

if a force is spread over a large area it creates a small pressure

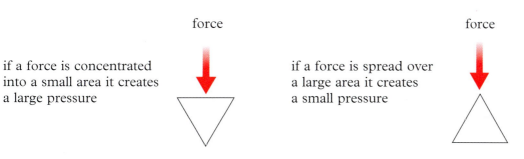

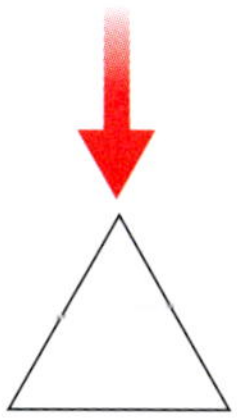

Examples of large and small pressures

snowshoes reduce the pressure due to your weight

if you wear snowshoes whilst walking on ice your weight will be spread over a large area and you won't sink

Without the snow shoes the force is more concentrated, i.e. the pressure is higher, and you are likely to sink.

you should never walk across a frozen pond or lake

the pressure your weight creates may be sufficient to crack the ice

Rescuers can avoid cracking the ice by using a long ladder which spreads their weight over a larger area and so reduces the pressure they exert on the ice.

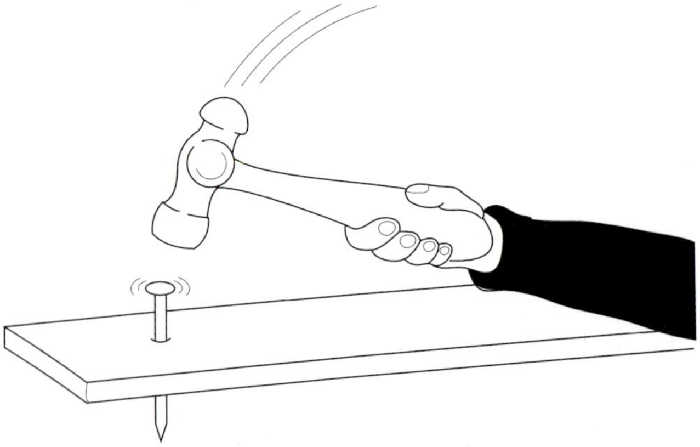

the pressure created at the point of this nail is large enough for the point to pierce the wood

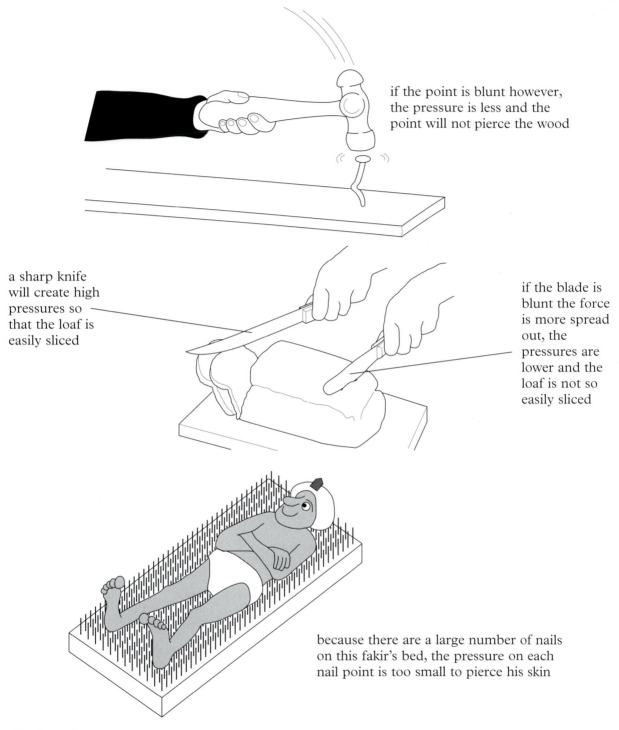

if the point is blunt however, the pressure is less and the point will not pierce the wood

a sharp knife will create high pressures so that the loaf is easily sliced
if the blade is blunt the force is more spread out, the pressures are lower and the loaf is not so easily sliced

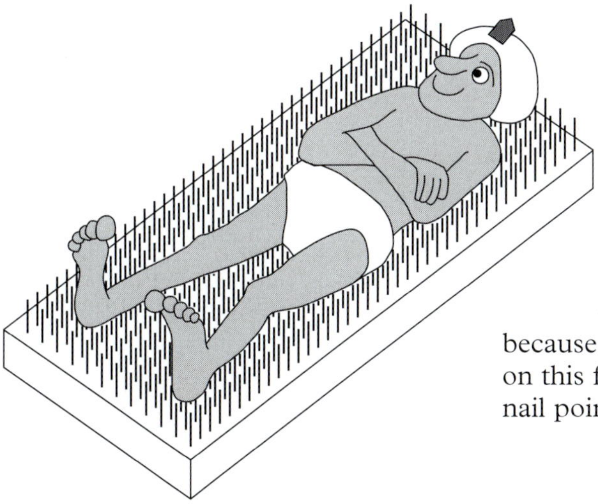
because there are a large number of nails on this fakir's bed, the pressure on each nail point is too small to pierce his skin

Calculating pressure

The pressure created by a force can be calculated using the equation

$$\textbf{pressure (p)} = \frac{\textbf{force (F)}}{\textbf{area (A)}}$$

Pressure is measured in **pascals (Pa)**. 1 Pa is the same as 1 N / m^2.

Example

A 1000 N weight is placed on a flat board whose area is 5 m^2. Calculate the pressure exerted on the floor by the board.

Using pressure = force / area

pressure = 1000 / 5

pressure = 200 Pa

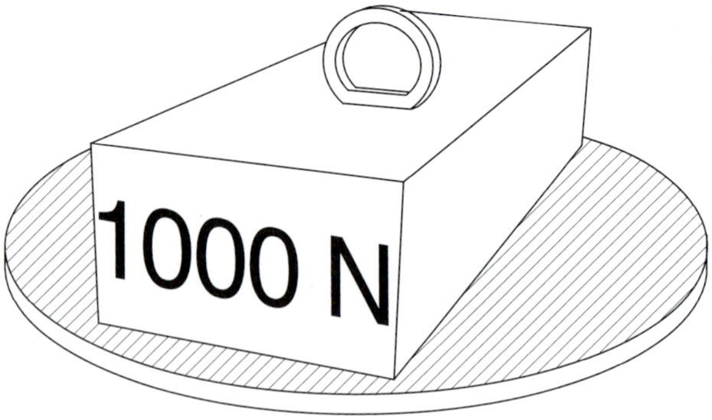

this skier's weight is 750 N

the total surface area of his skis is 0.25 m^2

p = F/A

p = 750/0.25

p = 3000 Pa

this elephant's weight is 20 000 N

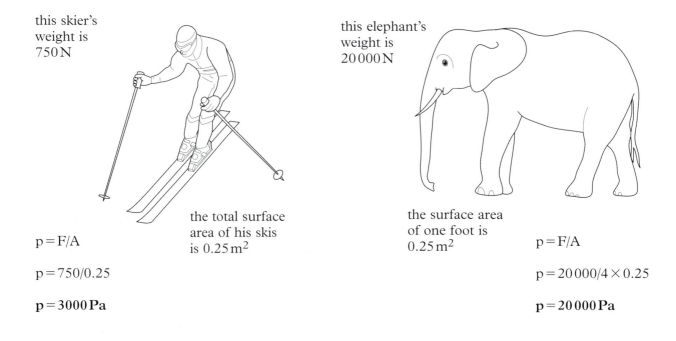

the surface area of one foot is 0.25 m^2

p = F/A

p = 20 000/4 × 0.25

p = 20 000 Pa

Questions

1. Explain in your own words the difference between force and pressure.
2. Why is a hiker walking over boggy ground more likely to sink if he stands on one foot rather than two?
3. Why is it easier to walk on an icy surface with spiked shoes rather than shoes with flat soles?
4. A crate weighing 5000 N is placed on a flat base which has an area of 2 m^2. Calculate the pressure exerted on the floor under the base.
5. A building weighing nine thousand million newtons (9 000 000 000 N) rests on foundations which are square and measure 30 m by 30 m. Calculate the average pressure the building exerts on the foundations.

Pressure in Liquids

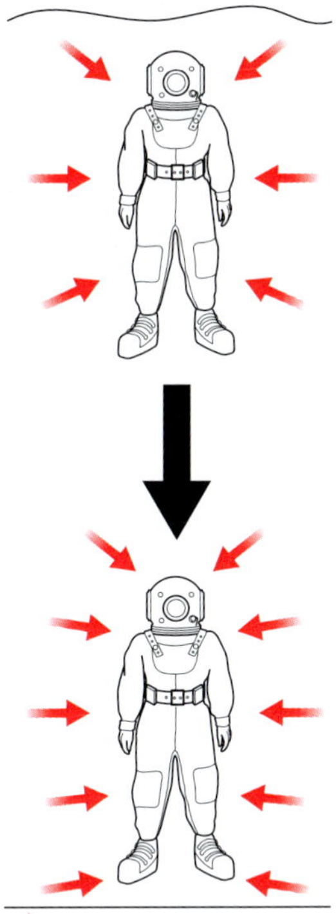
this deep sea diver
can feel the pressure
of the water around him
as he goes deeper
there is more water
above him so this
pressure increases

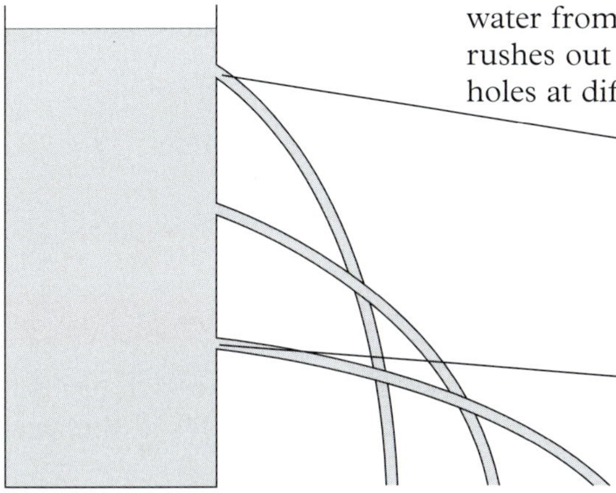
water from this can
rushes out of the three
holes at different rates
low pressure
where the pressure
is low, the water
flows out slowly
high pressure
where the pressure is
high, the water
flows out quickly

This simple experiment clearly shows that the pressure in a liquid increases with depth.

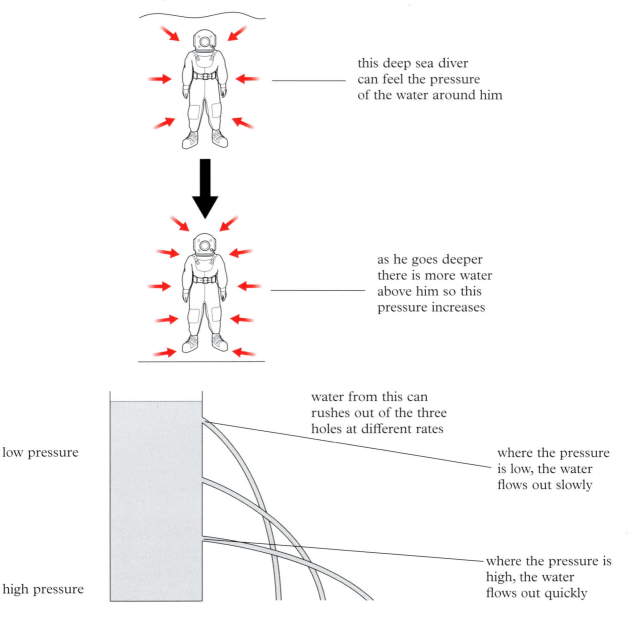

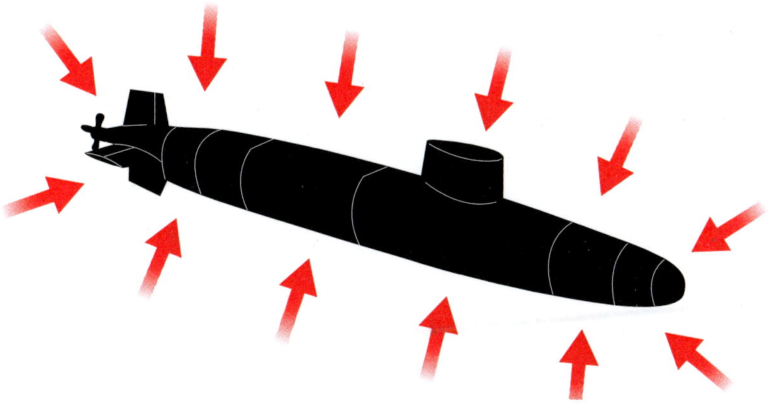
submarines
are designed
to go deep
below the
surface of
the sea
if a submarine goes
too deep there is a
danger that it will
be crushed by the
very high water
pressure

The pressure in a liquid acts in all directions. The water around the previous submarine tries to squash it from all sides and not just from above. Imagine taking a spherical balloon to the bottom of a deep swimming pool.

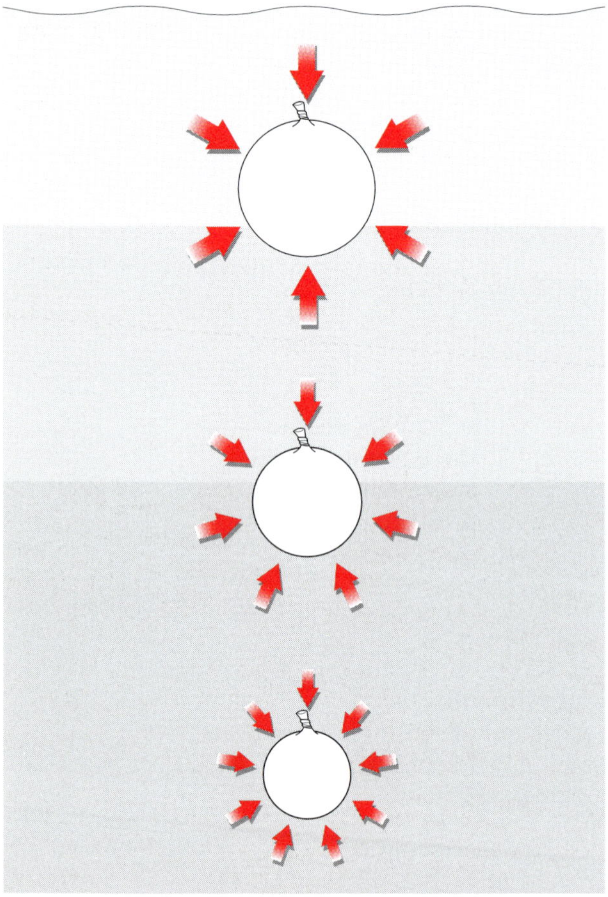

this spherical balloon becomes squashed as it is taken deeper underwater

the balloon remains spherical no matter how deep it goes

the balloon remains spherical because the pressure in a liquid is the same in all directions

Questions

1 Why is a submarine built to withstand large pressures?

2 Why are the walls of a dam thicker at the bottom than the top?

3 What happens to the shape of a spherical balloon as it is taken deeper and deeper in a swimming pool? Explain why this happens.

4 Why does water from this can rush out of all the holes at the same rate?

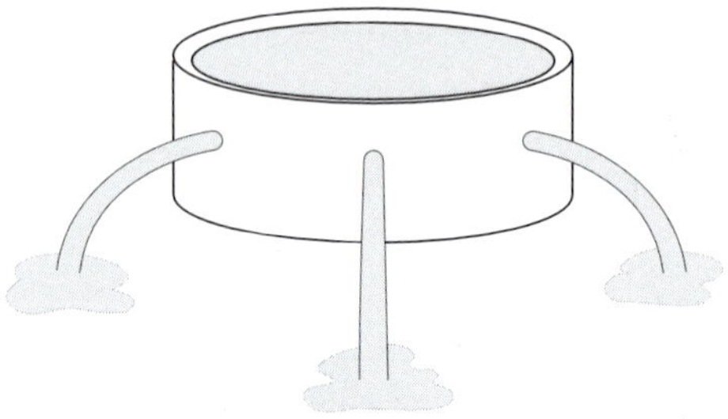

Making use of the pressure in a liquid

Liquids can be used to transmit a force over a distance. Liquids can also alter the size of this force. A machine or system which does this is called a **hydraulic machine** or a **hydraulic system**.

The Hydraulic Jack

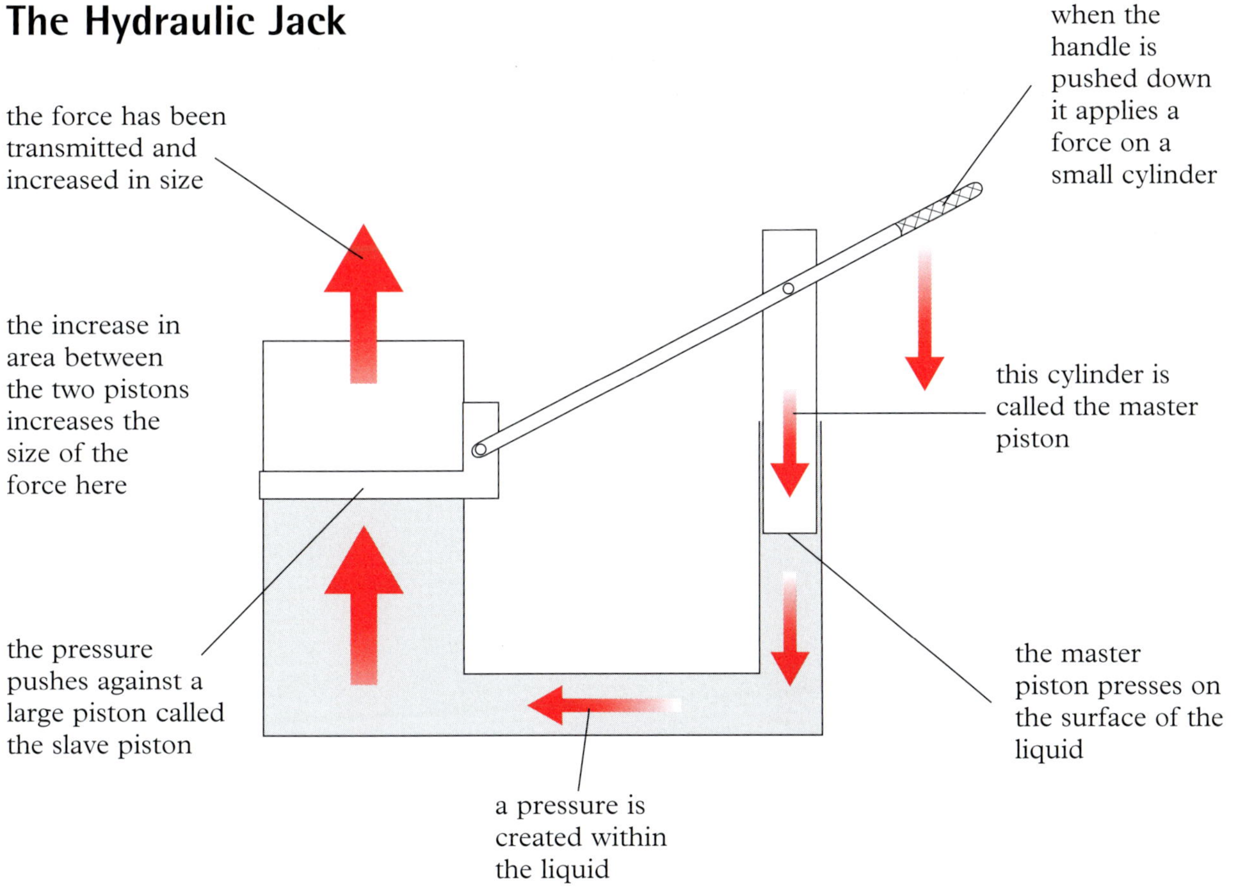
when the handle is pushed down it applies a force on a small cylinder
the force has been transmitted and increased in size
the increase in area between the two pistons increases the size of the force here
this cylinder is called the master piston
the pressure pushes against a large piston called the slave piston
the master piston presses on the surface of the liquid
a pressure is created within the liquid

The larger force created here can be used to lift very heavy objects, like cars. The jack increases the size of the applied force. It is a **force multiplier**.

Hydraulic Brakes

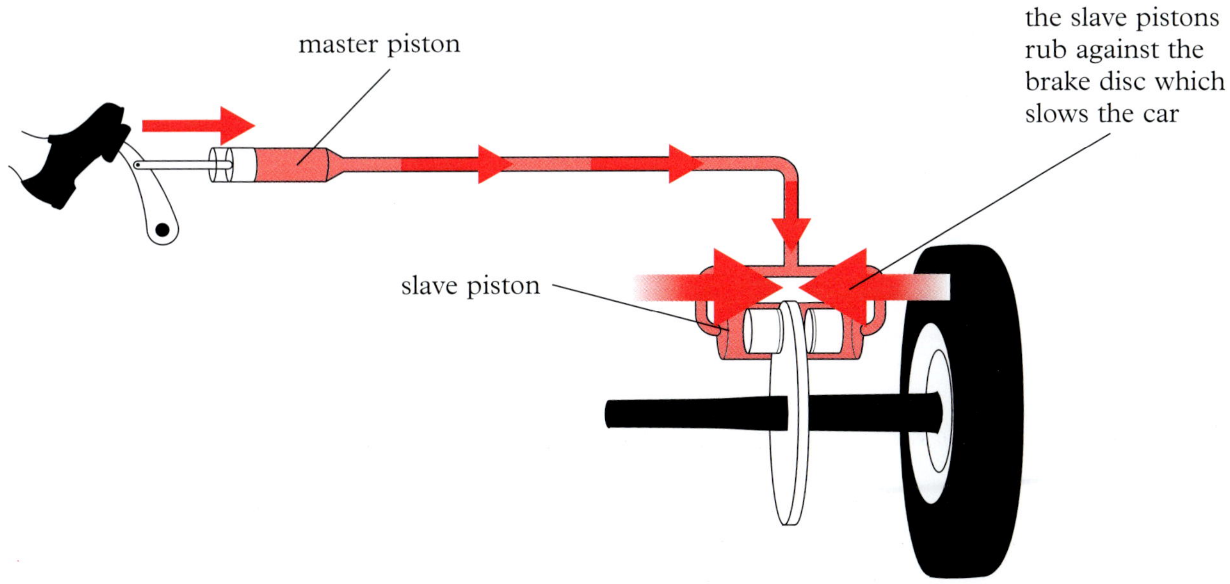
master piston
the slave pistons rub against the brake disc which slows the car
slave piston

The hydraulic brakes of a car work in a very similar way. When the driver presses the brake pedal a force is sent to where it is needed (to the four wheels) and the size of the force is increased to make braking easier for the driver.

Pressure in gases

Like liquids, gases exert a pressure. The air particles around us are continually bumping into us and exerting a pressure we call **atmospheric pressure**.

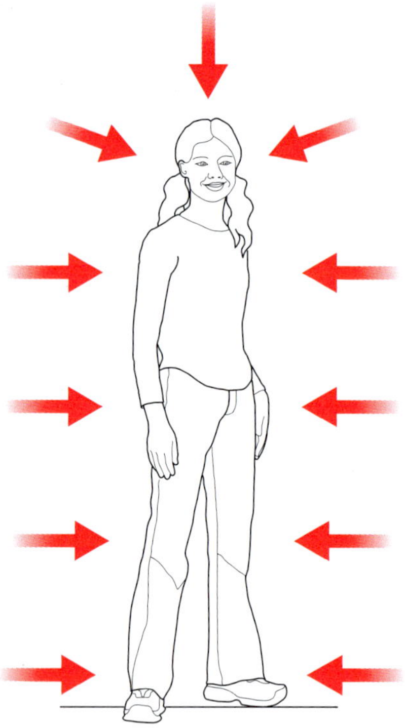
air pressure surrounds every part of our body, all the time

We are so used to this pressure that we are rarely aware it is there but it is easy to show that it exists.

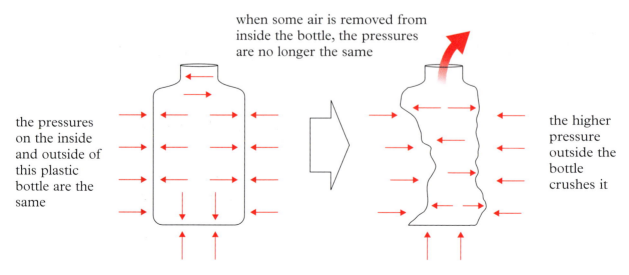

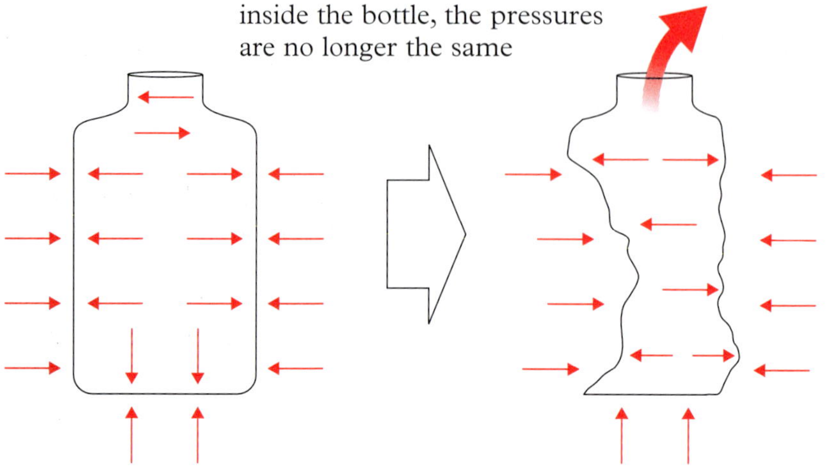
when some air is removed from inside the bottle, the pressures are no longer the same
the pressures on the inside and outside of this plastic bottle are the same
the higher pressure outside the bottle crushes it

Heating gases

The pressure of a gas is related to how warm it is.

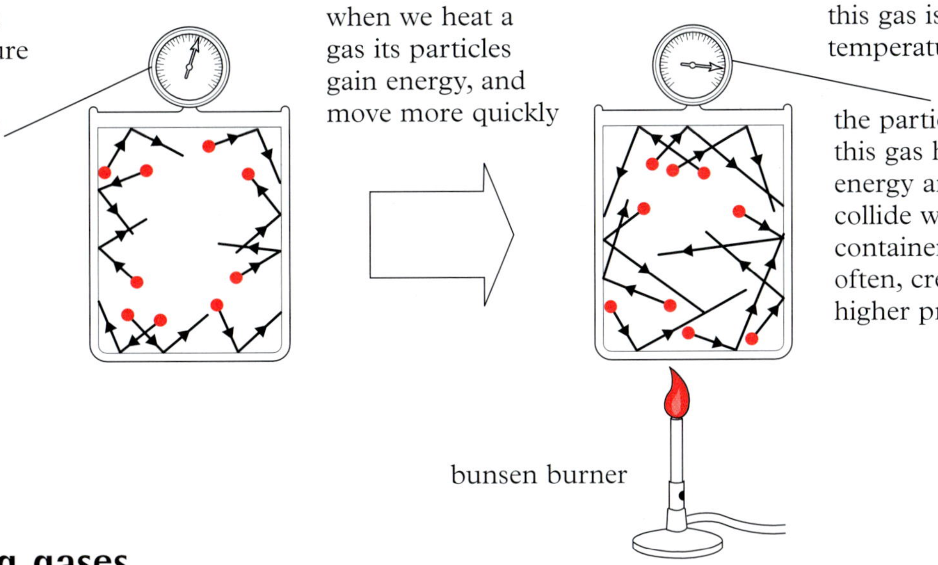
this gas is at a low temperature
this gas is at a low pressure
when we heat a gas its particles gain energy, and move more quickly
this gas is at a high temperature
the particles in this gas have more energy and so collide with the container more often, creating a higher pressure
bunsen burner

Squashing gases

If we squash a gas its pressure increases.

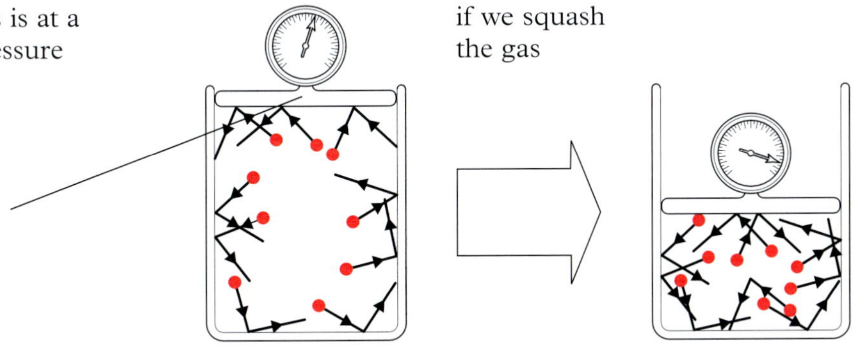
this gas is at a low pressure
piston
if we squash the gas
there is less room in the container so the gas particles collide with the container more often
because there are more collisions the gas is at a higher pressure

Questions

1 a) What is atmospheric pressure?
b) How is it created?

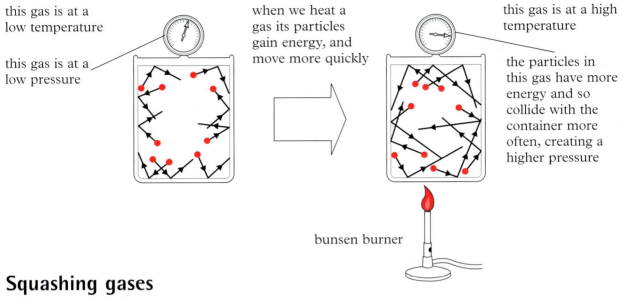

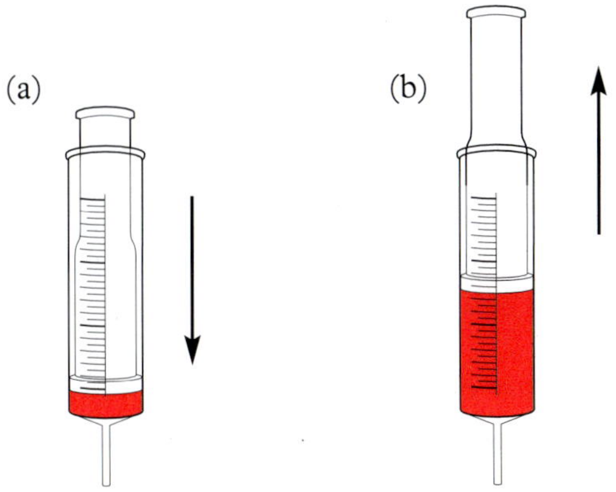
(a)
(b)

2 What would happen to the pressure of the gas in the syringe in the diagram if the piston was
a) pushed in
b) pulled out?

3 What would happen to the pressure in a cylinder of gas at room temperature if it was
a) placed in hot water
b) placed in melting ice?

Summary

When a force is applied to an object it may cause it to turn or rotate. This turning effect of a force is called a moment. If several moments act on an object they may create opposing moments. These will be in clockwise and anticlockwise directions. The Principle of Moments tells us that if the clockwise moments are equal to the anticlockwise moments the object will not rotate.

If a force is applied over a small area it creates a large pressure. If the same force is applied to a much larger area it creates a smaller pressure. We can calculate the size of the pressure created by a force using the equation

pressure = force / area.

The particles of a liquid apply pressure to any object which is placed in it. This pressure is exerted in all direction and increases with depth.

Gas particles also exert pressure. This pressure increases if the gas is heated or squashed.

Key words

atmospheric pressure	The pressure exerted on an object by air molecules.
force multiplier	A machine which increases the size of the applied force.
hydraulic brakes	A hydraulic system which is used to apply braking forces to all wheels of a car when the brake pedal is pressed.
hydraulic jack	An example of a hydraulic system which increases the size of the applied force and is used to lift heavy objects.
hydraulic system	A machine which uses liquids to send forces where they are needed.
moment	The turning effect caused by a force.
pivot	The point around which a force turns.
pressure	A measure of the force which is being applied to each square metre.
principle of moments	An object will not turn or rotate if the clockwise moments and the anticlockwise moments are equal.

End of Chapter 8 Questions

1 a) What is meant by the moment of a force?
   b) The diagram on the next page shows two children who have just climbed onto a see-saw.
      i) Calculate the moment trying to turn this see-saw clockwise.
      ii) Calculate the moment trying to turn the see-saw anticlockwise.
      iii) Will the see-saw stay balanced, tilt down to the left or tilt down to the right? Explain your answer.

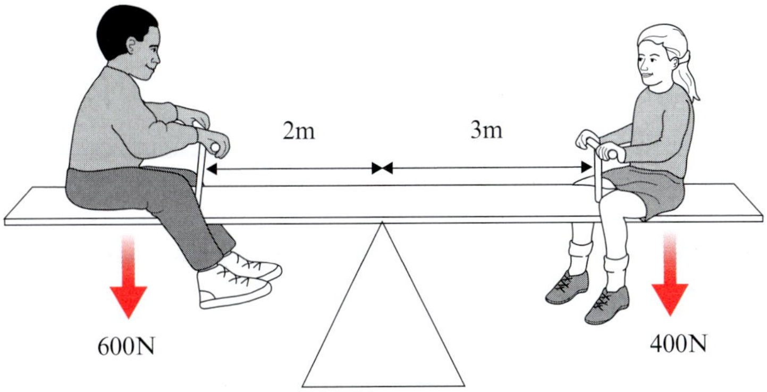

2 a) What is likely to happen if you press down on a drawing pin as shown in the diagram below?

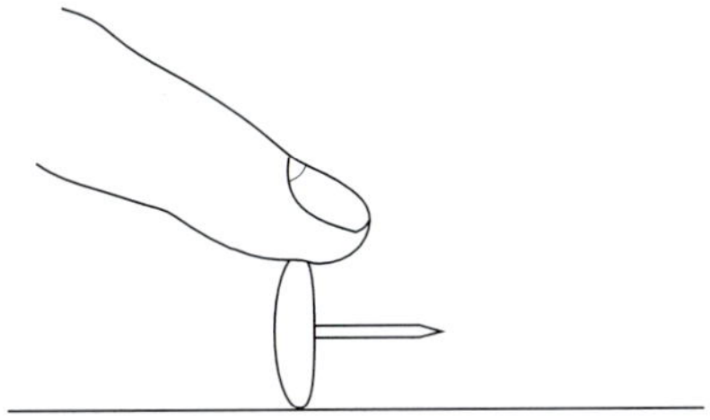

b) What is likely to happen if you press down on the drawing pin as shown in the diagram below?

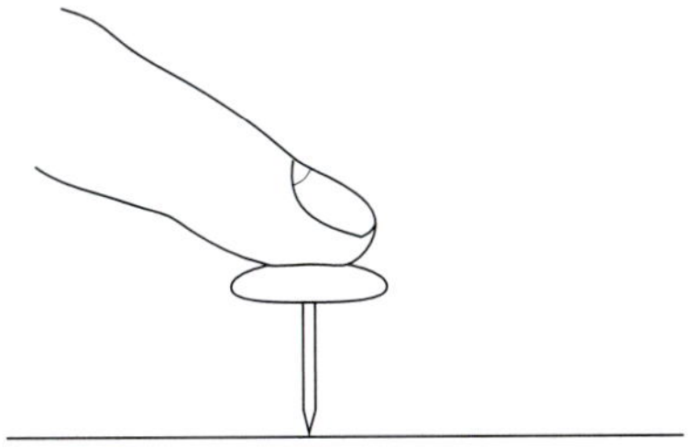

c) Explain the difference between these two situations.

3 The crate drawn below is 5 m high, 3 m wide and 2 m deep. It weighs 120 N.

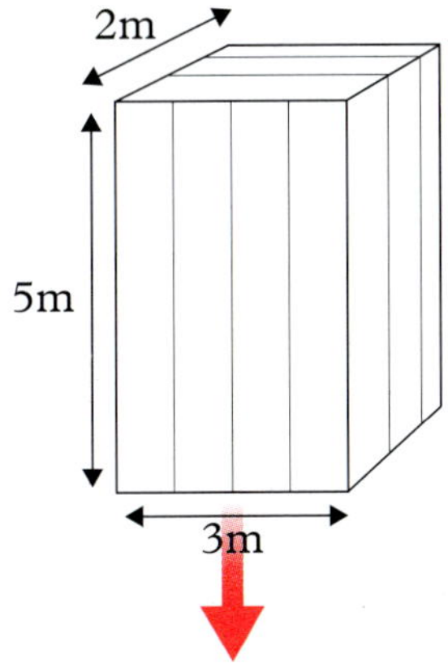

a) i) On which face should the crate stand if the pressure it exerts on the floor is to be kept to a minimum?
ii) Calculate the size of this pressure.

b) i) On which face should the crate stand if the pressure it is to exert on the floor is to be as large as possible?
ii) Calculate the size of this pressure.

4 A car weighing 9000 N is stationary. Its weight is evenly distributed on each of its four tyres. The area of each of the four tyres which is in contact with the road is $0.025\,m^2$.

a) Calculate the total area of tyre in contact with the road.

b) Calculate the pressure on the road due to the weight of the car.

c) What would happen to the pressure of the air in the tyre if the tyre became warm?

d) What would happen to the pressure in each of the tyres if a very heavy load was put into the car?

5 The diagram below shows the hydraulic brakes of a car

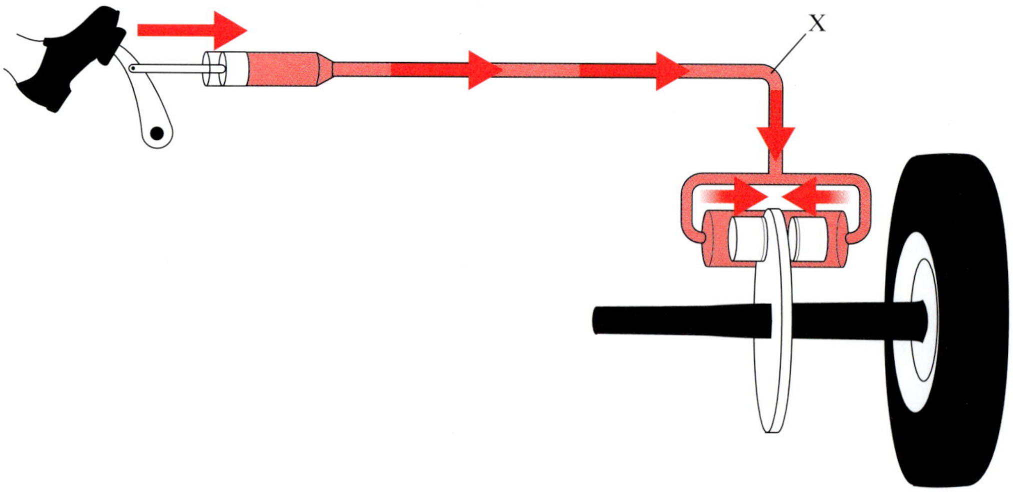

a) When a force is applied to the brake pedal what happens inside the brake fluid?

b) What would happen if there was a blockage in the pipe at X?

c) What would happen if the brake fluid was replaced by a gas?

d) Why is it an advantage to the driver to have large pistons next to the brakes in the wheels and a smaller cylinder next to the brake pedal?

9 Waves and Sound Waves

Waves are very important to us. We use them every day of our lives.

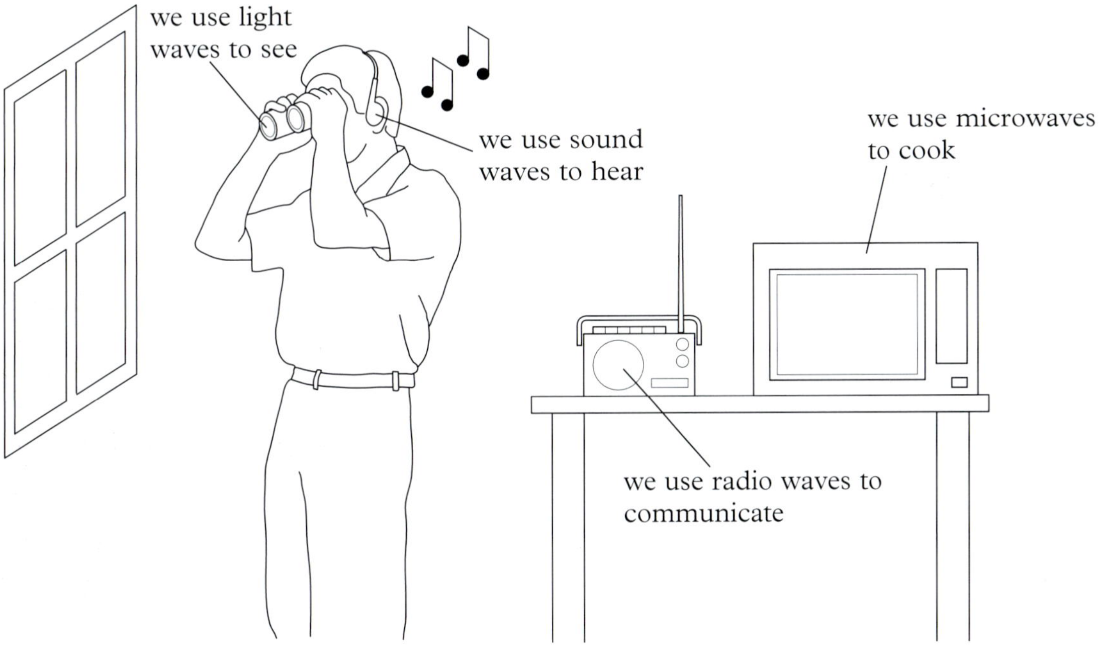
we use light waves to see
we use sound waves to hear
we use microwaves to cook
we use radio waves to communicate

Most waves carry energy from place to place.

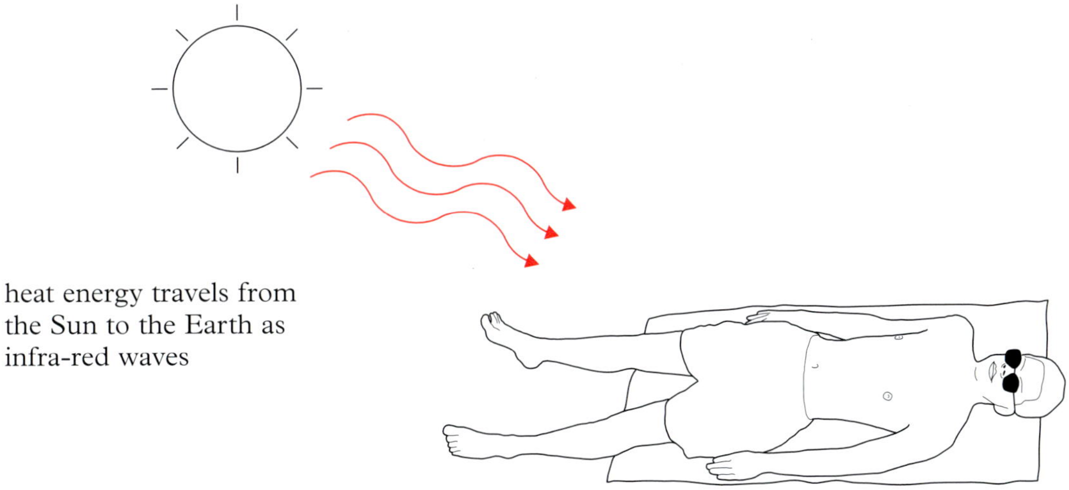
heat energy travels from the Sun to the Earth as infra-red waves

There are two main groups of waves. These are **transverse waves** and **longitudinal waves**.

Transverse waves

If we move a slinky spring up and down we will see a wave move along the spring.

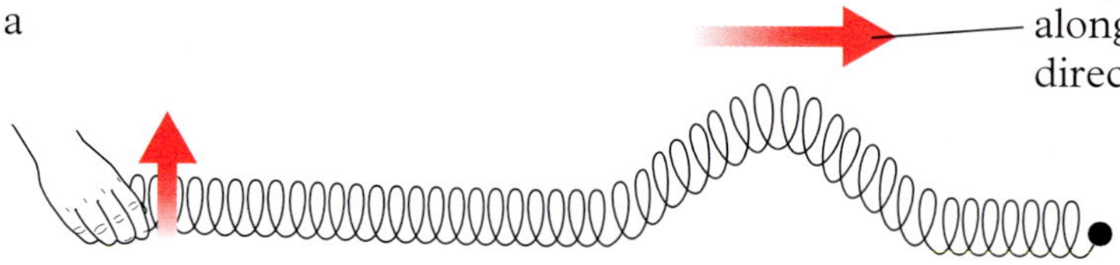
a single upwards movement produces a pulse in the spring
the pulse is moving along the spring in this direction

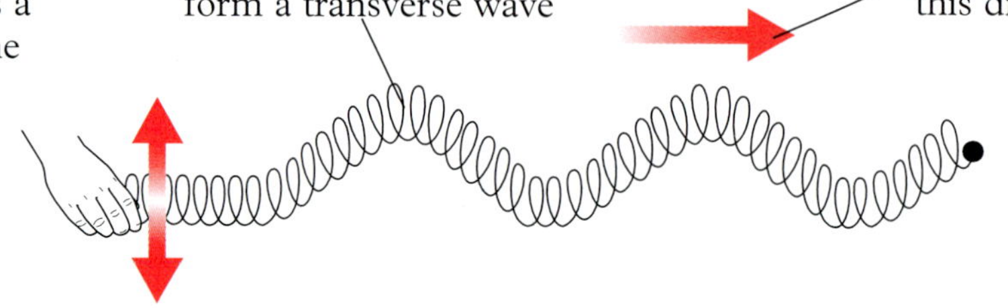
repeated up and down movement produces a series of pulses in the spring
several of these pulses form a transverse wave
the wave is moving in this direction

The coils of the slinky vibrate up and down, yet the wave itself is moving from left to right. This is an example of a transverse wave.

A transverse wave is one which vibrates at right angles to the direction in which the wave is moving.

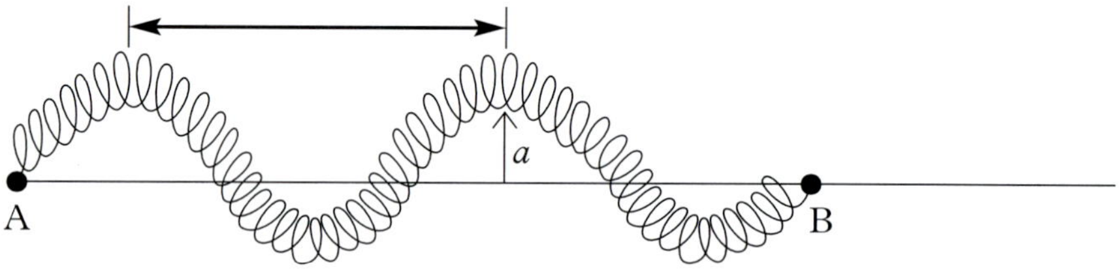
the distance between one peak and the next is called the wavelength
a
A
B
the maximum height or displacement of the wave is called the wave amplitude, a

Light waves and water waves are examples of transverse waves.

Longitudinal waves

If we push a slinky spring back and forth we will create a longitudinal wave which moves along the spring.

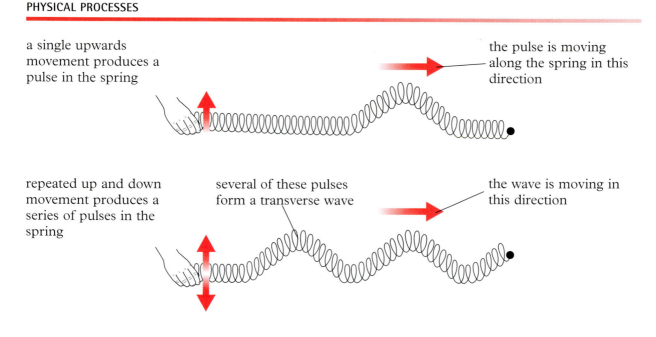

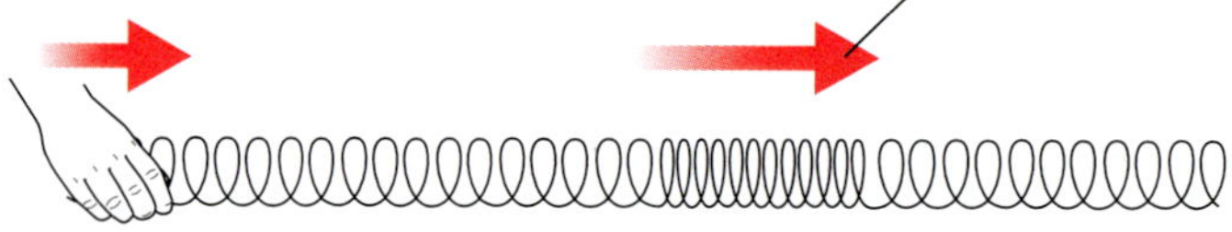
a single push produces a single pulse in the spring
the pulse is moving along the spring in this direction

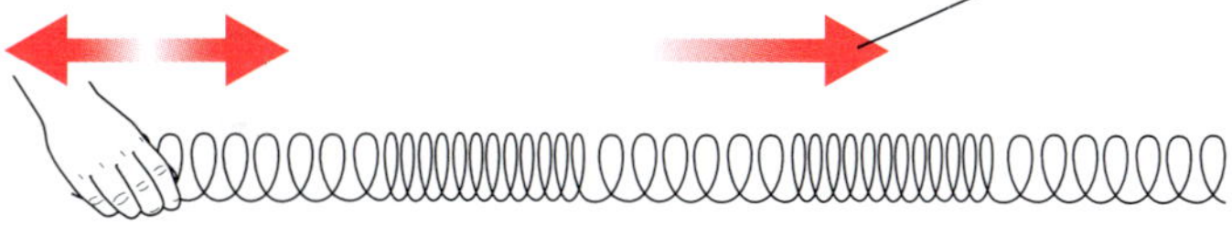
repeatedly pushing back and forth produces a series of pulses in the spring
several of these pulses form a longitudinal wave
the wave is moving in this direction

The coils of the slinky are vibrating from left to right and back again. The wave itself is moving left to right. This is an example of a longitudinal wave.

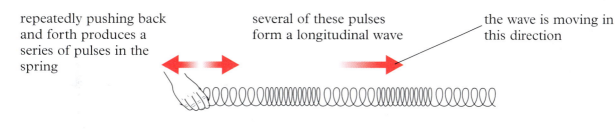

A longitudinal wave has vibrations which are along the direction in which the wave is moving.

Properties of waves

Water waves are transverse waves. The water molecules move up and down whilst the energy moves along the surface with the ripples. Water waves are easy to make and to study. By watching how they behave we can learn a lot about other types of waves.

Creating water waves using a ripple tank

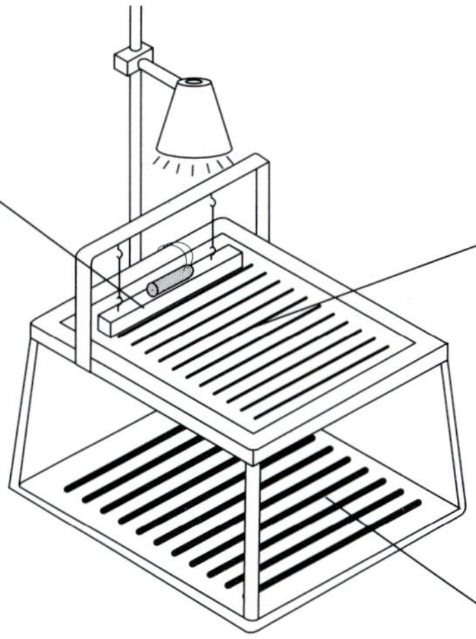
when the electric motor is turned on it makes the wooden bar vibrate
as the wooden bar vibrates it creates ripples (water waves) on the surface of the water
the lamp casts a shadow of the ripples onto a screen making it easier to see the movement of the water waves

Plane and point sources of waves

(a)

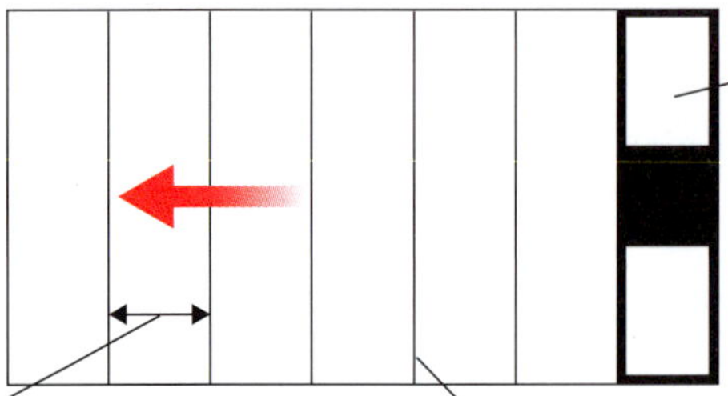
a vibrating wooden bar is a plane wave source
a plane wave source creates flat, straight, parallel wavefronts
the distance between neighbouring wavefronts is the wavelength of the waves
wavefronts like these are called plane wavefronts

(b)

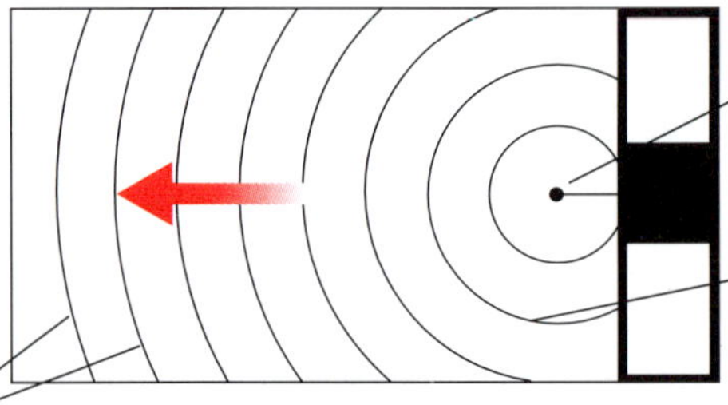
a vibrating point is called a point source
a point source creates circular wavefronts
the distance between neighbouring wavefronts is the wavelength of the waves

Frequency and wavelength

If the speed at which the motor turns is changed, the wave pattern seen on the screen also changes.

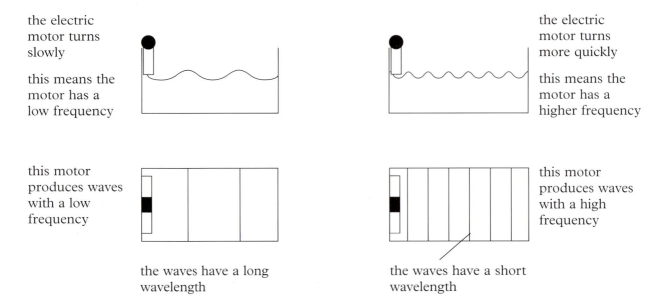

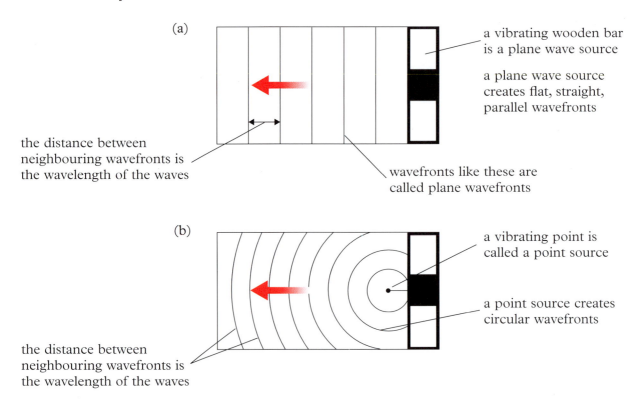

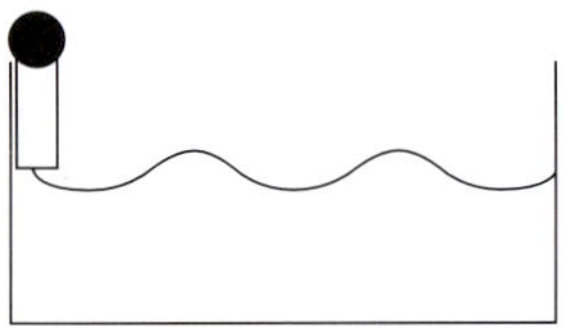
the electric motor turns slowly
this means the motor has a low frequency

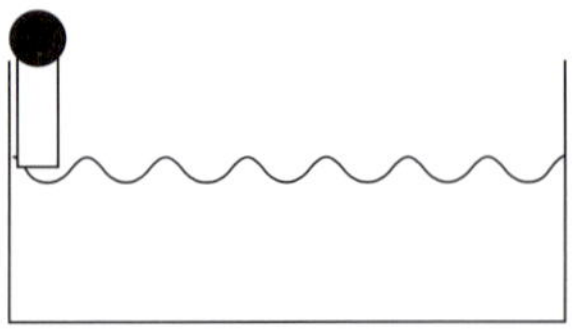
the electric motor turns more quickly
this means the motor has a higher frequency

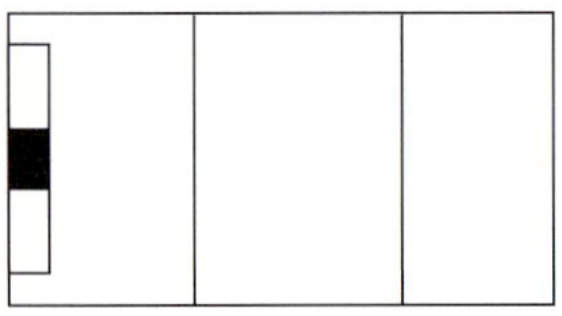
this motor produces waves with a low frequency
the waves have a long wavelength

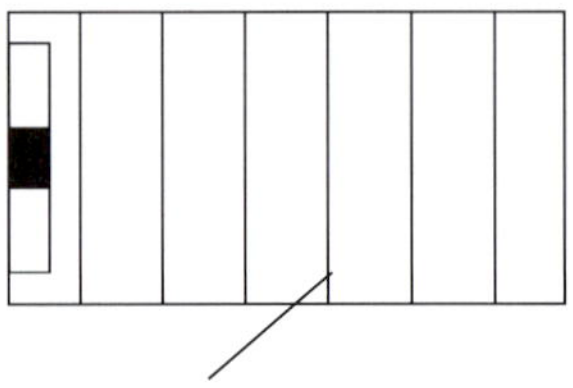
this motor produces waves with a high frequency
the waves have a short wavelength

The number of waves the vibrating bar makes on the surface of the water is called the **frequency** of the waves. Frequency is measured in hertz (Hz). If the bar makes five waves each second the frequency of the waves is 5 Hz.

Questions

1 What do most waves carry?

2 a) What is a transverse wave?
b) Give one example of a transverse wave.

3 a) What is a longitudinal wave?
b) Give one example of a longitudinal wave.

4 a) Draw a diagram showing two full waves.
b) Mark on the diagram the amplitude and the wavelength of the wave.

5 Draw two diagrams to show the differences between the waves produced by a vibrating point source and a vibrating plane source.

6 A vibrating bar produces the following:
a) 20 waves in one second
b) 40 waves in one second
c) 120 waves in one minute.
What is the frequency of each of these waves?

Reflection of water waves

When waves strike a surface or barrier they are often reflected. Barriers with different shapes can make them reflect in different ways.

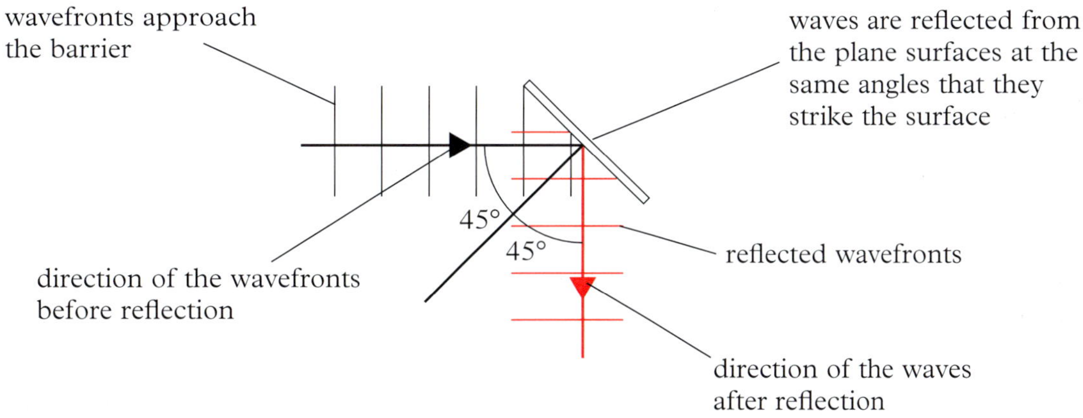
wavefronts approach the barrier
waves are reflected from the plane surfaces at the same angles that they strike the surface
45°
45°
reflected wavefronts
direction of the wavefronts before reflection
direction of the waves after reflection

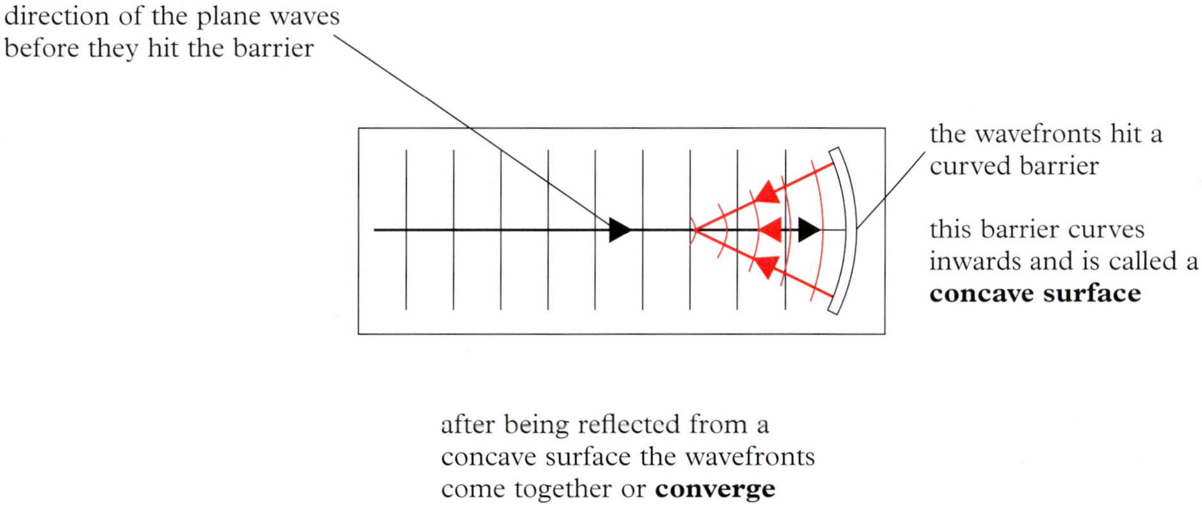
direction of the plane waves before they hit the barrier
the wavefronts hit a curved barrier
this barrier curves inwards and is called a **concave surface**
after being reflected from a concave surface the wavefronts come together or **converge**

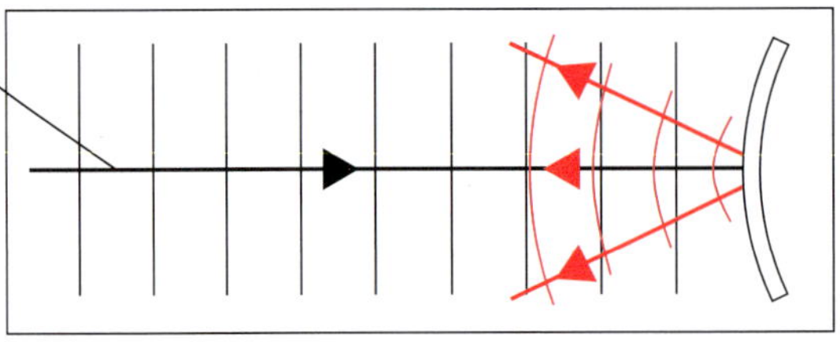
direction of the plane waves before they hit the barrier
the wavefronts hit a curved barrier
this barrier curves outward and is called a **convex surface**
after being reflected from a convex surface the wavefronts spread out or **diverge**

Refraction of water waves

When water waves travel from deep to shallow water they slow down and decrease their wavelength. The waves, as they cross the boundary, may change direction. This change in direction is called **refraction**.

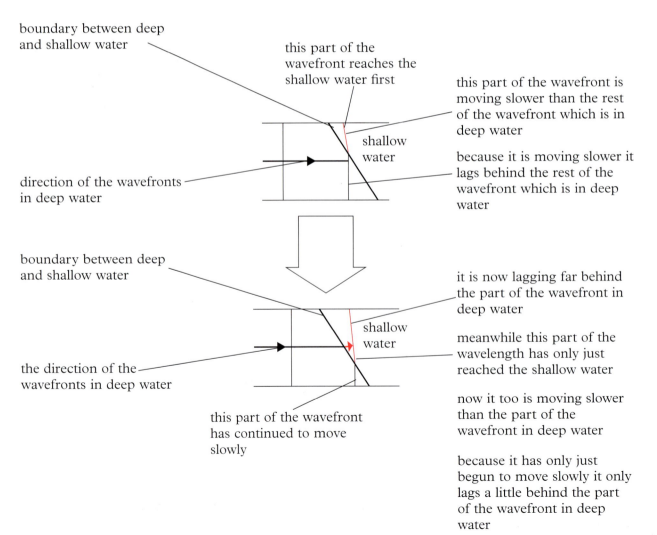

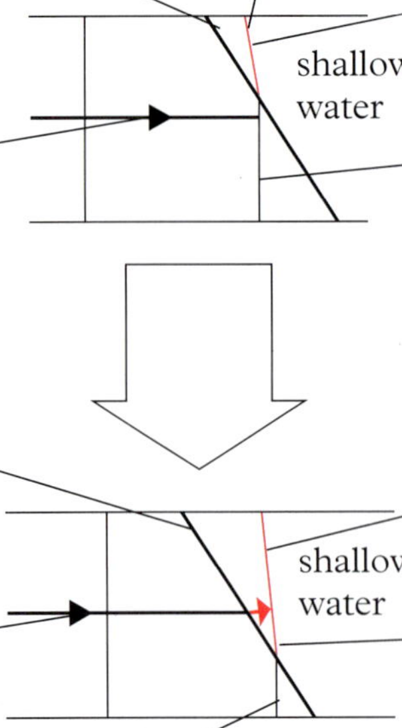
boundary between deep and shallow water
this part of the wavefront reaches the shallow water first
this part of the wavefront is moving slower than the rest of the wavefront which is in deep water
shallow water
because it is moving slower it lags behind the rest of the wavefront which is in deep water
direction of the wavefronts in deep water
boundary between deep and shallow water
it is now lagging far behind the part of the wavefront in deep water
shallow water
meanwhile this part of the wavelength has only just reached the shallow water
the direction of the wavefronts in deep water
now it too is moving slower than the part of the wavefront in deep water
this part of the wavefront has continued to move slowly
because it has only just begun to move slowly it only lags a little behind the part of the wavefront in deep water

Because different parts of the wavefront lag behind by different amounts, the entire wavefront eventually changes direction.

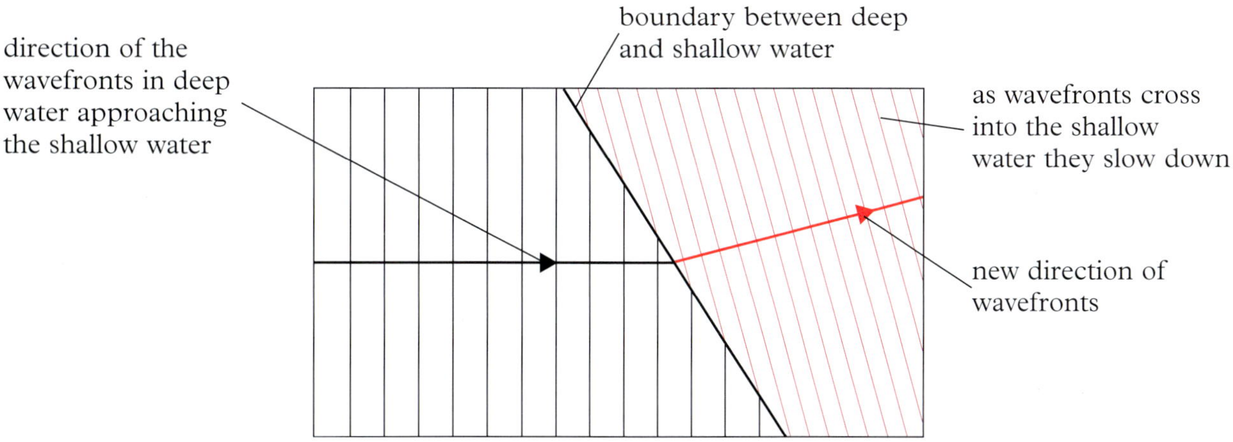
boundary between deep and shallow water
direction of the wavefronts in deep water approaching the shallow water
as wavefronts cross into the shallow water they slow down
new direction of wavefronts

Diffraction of water waves

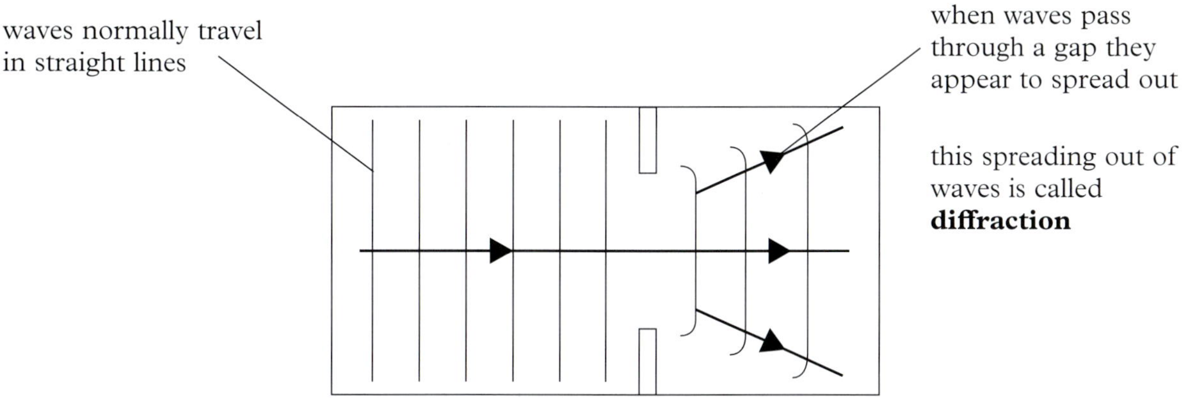
waves normally travel in straight lines
when waves pass through a gap they appear to spread out
this spreading out of waves is called **diffraction**

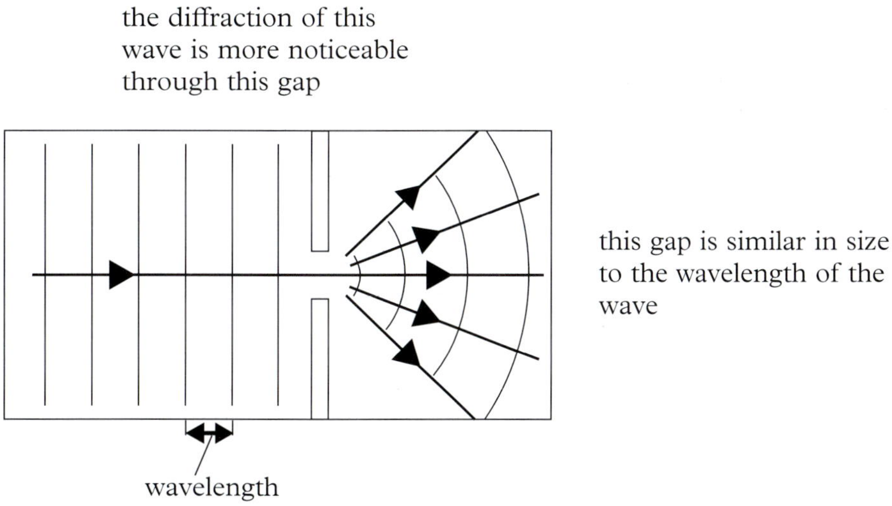
the diffraction of this wave is more noticeable through this gap
this gap is similar in size to the wavelength of the wave
wavelength

The diffraction of a wave is most noticeable when the gap through which it is travelling is approximately the same size as its wavelength.

Interference between water waves

when waves overlap like this they produce a taller wave

the new wave has a larger amplitude

this is called **constructive interference**

when waves overlap like this they cancel each other out or produce a smaller wave

the two waves have overlapped in such a way that the new wave has no amplitude – it is just a line

this is called **destructive interference**

When two or more waves overlap, they may combine to make a new wave. This is called **interference**.

The pattern created when waves overlap in this way is called an **interference pattern**.

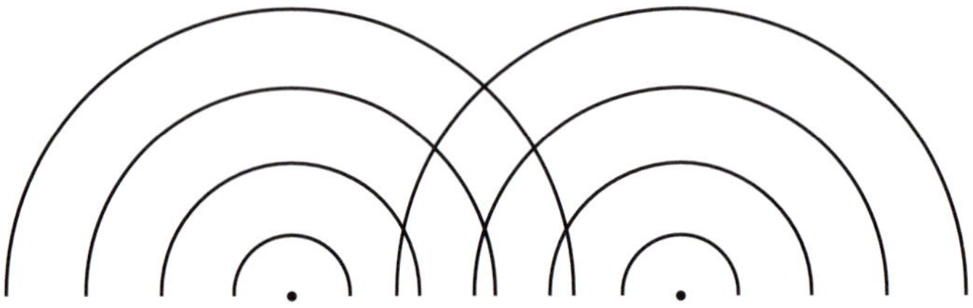

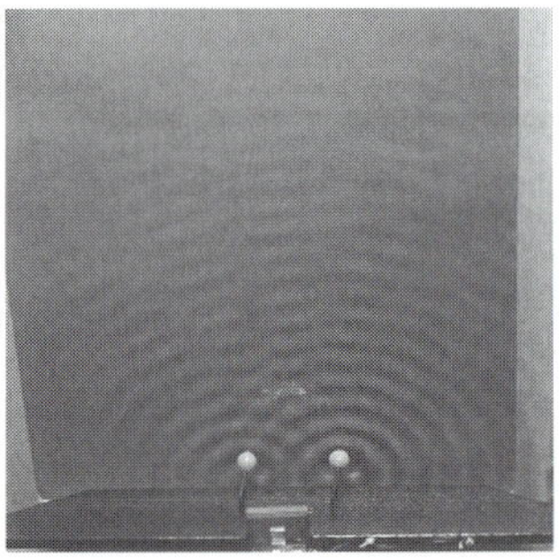

The Wave Equation

The speed at which a wave moves can be calculated using the equation:

speed = frequency × wavelength

Example
A sound wave has a frequency of 170 Hz and a wavelength of 2 m. Calculate the speed of sound.

Using speed = frequency × wavelength

speed = 170 Hz × 2 m

speed = 340 m/s

Questions

1 A water wave strikes a plane surface at 60°. At what angle will the wave be reflected from this surface?

2 What causes a water wave to refract as it passes from deep water into shallow water?

3 Water waves with a wavelength of 2 cm pass through a gap and are diffracted. What size should this gap be for the diffraction to be most noticeable?

4 Explain the following terms:
a) constructive interference
b) destructive interference.

5 Calculate the speed of a water wave which has a frequency of 50 Hz and a wavelength of 3 cm.

6 Calculate the speed of a sound wave which has a frequency of 250 Hz and a wavelength of 1.5 m

Sound

All sounds begin with objects that are vibrating

when you speak your vocal chords vibrate producing the sounds you want to make

when this bee flies its wings beat so quickly that they produce a buzzing sound

when the drum skin is hit it vibrates, producing a sound

when the strings on this guitar are plucked they vibrate, producing musical sounds

when the cone of this speaker vibrates it produces sound

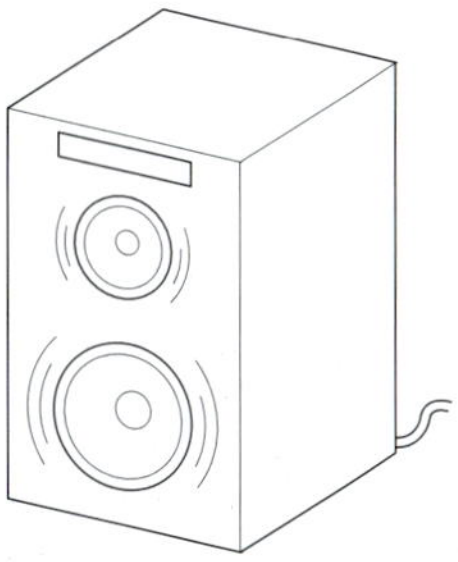

Frequency of vibration and pitch

if an object is large, like one of the strings of this double bass, it will vibrate slowly when you pluck it and you will hear a **low pitched** note

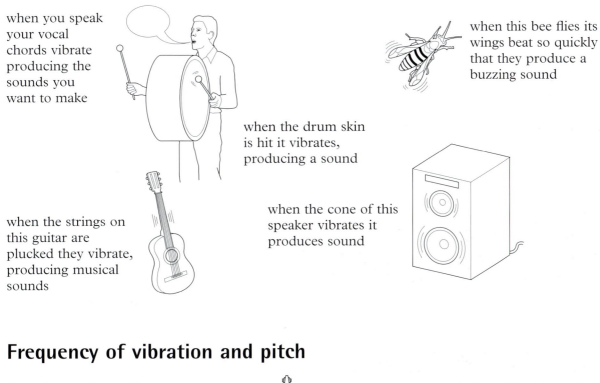

if an object is small, like one of the strings of this violin, it will vibrate quickly when you pluck it and you will hear a **high pitched** note

Small objects vibrate quickly producing high pitched sounds.

Large objects vibrate slowly producing low pitched sounds.

We can show that these two statements are true by carrying out the 'ruler experiment'.

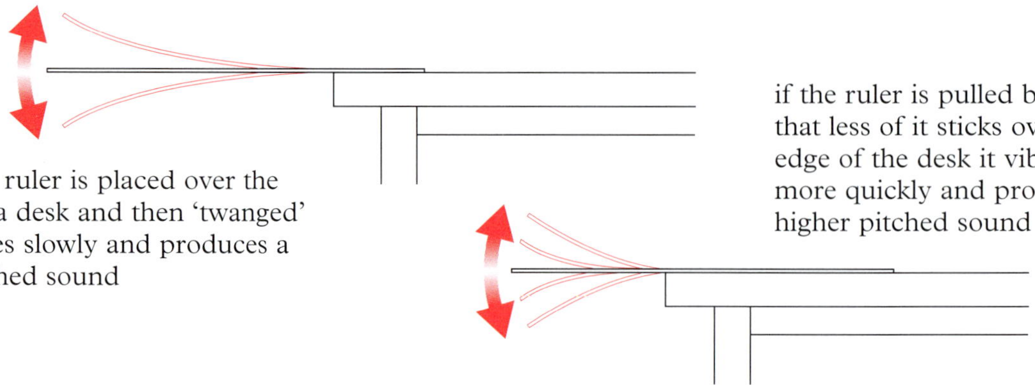
if a long ruler is placed over the edge of a desk and then 'twanged' it vibrates slowly and produces a low pitched sound
if the ruler is pulled back so that less of it sticks over the edge of the desk it vibrates more quickly and produces a higher pitched sound

Questions

1 a) Write down the names of four musical instruments you would find in a band or orchestra.

b) Which of these instruments produces high pitched notes?

c) Which of these instruments produces low pitched notes?

d) Explain in each case what vibrates to produce the sound.

2 a) Explain how a flying mosquito produces a sound.

b) Why is this sound always high pitched?

Sound waves

The sound an object makes travels to our ears by means of sound waves. Although we can't see sound waves the diagrams below show how they move through the air.

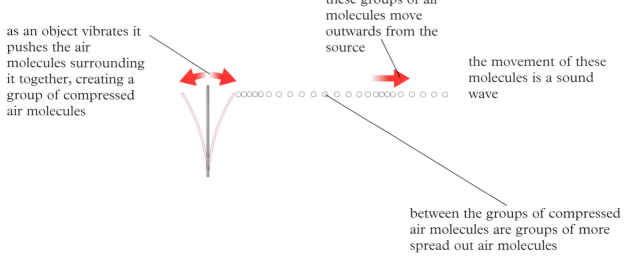

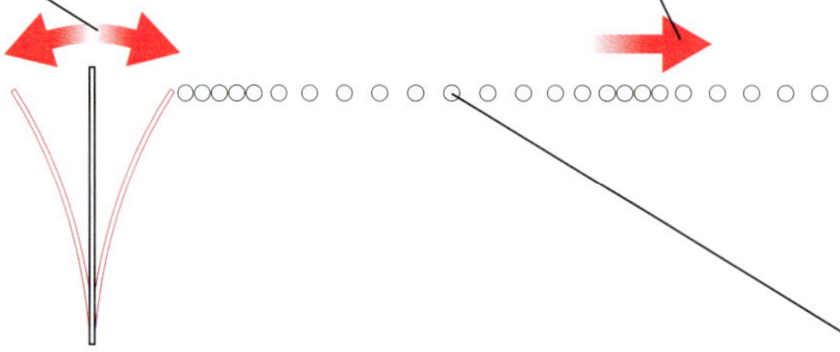
as an object vibrates it pushes the air molecules surrounding it together, creating a group of compressed air molecules
these groups of air molecules move outwards from the source
the movement of these molecules is a sound wave
between the groups of compressed air molecules are groups of more spread out air molecules

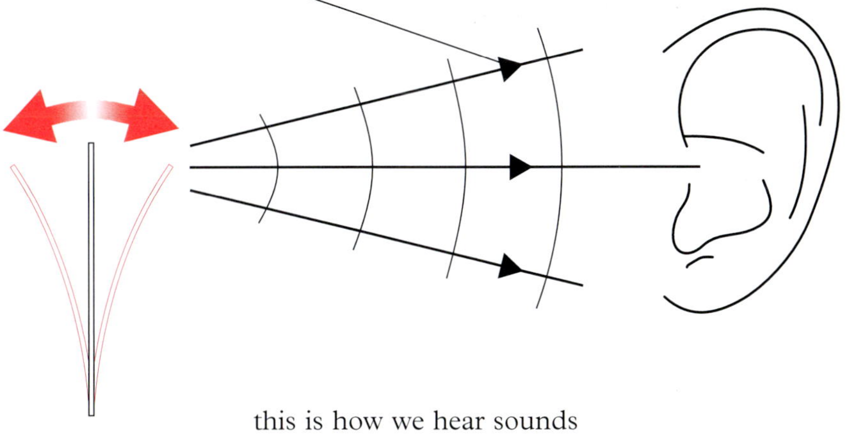
movement of the
sound wave
when a
sound wave
enters the ear
the air
molecules
strike the ear
drum

this is how we hear sounds

Sound in a vacuum

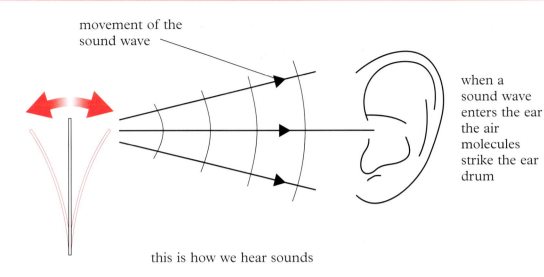

Sound waves must have something to travel through. They can travel through solids – that's why your mum and dad can hear your music through the walls and floor of your room if you have it turned up too loud! They can travel through liquids – that's how whales are able to 'talk' to each other deep in the ocean. They can travel through gases such as air – that is how we are able to hear all the sounds around us. But sounds cannot travel through a **vacuum**. A vacuum is empty space where there are no particles.

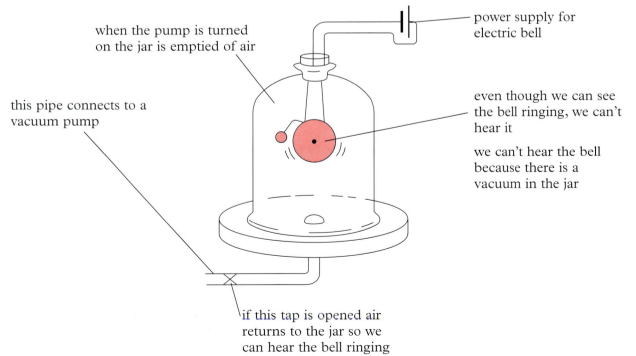

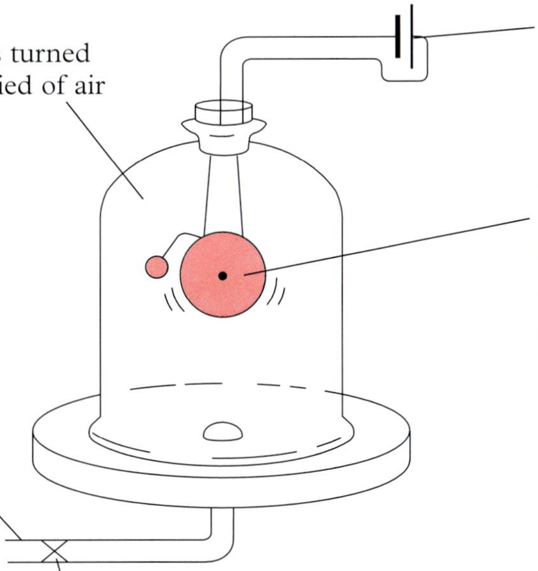
when the pump is turned
on the jar is emptied of air
power supply for
electric bell
this pipe connects to a
vacuum pump
even though we can see
the bell ringing, we can't
hear it
we can't hear the bell
because there is a
vacuum in the jar
if this tap is opened air
returns to the jar so we
can hear the bell ringing

Loudness

We hear sounds when molecules (usually air molecules) strike our ear drums making them vibrate. If lots of air molecules hit our ear drums the sound we hear is loud. If just a few air molecules strike our ear drums the sound we hear will be quiet.

this tuning fork only pushes a few air molecules as it vibrates

it makes a quiet sound

when the tuning fork is held against a desk it makes the desk vibrate
the desk pushes against lots of molecules as it vibrates

it makes a loud sound

Very loud sounds such as those made by aircraft, machinery in factories or personal stereos with their volume turned up high can damage your hearing.

this road worker is wearing ear protectors
constant exposure to the very loud noises produced by his drill might permanently damage his hearing

The loudness of a sound is measured on the **decibel scale.** The following gives some decibel values for sounds you may be familiar with.

no sound 0dB

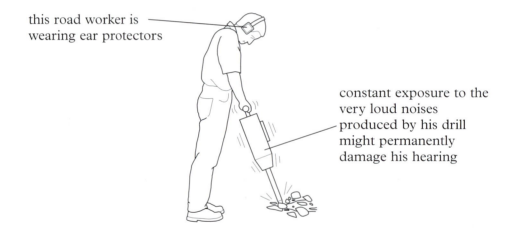

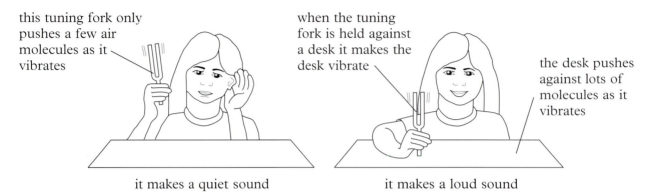

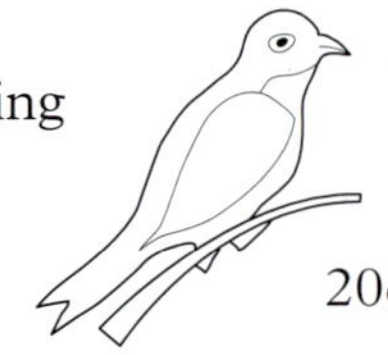
a bird singing quietly
20dB

a dog howling (close by)
80dB

people talking reasonably quietly
60dB

a noisy factory
90dB

someone playing the trumpet

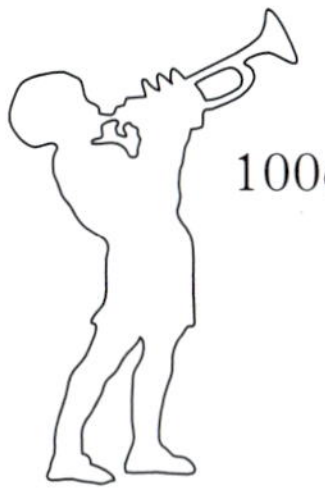

100dB

a jet aircraft taking off

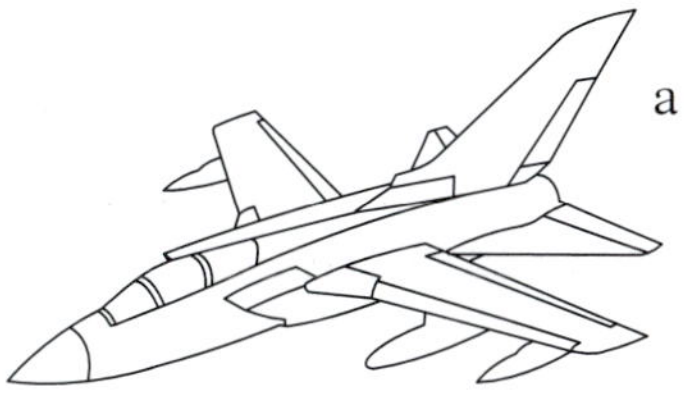

130dB

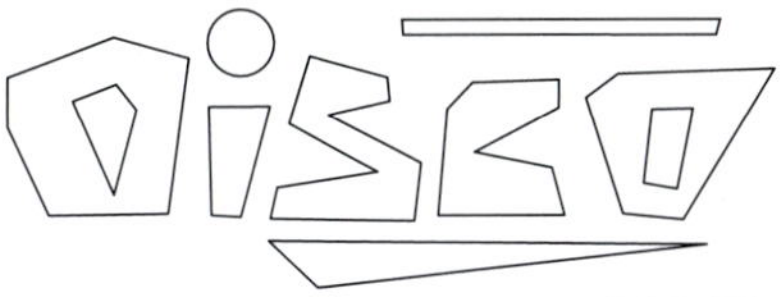

a noisy disco

110dB

a bomb exploding

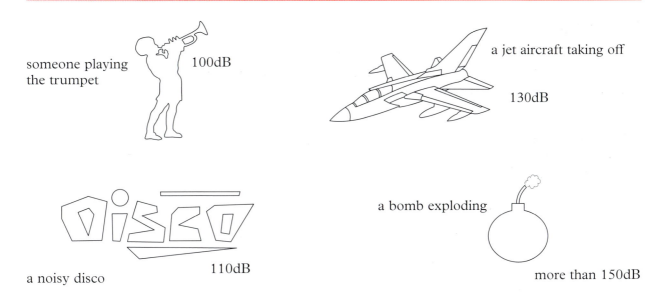

more than 150dB

Noise and noise pollution

Unwanted sound is called **noise**. Noise pollution can be a real problem. It can cause stress and make it more difficult to concentrate. There are several ways in which planners, builders and architects try to reduce this problem.

1 Although sound waves can be reflected most will travel in a straight line from the source to the listener. If a barrier is placed between the two much of the sound will be blocked off. This means that the noise heard from a nearby motorway can be reduced by planting trees or building a wall.

2 Double glazing is an excellent barrier for keeping out unwanted sounds. Because there are very few particles between the two sheets of glass it is very difficult for sound waves to pass through.

3 Advances in technology mean that many machines such as cars and aeroplanes are much quieter than they used to be. This work is continuing and should result in even quieter machines in the future.

Questions

1 Explain using a labelled diagram how the sounds produced by a drum are heard by a listener.

2 Why can sound waves not travel through a vacuum?

3 a) Explain how you would make the sound produced by a drum louder.
b) What effect does this have on the number of air molecules striking the ear drum of the listener?

4 Estimate on the decibel scale the noise level of the following:
a) shouting
b) ringing a bicycle bell
c) popping a balloon
d) whispering.

Frequency, amplitude and tone

When we listen to some music or hear someone speak different kinds of sound waves enter our ears. Some of them will carry high pitched sounds, some low pitched sounds. Some will be loud sounds, others will be quiet. What do each of these waves look like? We can 'see' the differences between these waves using a piece of equipment called a **C**athode **R**ay **O**scilloscope or **CRO**.

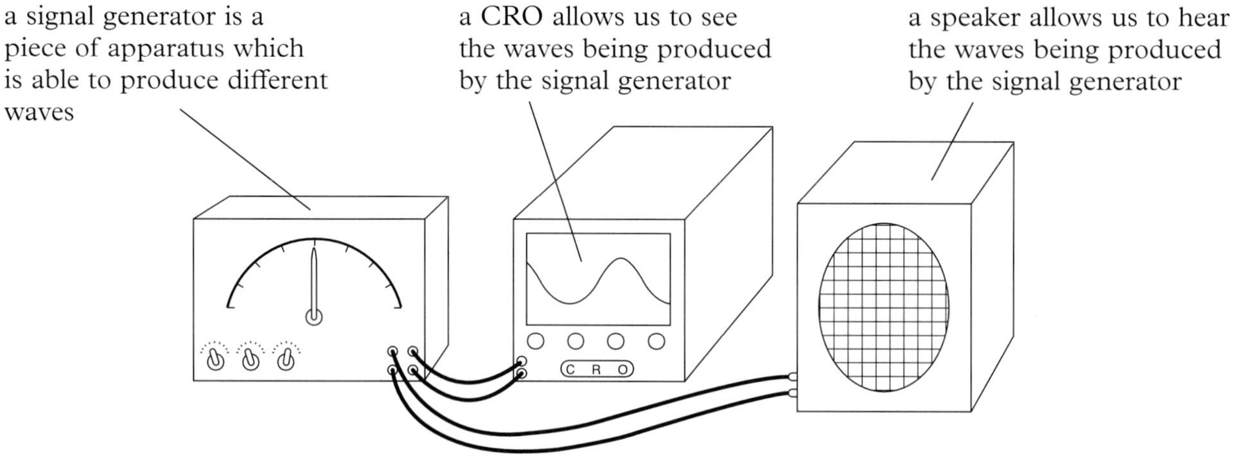
a signal generator is a piece of apparatus which is able to produce different waves
a CRO allows us to see the waves being produced by the signal generator
a speaker allows us to hear the waves being produced by the signal generator
C R O

Frequency and Pitch

high frequency or high pitched sounds have a short wavelength

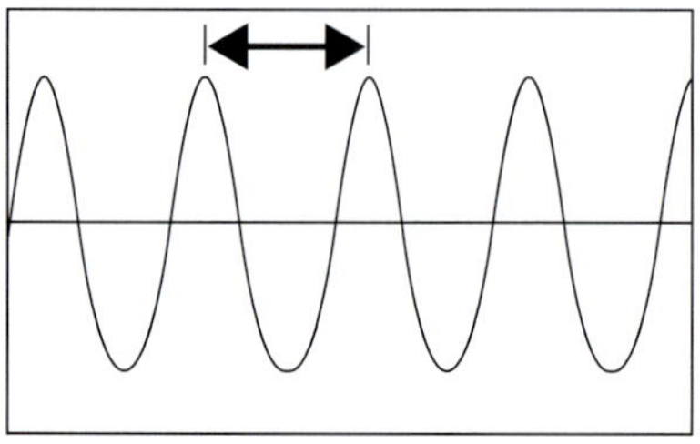

low frequency or low pitched sounds have a long wavelength

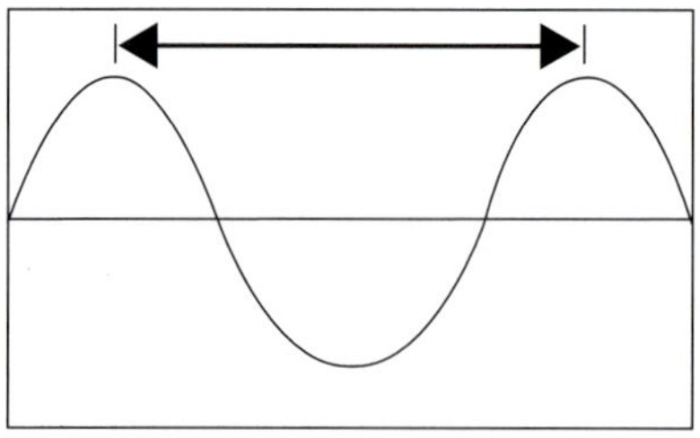

Loudness and amplitude

sounds which are loud have a large amplitude

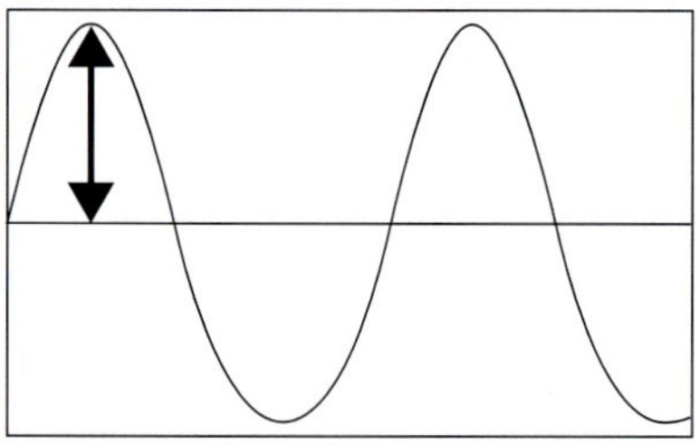

sounds which are quiet have a small amplitude

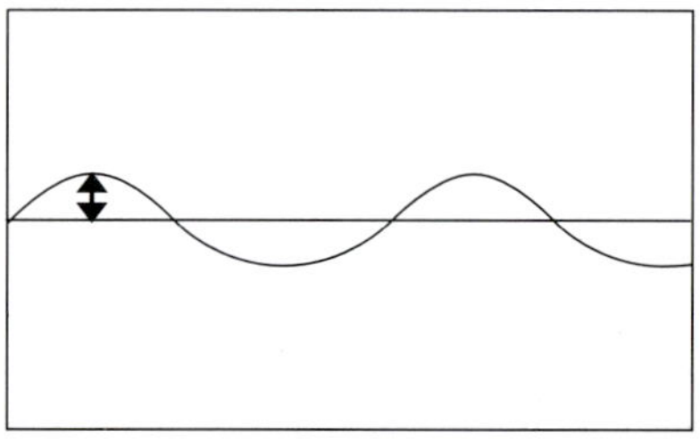

Tone

If we close our eyes and someone played the same note on a piano, violin and trumpet we would have no difficulty in deciding which instrument made which sound. This is because the sounds have different **tones**. Notes of different tones are shown on the CRO as waves with different shapes.

these two waves have the same pitch and loudness but we hear a difference because they have different tones

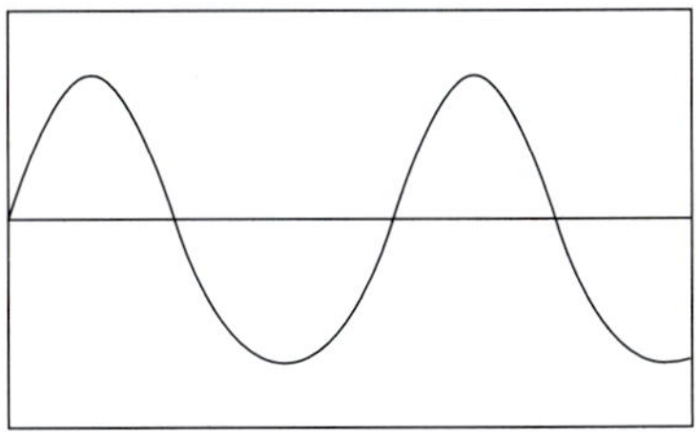

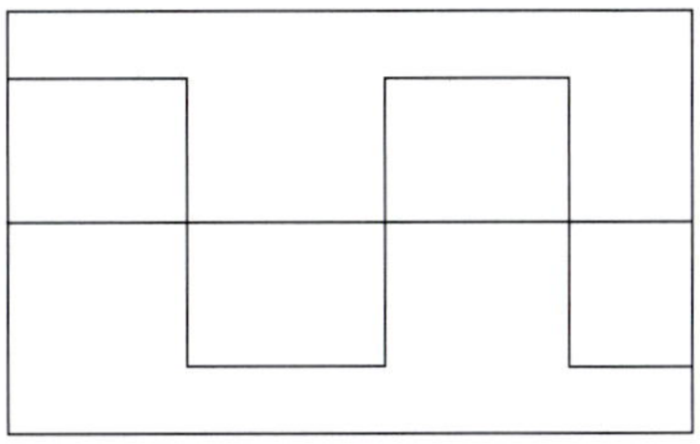

altering the tone of a sound wave changes its shape

On our modern stereo systems we have 'graphic equalizers' or bass and treble controls which allow us to alter sounds to our taste. We can change the sound so that we hear more of the bass notes, or perhaps add a little more emphasis on the high notes.

Echoes

An echo is a reflected sound wave. We often hear echoes in caves, tunnels and in mountain valleys.

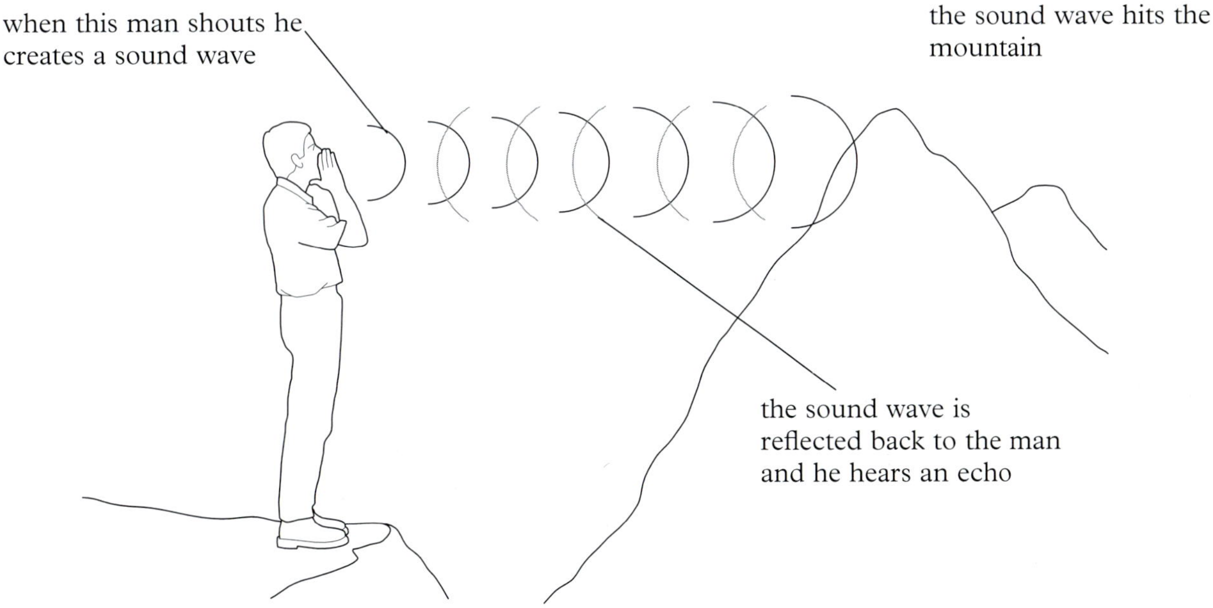
when this man shouts he creates a sound wave
the sound wave hits the mountain
the sound wave is reflected back to the man and he hears an echo

Echo-sounding

Echoes can be used by fishing boats to search for shoals of fish, and by ships to check the depth of the water beneath them.

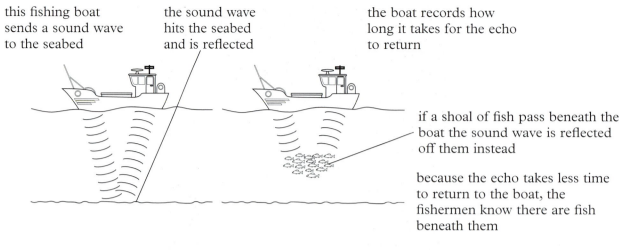

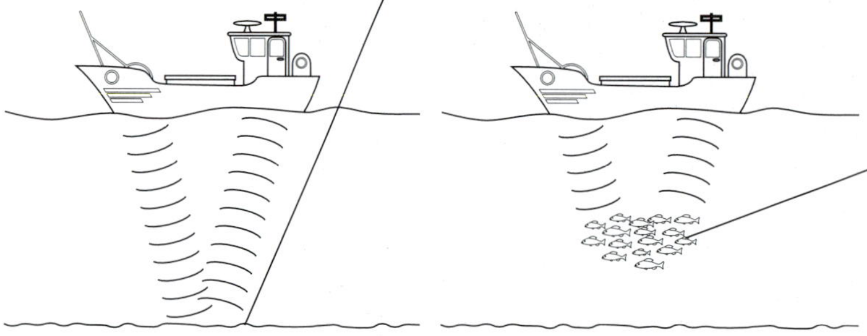
this fishing boat sends a sound wave to the seabed
the sound wave hits the seabed and is reflected
the boat records how long it takes for the echo to return
if a shoal of fish pass beneath the boat the sound wave is reflected off them instead
because the echo takes less time to return to the boat, the fishermen know there are fish beneath them

Ultra sounds

Some sound waves have frequencies that are so high we are unable to hear them. These sounds are called **ultra sounds**. Bats use ultrasound to 'see'.

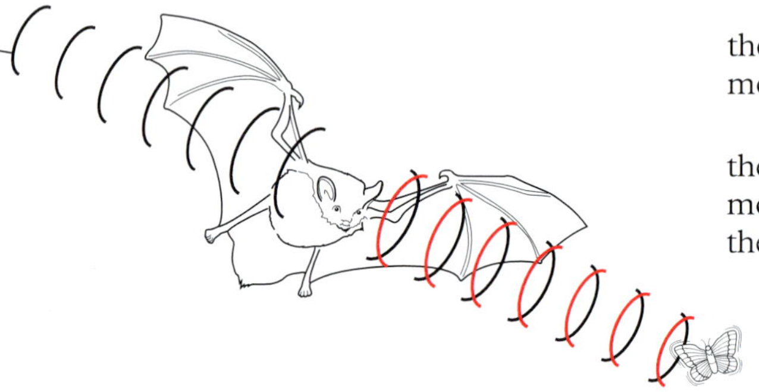
bats emit high pitched sounds
the sound wave hits this moth and is reflected back
the bat knows where the moth is by listening for the echo

By using ultrasound echoes, bats are not only able to navigate but also to detect, chase and catch prey in mid flight.

Ultra-scans

Normally if a doctor wanted to see inside your body to see if there are any problems such as broken bones he would use X-rays. If however he wanted to check on the growth of an unborn baby in the mother's womb, he would be more likely to use ultrasound. These waves are less likely to harm the baby than X-rays and therefore are safer to use.

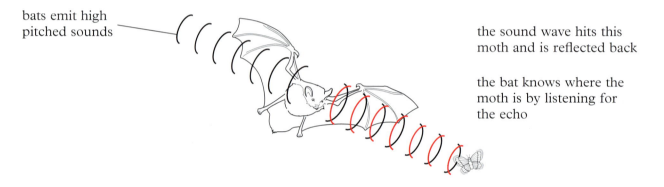

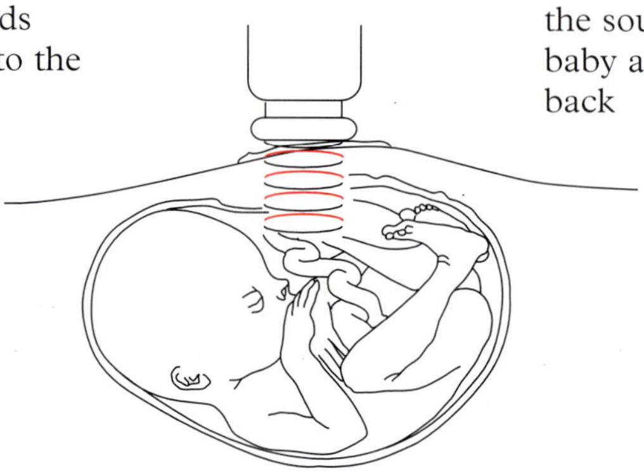
the probe sends ultrasound into the womb
the soundwaves hit the baby and are reflected back

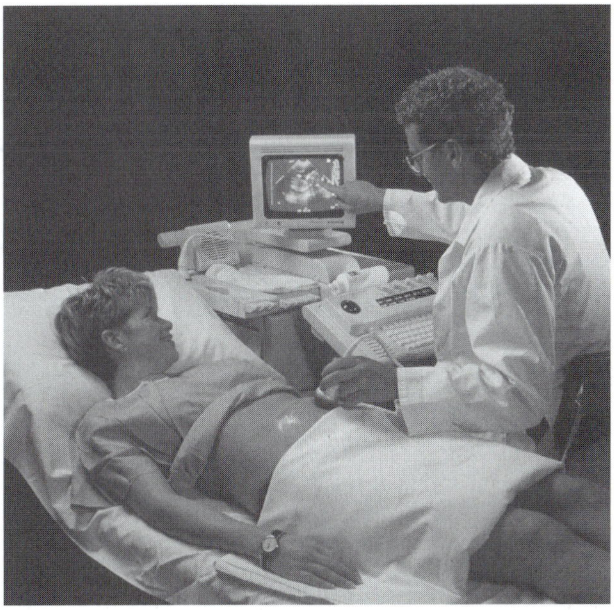

The reflected signals are detected by the probe. The signals are then processed by a computer and an image of the baby can be seen on the monitor.

Speed of sound

Sound waves travel much more slowly than light waves. This is why we always see a flash of lightning before we hear the thunder. The speed of light in air is approximately 300 000 000 m/s. The speed of sound in air is approximately 340 m/s. Aeroplanes which fly at speeds faster than the speed of sound are called **supersonic** aircraft.

Questions

1 Draw a diagram to show how the following waves might appear on a CRO:
a) a loud, high pitched sound
b) a quiet low pitched sound
c) two sounds which have the same pitch and loudness but have different tones.

2 Explain using diagrams how a ship might use sound waves to discover the depth of the ocean directly beneath it.

3 Why would a doctor use an ultrasonic scan to check the condition of an unborn baby and not X-rays?

Seismic Waves

When an earthquake occurs, shock waves travel outwards from the centre of the earthquake. The centre of the earthquake is called the **epicentre**. The shockwaves are called **seismic waves**.

Seismic waves can cause great damage and loss of life.

There are two main types of seismic waves which travel through the Earth.

1 **Primary waves** or **p-waves**. These are longitudinal waves and are able to travel through solids and liquids. They travel quickly through dense rock but more slowly through less dense rock.

(i)

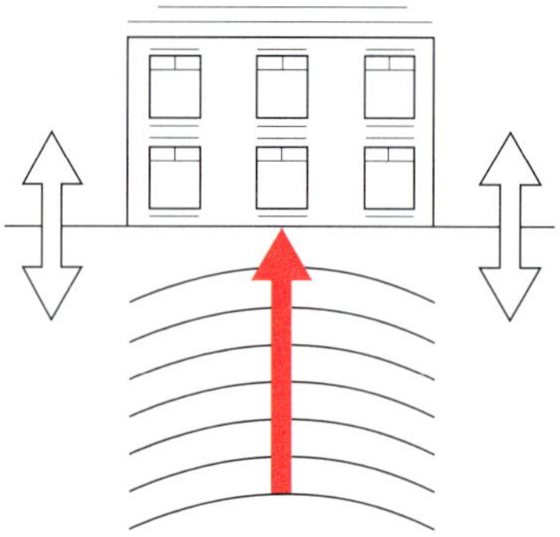
as the waves reach the surface of the Earth the rocks are stretched and compressed
this causes the ground to vibrate up and down

2 **Secondary waves** or **s-waves**. These are transverse waves. They are able to travel through solid rocks but not through liquids such as molten rock. They travel more slowly than p-waves.

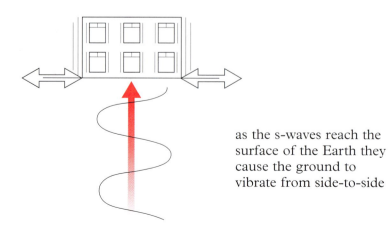

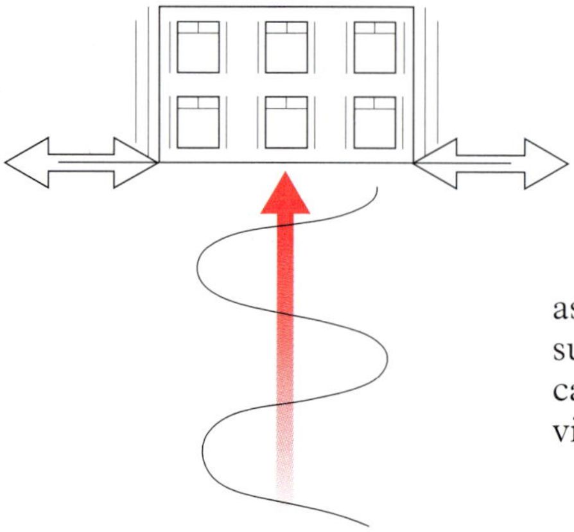
as the s-waves reach the surface of the Earth they cause the ground to vibrate from side-to-side

Seismic waves can be detected and measured using a piece of equipment called a **seismometer**. By analysing the readings from seismometers after an earthquake has taken place, scientists have been able to gain knowledge about the structure of the Earth.

The diagram below shows the kinds of observations made and the conclusions that have been drawn.

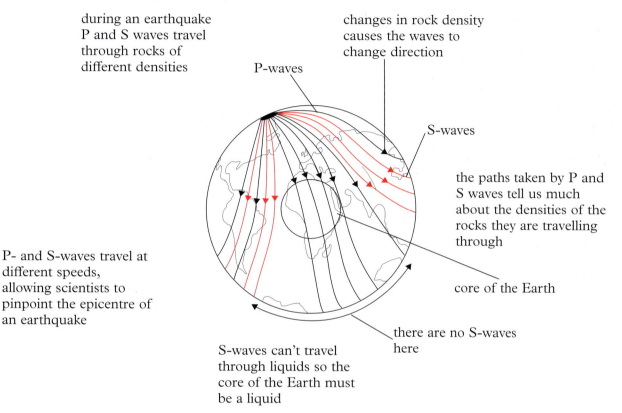

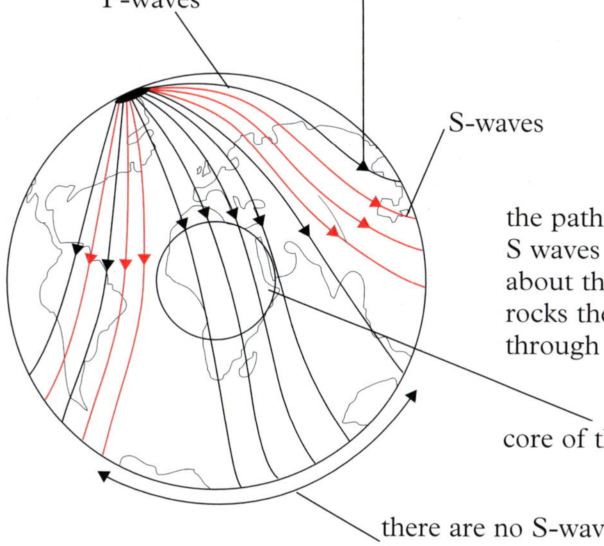
during an earthquake P and S waves travel through rocks of different densities
changes in rock density causes the waves to change direction
P-waves
S-waves
the paths taken by P and S waves tell us much about the densities of the rocks they are travelling through
P- and S-waves travel at different speeds, allowing scientists to pinpoint the epicentre of an earthquake
core of the Earth
there are no S-waves here
S-waves can't travel through liquids so the core of the Earth must be a liquid

Summary

Light and sound are examples of wave motion. Waves transfer energy away from a source. There are several properties which all waves share. They can all be reflected, refracted, diffracted and form interference patterns.

All sound waves begin with an object which vibrates. If the object vibrates quickly it produces high pitched sounds. If it vibrates slowly it produces low pitched sounds. The larger the amplitude of vibration the louder the sound. We measure the loudness of sounds on the decibel scale. Constant exposure to loud sounds can permanently damage our hearing.

Sound waves can travel through solids, liquids and gases but not through a vacuum. They travel much more slowly than light waves which is why there is sometimes a short delay between seeing and hearing an event. This can be seen when watching thunder and lightning or exploding rockets on bonfire night.

When an earthquake occurs seismic waves spread outwards from the epicentre. Analysis of these waves gives us clues about the inner structure of the Earth.

Key words

amplitude	The height of the wave from the undisturbed position to the peak of the crest.
decibel scale	A scale used to measure the loudness of sounds.
diffraction	The spreading out of waves as they pass through narrow gaps.
echo	A reflected sound.
epicentre	The centre of an earthquake.
interference	The overlapping of waves to produce new waves.
longitudinal wave	A wave whose vibrations are in the same direction as the wave is moving.
noise / noise pollution	Unwanted sounds.
p-waves (primary waves)	Longitudinal seismic waves.
reflection	The 'bouncing off' of a wave as it hits a barrier.
refraction	The change in direction of a wave caused by a change in its speed.
seismic waves	Waves that carry energy away from the centre of an earthquake.
seismometer	Device for detecting and measuring seismic waves.
s-waves (secondary waves)	Transverse seismic waves.
transverse wave	A wave whose vibrations are at 90° to the direction in which the wave is moving.
ultrasound	Sound which has such a high frequency it cannot be detected by the human ear.
vacuum	A volume of space which contains no particles.
wave frequency	The number of waves produced each second by the source.
wavelength	The distance from the crest of one wave to the crest of the next wave.

End of Chapter 9 Questions

1 State four properties which all waves have in common.

2 Draw a diagram of the wave pattern produced in a ripple tank by the following:
a) a plane source vibrating slowly
b) a point source vibrating with twice the frequency of **a**).

3 Draw a diagram of a wave.
  a) Label both its amplitude and its wavelength.
  b) If your diagram represented a sound wave what would you hear when the following happened?
    i) the wavelength got shorter,
    ii) the amplitude got smaller,
    iii) the shape of the wave changed.

4 A ship searching for fish emits sound waves which are reflected from the sea-bed.
  a) How will the operator know if a shoal of fish swims under the ship?
  b) Suggest one way in which the operator might receive a false signal that tells him there are fish present when in fact there are none actually under the ship.
  c) If the waves emitted by the ship have a wavelength of 1.4 m and a frequency of 1000 Hz calculate their speed through the water.

5 Explain why a spectator might see a batsman strike a cricket ball before he hears the sound of the shot.

6 Look at the musical instruments drawn below.

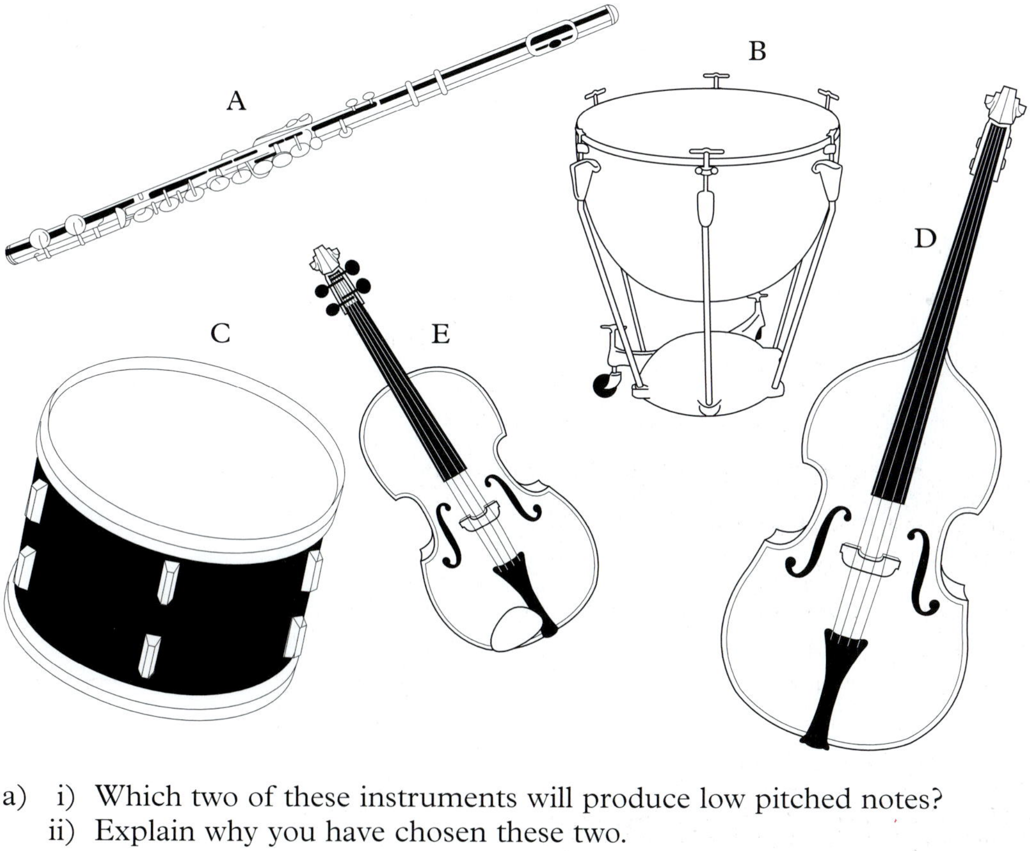

  a) i) Which two of these instruments will produce low pitched notes?
    ii) Explain why you have chosen these two.
  b) Draw two diagrams to show the wave pattern instrument A would make if played quietly and instrument D would make if it was played loudly.

7 a) What is a seismic wave?
  b) Name two types of seismic wave which travel through the Earth.
  c) Explain how scientists have been able to gain evidence about the structure of the Earth by studying seismic waves.

10 Light Waves: Reflection

Seeing with light

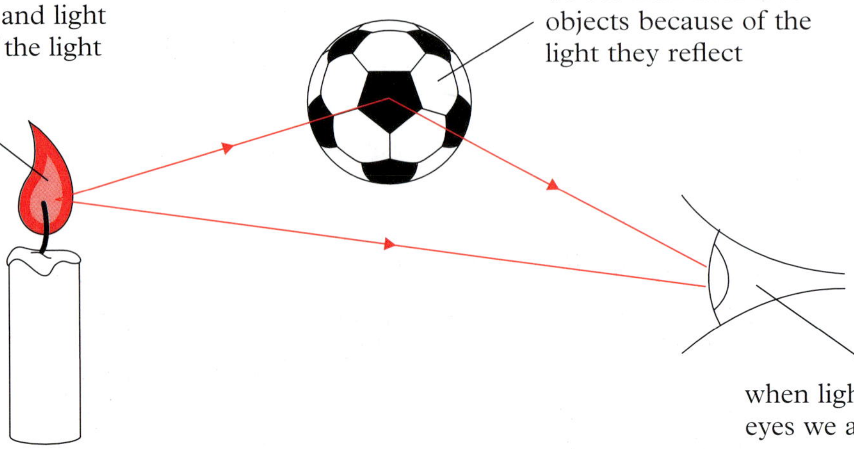

we see luminous objects (like the Sun, fire and light bulbs) because of the light they emit
we see non-luminous objects because of the light they reflect
when light enters our eyes we are able to see

If there is no light entering our eyes we are unable to see.

How does light travel?

On a bright day when the Sun is shining through the clouds we can see that light travels in beams or straight lines. Light travels very, very quickly – at 300 000 000 metres per second. It takes just 8 minutes for light to travel from the Sun to the Earth.

Questions

1 Draw a diagram to show how a pinhole camera produces an image.

2 Describe the image produced by a pinhole camera.

3 Give one advantage and one disadvantage of making the pinhole larger.

4 What extra device do we have in all modern cameras to help create a focused image, which we do not have in pinhole cameras?

Reflection of Light

When a ray of light strikes a flat or plane surface it is easy to predict the direction of the reflected ray.

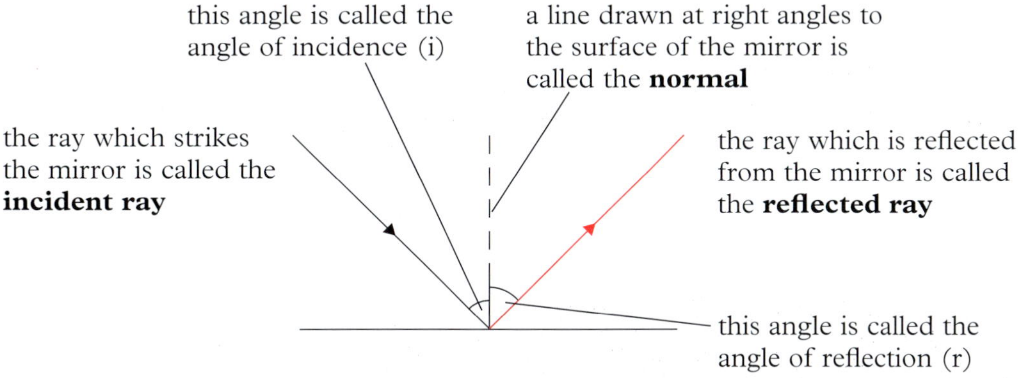
this angle is called the angle of incidence (i)
a line drawn at right angles to the surface of the mirror is called the **normal**
the ray which strikes the mirror is called the **incident ray**
the ray which is reflected from the mirror is called the **reflected ray**
this angle is called the angle of reflection (r)

When a ray of light strikes a mirror it is always reflected such that the angle of incidence equals the angle of reflection.

From the diagram above we can see that if a ray of light strikes a mirror at 45° to the normal it will be reflected at 45° to the normal. After striking the mirror, the ray has turned through 90°. This idea is used in the simple periscope. It is a useful instrument for seeing over high objects or looking round corners.

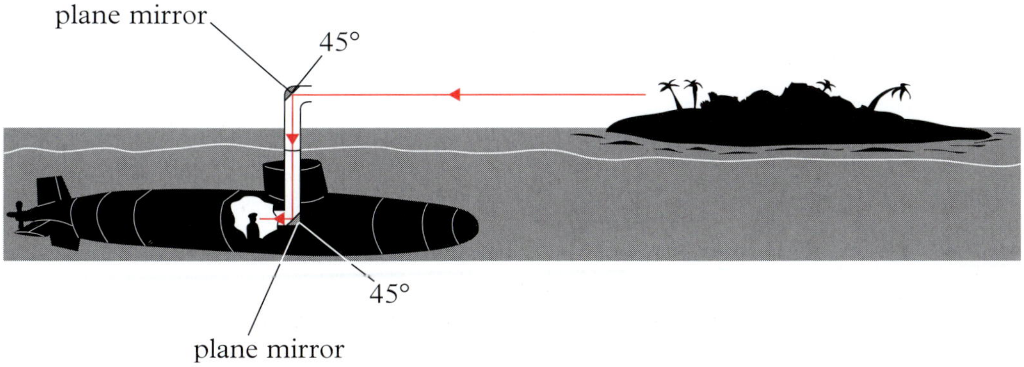
plane mirror
45°
45°
plane mirror

Shiny and dull surfaces

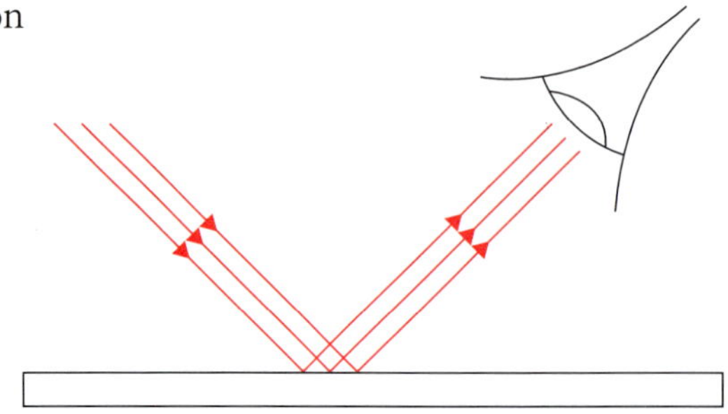
if a parallel beam of light strikes a flat surface all the rays are reflected in the same direction
to an observer the surface therefore looks shiny

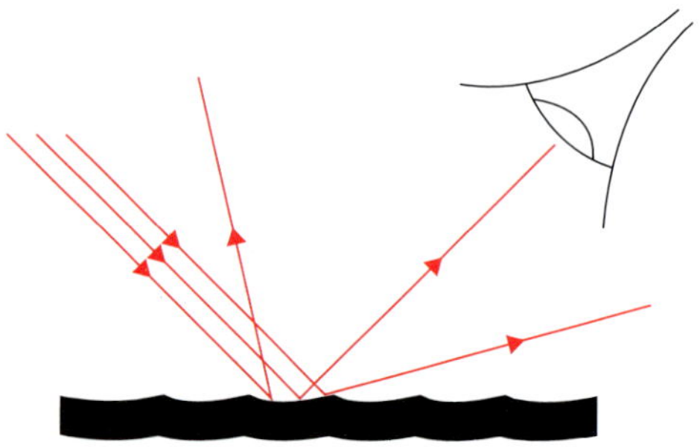
if a parallel beam of light strikes a rough surface the rays are reflected in lots of different directions
only a few of the rays enter the eyes of the observer so the surface looks dull

The image created by a plane mirror

When we look in a plane mirror we see images of objects which are in front of the mirror. The diagram below shows how these images are created.

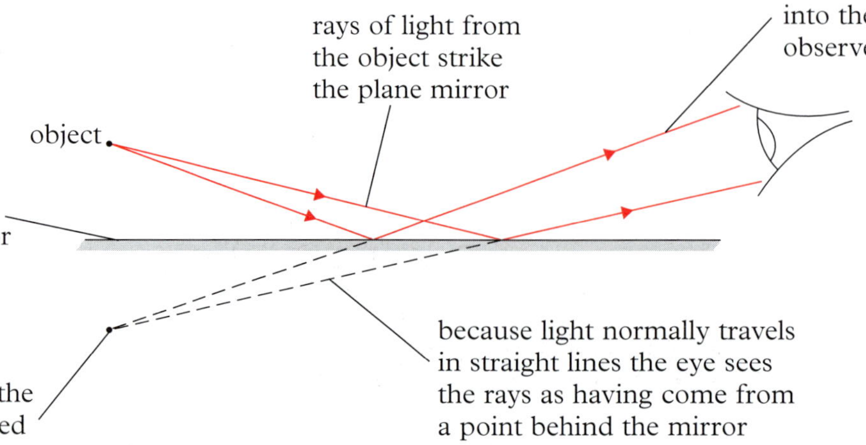
the rays are reflected into the eyes of the observer
rays of light from the object strike the plane mirror
object
plane mirror
because light normally travels in straight lines the eye sees the rays as having come from a point behind the mirror
position of the image created by the mirror

The images created by a plane mirror are always

- upright
- the same size as the object
- the same distance behind the mirror as the object is in front
- back to front
- virtual – not real

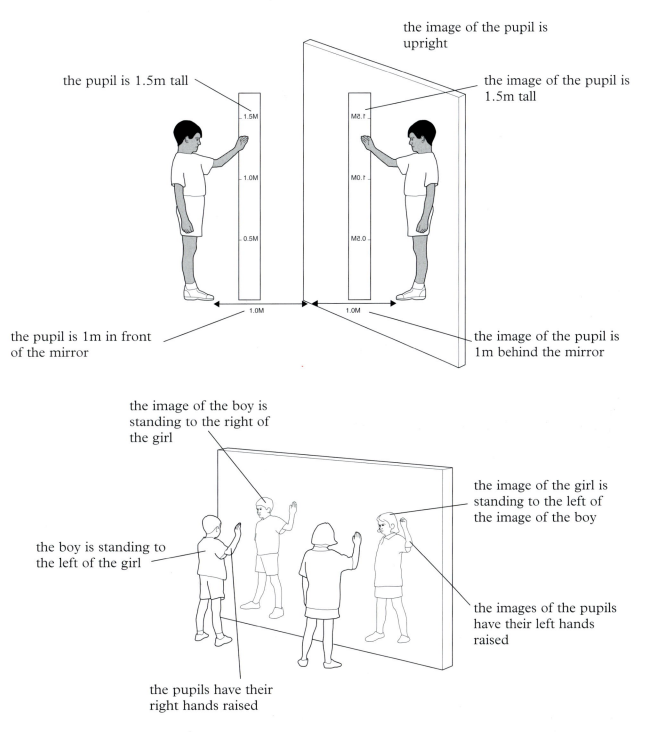

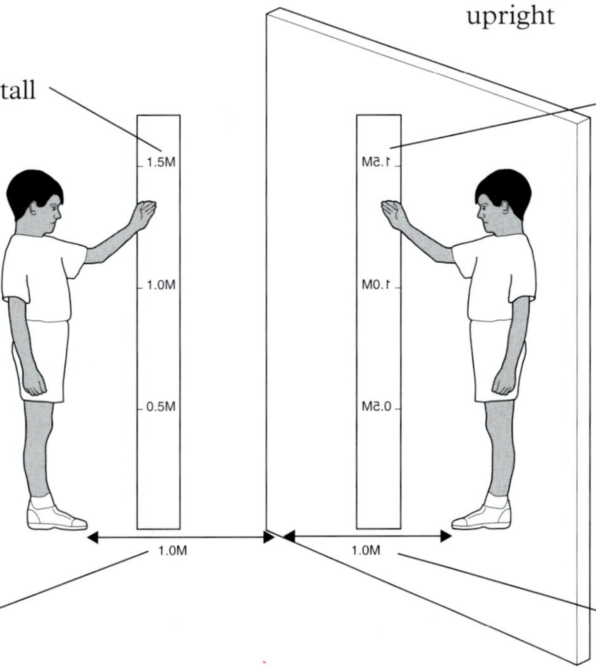
the image of the pupil is upright
the pupil is 1.5m tall
the image of the pupil is 1.5m tall
1.5M
1.0M
0.5M
1.0M
1.0M
the pupil is 1m in front of the mirror
the image of the pupil is 1m behind the mirror

the image of the boy is standing to the right of the girl
the image of the girl is standing to the left of the image of the boy
the boy is standing to the left of the girl
the images of the pupils have their left hands raised
the pupils have their right hands raised

The plane mirror has created an image that is back to front. It is **laterally inverted**.

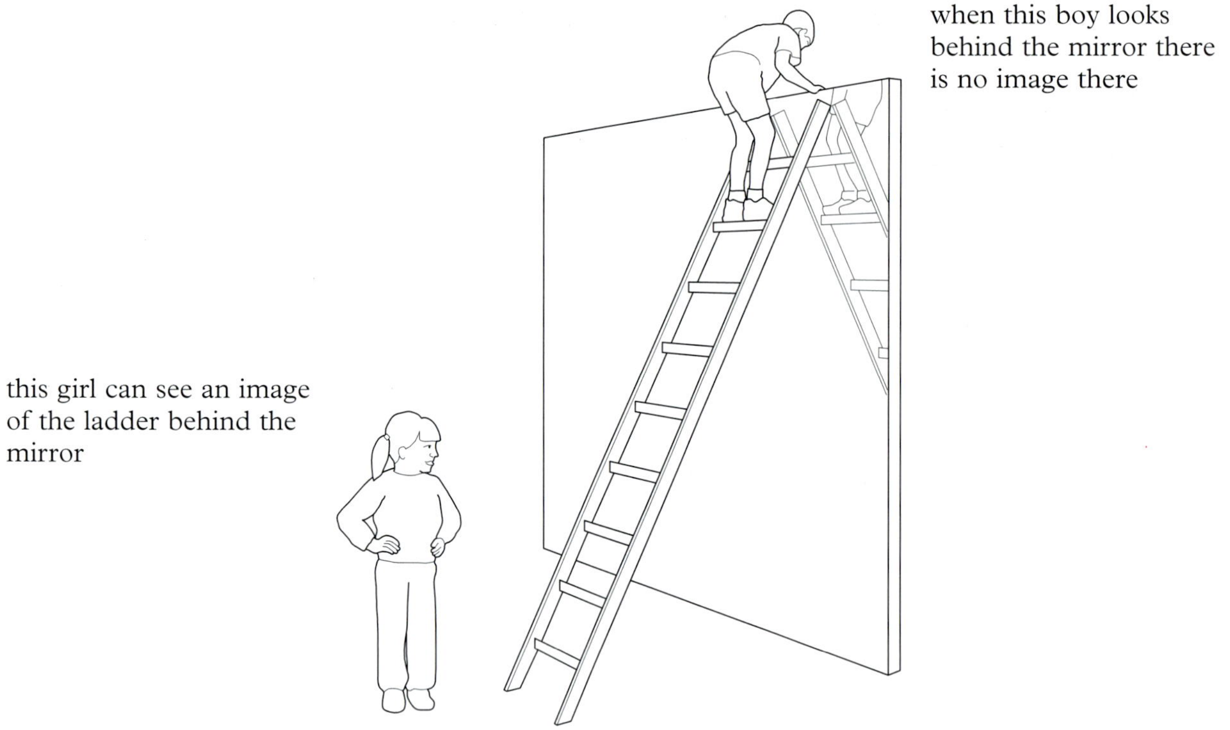
when this boy looks behind the mirror there is no image there
this girl can see an image of the ladder behind the mirror

The eye is tricked into seeing an image which isn't really there. It is a **virtual image**. Mirrors always produce virtual images. An image which really exists, like the image created on a screen by a projector, is called a **real image**.

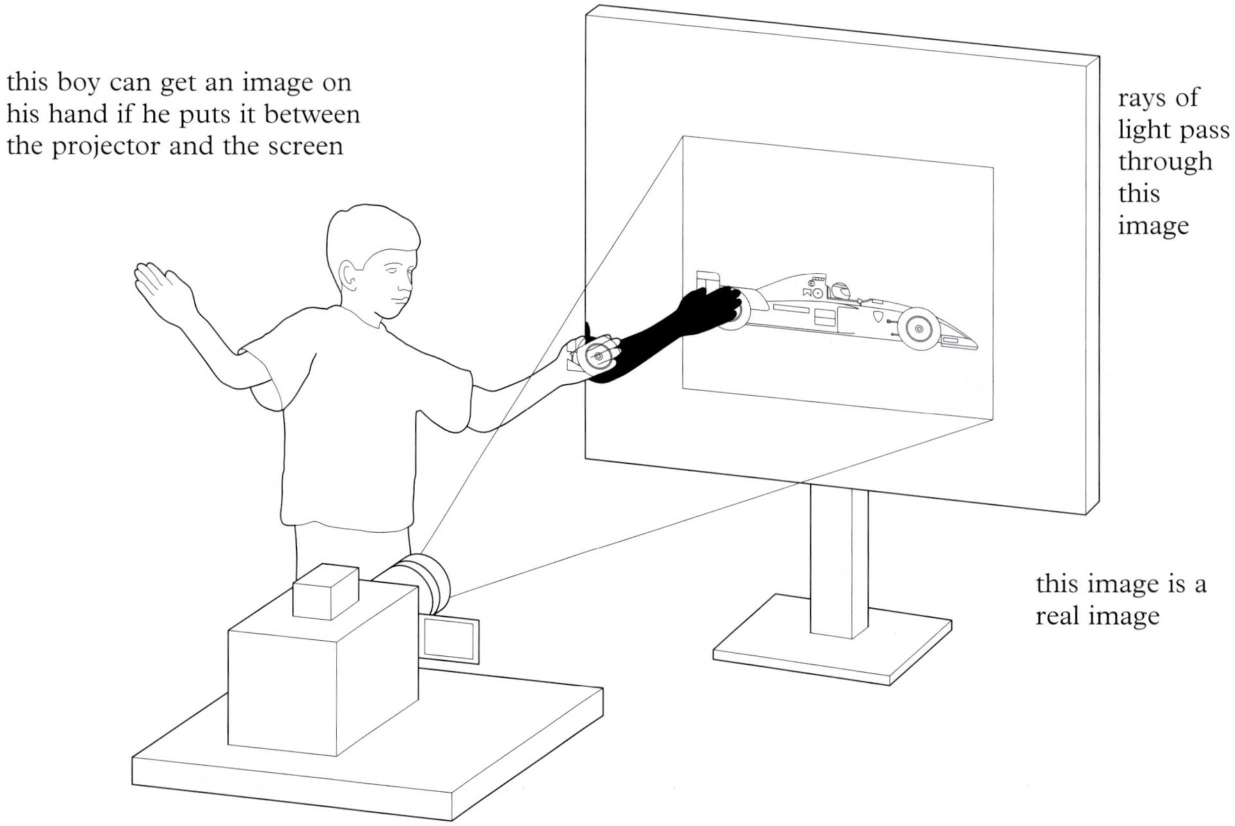
this boy can get an image on his hand if he puts it between the projector and the screen
rays of light pass through this image
this image is a real image

Some mirrors have surfaces which are not flat, instead they are curved.

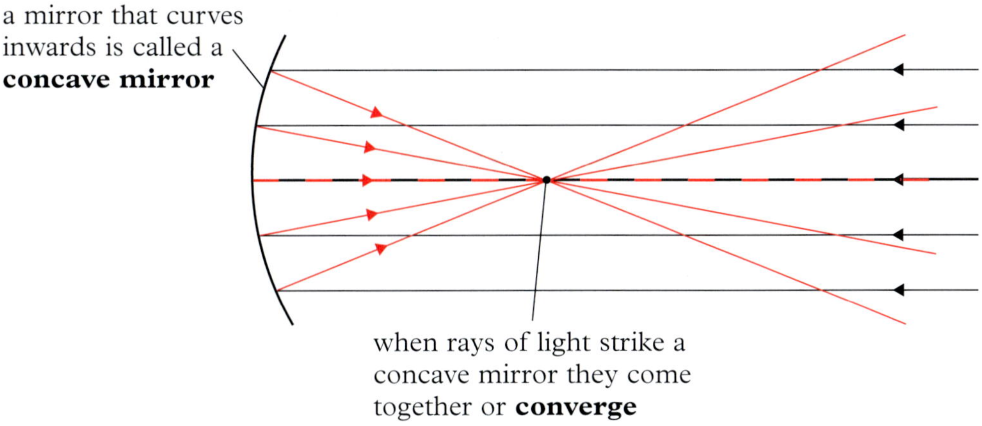
a mirror that curves inwards is called a **concave mirror**
when rays of light strike a concave mirror they come together or **converge**

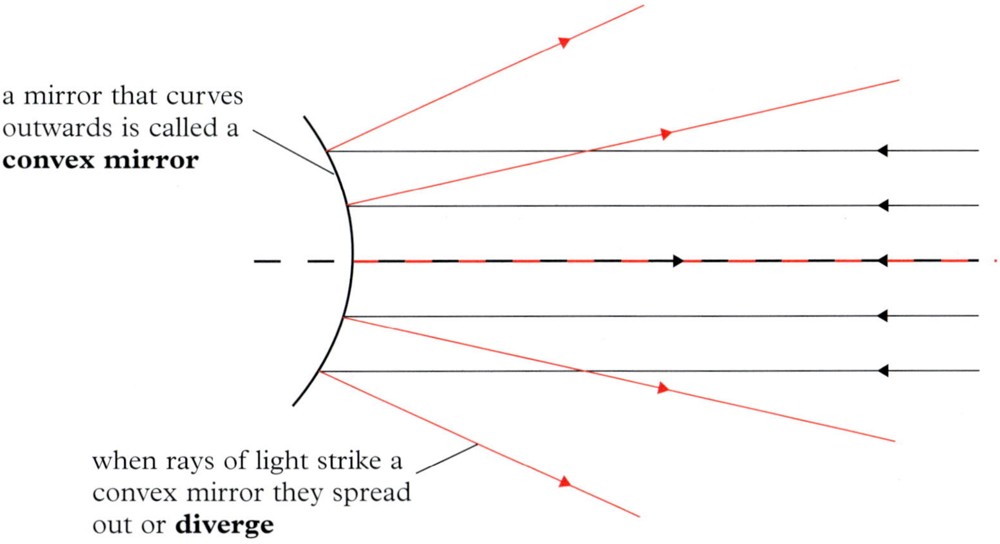
a mirror that curves outwards is called a **convex mirror**
when rays of light strike a convex mirror they spread out or **diverge**

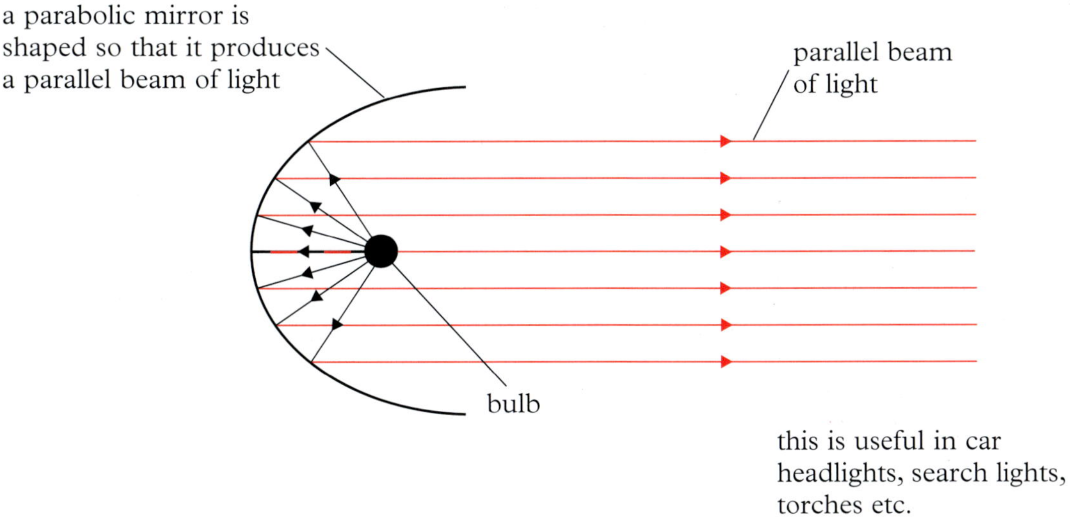
a parabolic mirror is shaped so that it produces a parallel beam of light
parallel beam of light
bulb
this is useful in car headlights, search lights, torches etc.

Questions

1 Draw a fully labelled diagram to show how a ray of light is reflected from a plane mirror. Then complete the following sentence.

If a ray of light strikes a plane mirror at an angle of 30° to the normal it will ...

2 a) Draw two diagrams to show why smooth surfaces look shiny but rough surfaces look dull.
b) Why do polished surfaces look shiny?

3 a) Draw a diagram to show how two mirrors can be used to make a simple periscope.
b) Give one use for a simple periscope.

5 Explain clearly what is meant by the following terms:
a) an inverted image,
b) a laterally inverted image,
c) a real image,
d) a virtual image.

Summary

Light normally travels in straight lines but when it strikes an object it may be reflected in a new direction. We see non-luminous objects because they reflect light into our eyes.

If a ray of light strikes a flat surface such as a mirror it is possible to predict the direction in which it will be reflected. The angle of incidence is equal to the angle of reflection. Reflected rays from a plane mirror create images. These images are always virtual, laterally inverted, the same size as the object and the same distance behind the mirror as the object is in front. Optical instruments such as periscopes use mirrors to change the direction in which a ray of light is travelling.

Key words

laterally inverted Something that has been changed so that the left side is now on the right and the right side is on the left.

luminous object An object which emits light, for instance, the Sun.

non-luminous object An object which does not emit light.

opaque material A material which does not allow light to pass through it.

real image An image which really exists and can be seen on a screen.

shadow An area of darkness created when an opaque object is placed in front of a light source.

transparent material A material which allows light to pass through it.

virtual image An image which cannot be seen on a screen.

End of Chapter 10 Questions

1 a) Draw a fully labelled diagram of a solar eclipse.
b) Name one luminous and one non-luminous object you have drawn in your diagram.
c) Explain the difference between a region where a total eclipse is seen and one where a partial eclipse is seen.

2 a) Draw an accurate and fully labelled diagram to show how a plane mirror creates an image of an object placed in front of it.
b) Describe the properties of the image created by the plane mirror.
c) Why is the word AMBULANCE often written on the front of the vehicle as ƎƆИA⅃UᙠMA?
d) Write down your name and address in mirror writing.
e) i) The diagrams below show the faces of three clocks seen in a mirror.
ii) What times are the clocks actually showing?

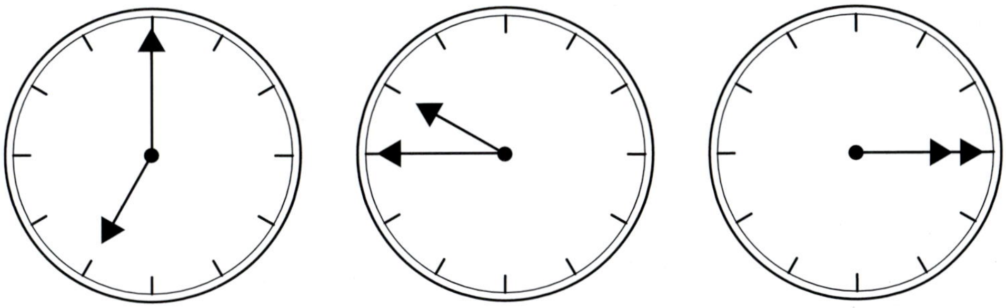

3 Draw a labelled diagram to show how you could use a) one mirror b) two mirrors to turn a ray of light through 180° (reflect it back in the direction from which it came).

11 Light Waves: Refraction and Total Internal Reflection

Light waves can travel through many different transparent materials such as air, water and glass. These different materials are called **media**. When light waves travel through a vacuum they move at a speed of 300 000 000 m/s. When they travel through other media they travel more slowly. For instance in glass light waves travel at 200 000 000 m/s. As a light wave travels across the boundary between two different media its speed has to change. This change in speed can cause the ray of light to change direction. This change in direction is called **refraction**.

A ray of light may travel from a less dense medium into a more dense medium.

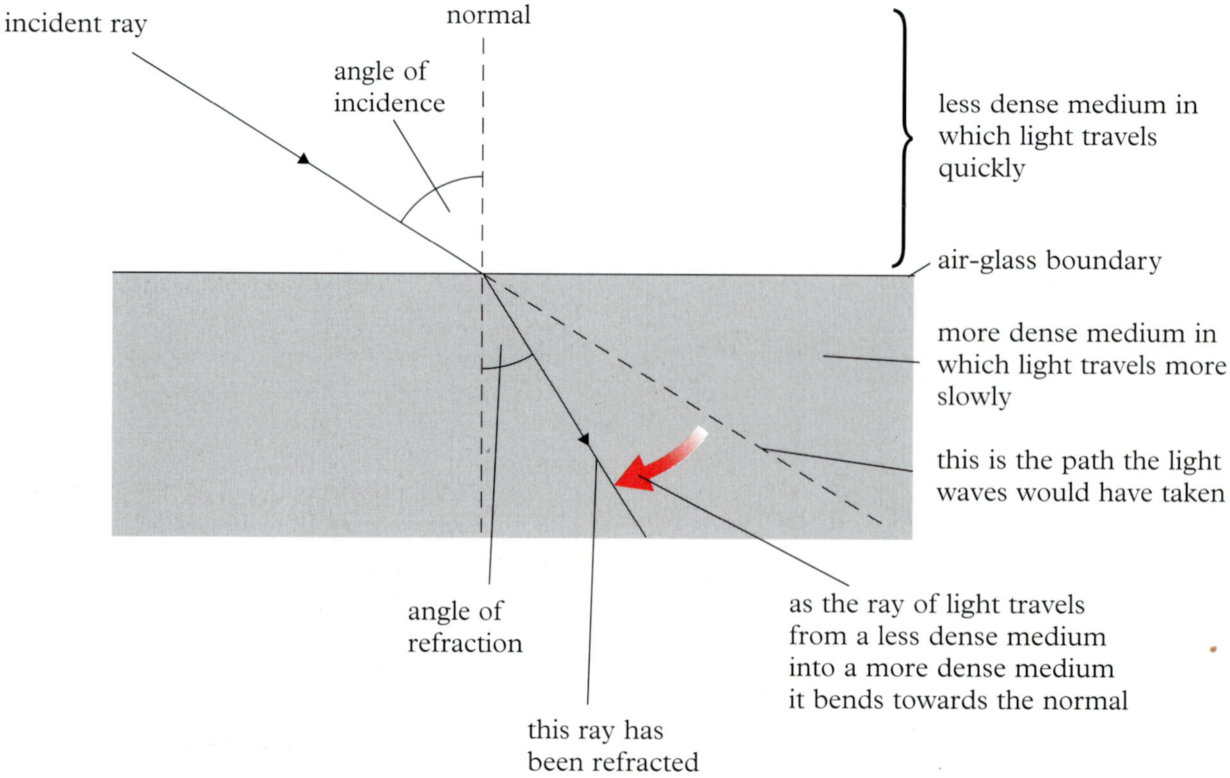

incident ray
normal
angle of incidence
less dense medium in which light travels quickly
air-glass boundary
more dense medium in which light travels more slowly
this is the path the light waves would have taken
angle of refraction
as the ray of light travels from a less dense medium into a more dense medium it bends towards the normal
this ray has been refracted

A ray of light may travel from a more dense medium into a less dense medium.

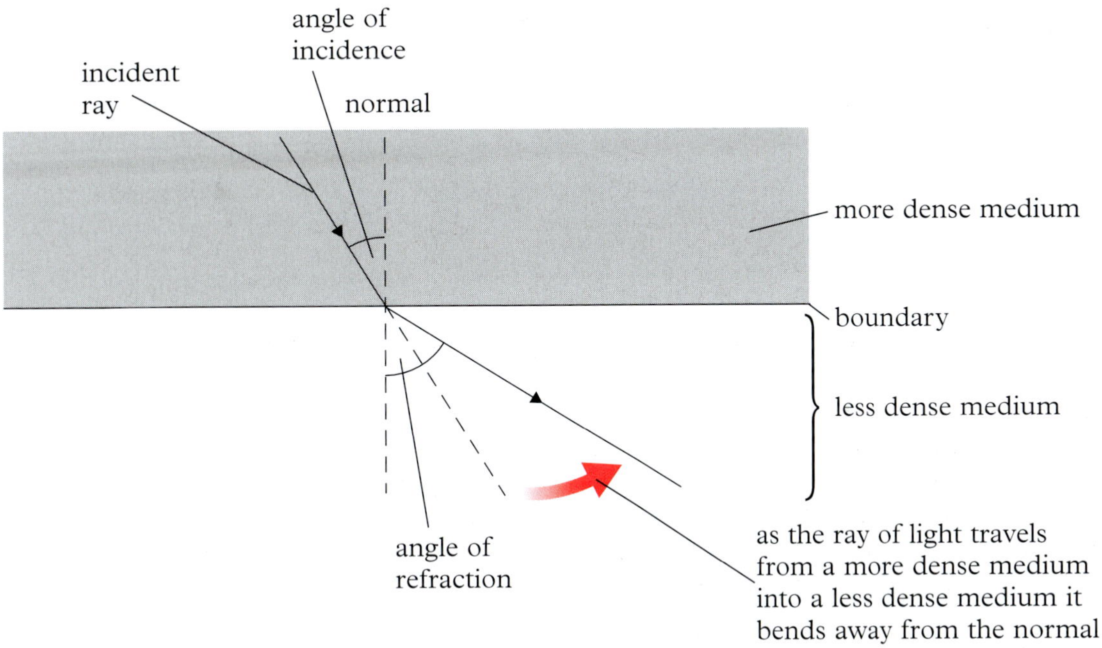
angle of
incidence
incident
ray
normal
more dense medium
boundary
less dense medium
angle of
refraction
as the ray of light travels
from a more dense medium
into a less dense medium it
bends away from the normal

A ray of light passing between media does not always change direction.

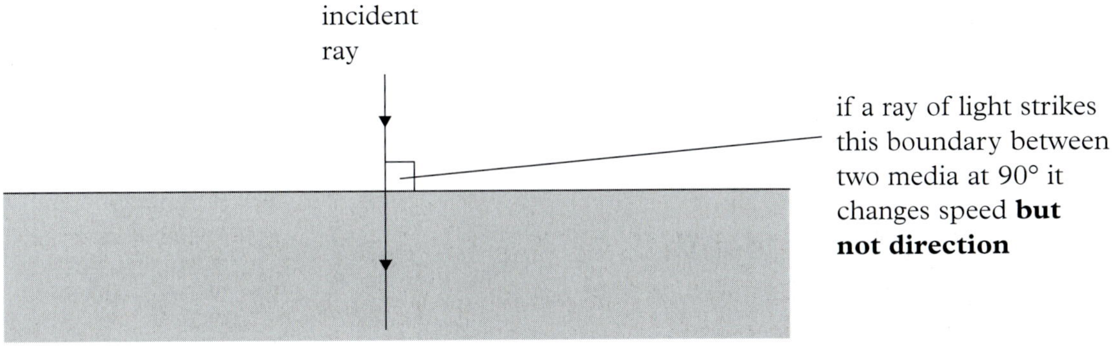
incident
ray
if a ray of light strikes
this boundary between
two media at 90° it
changes speed **but
not direction**

Questions

1 What happens to the speed of a ray of light as it passes from a less dense medium into a more dense medium?

2 What happens to the speed of a ray of light as it passes from a more dense medium into a less dense medium?

3 a) Draw a labelled diagram to show what happens to a ray of light when it crosses the boundary from air to glass at an angle of approximately 40° to the normal.
b) Do the same for light as it passes from glass to air.

4 Draw a diagram to show what happens to a ray of light which travels at 90° to the boundary between two media.

Everyday effects of refraction

We expect light to travel in straight lines. So when refraction takes place it can produce some strange effects.

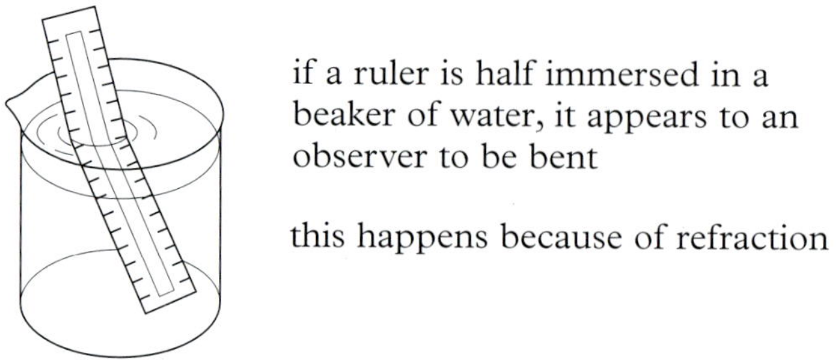
if a ruler is half immersed in a beaker of water, it appears to an observer to be bent
this happens because of refraction

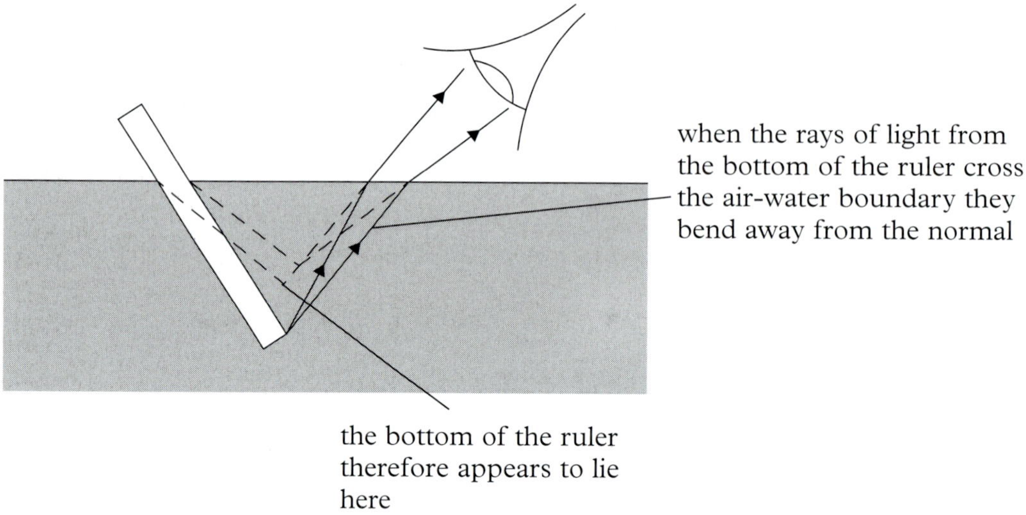
when the rays of light from the bottom of the ruler cross the air-water boundary they bend away from the normal
the bottom of the ruler therefore appears to lie here

Real and apparent depth

A river or a swimming pool always appears shallower than it really is. This again is the result of refraction.

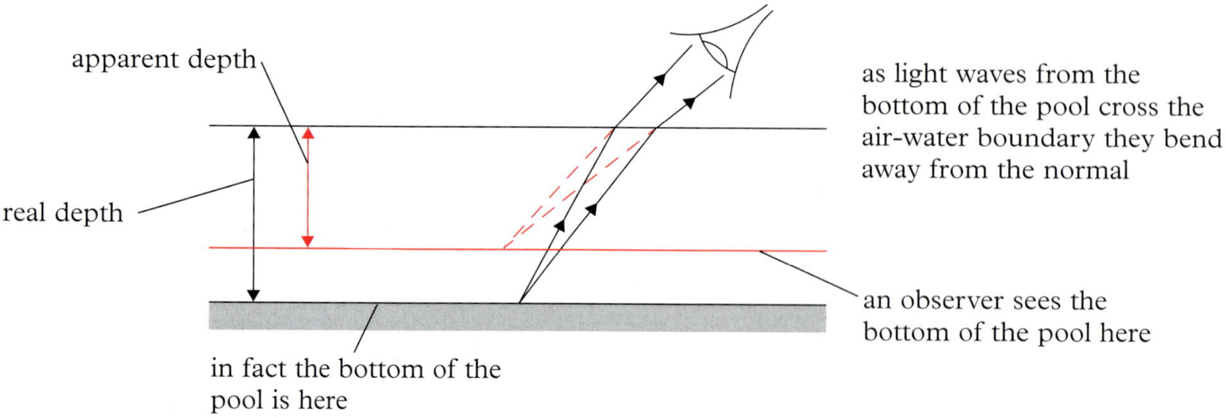
apparent depth
real depth
as light waves from the bottom of the pool cross the air-water boundary they bend away from the normal
an observer sees the bottom of the pool here
in fact the bottom of the pool is here

Water is usually about half as deep again as it appears. This means that if water appears to be about 2 m deep it is actually about 3 m deep!

Questions

1 Draw a labelled ray diagram to show why a swimming pool looks shallower than it really is.

2 Draw a labelled ray diagram to show why a pencil half immersed in water looks bent.

Total Internal Reflection

We expect a ray of light leaving a more dense medium as glass to be refracted away from the normal.

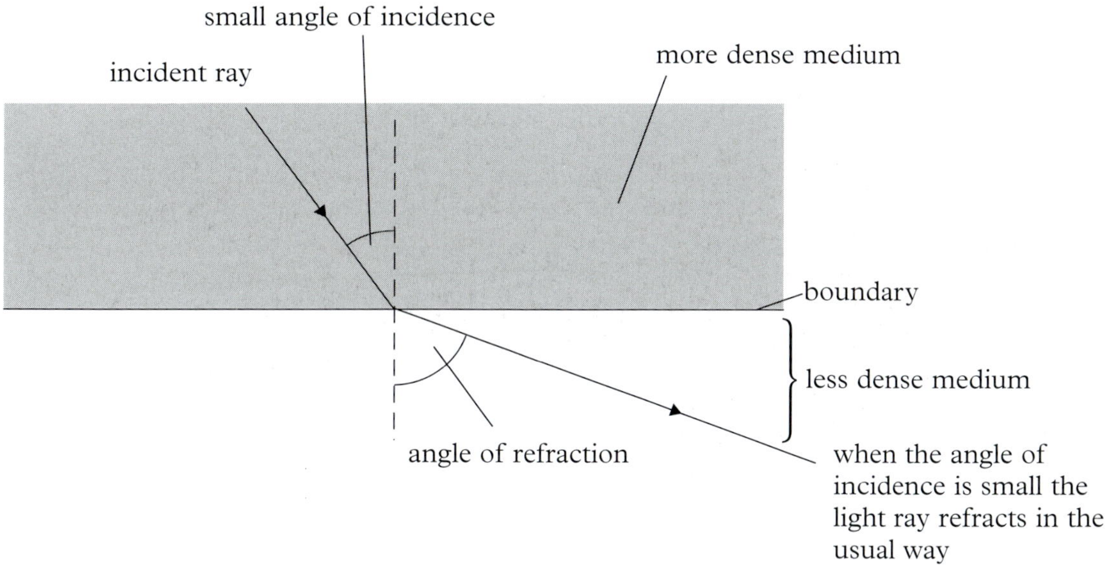
small angle of incidence
incident ray
more dense medium
boundary
less dense medium
angle of refraction
when the angle of incidence is small the light ray refracts in the usual way

Sometimes the ray is not refracted but is reflected back into the glass. This is called **total internal reflection**.

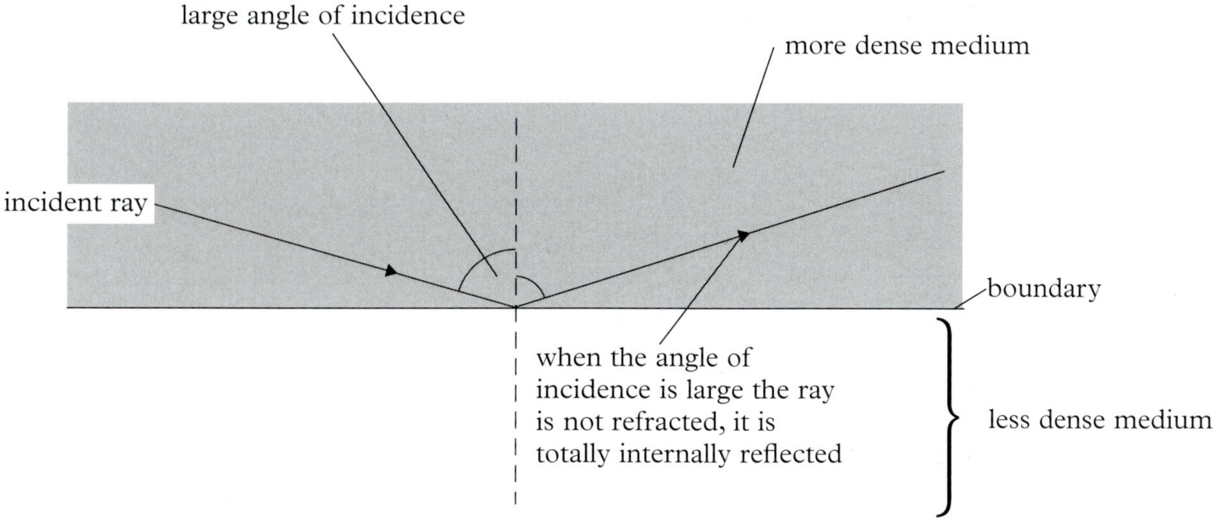
large angle of incidence
more dense medium
incident ray
boundary
when the angle of incidence is large the ray is not refracted, it is totally internally reflected
less dense medium

The boundary between the less dense and more dense media acts like a mirror during total internal reflection.

Critical angle

If the angle of incidence is equal to the **critical angle** the refracted ray travels along the boundary between the two media.

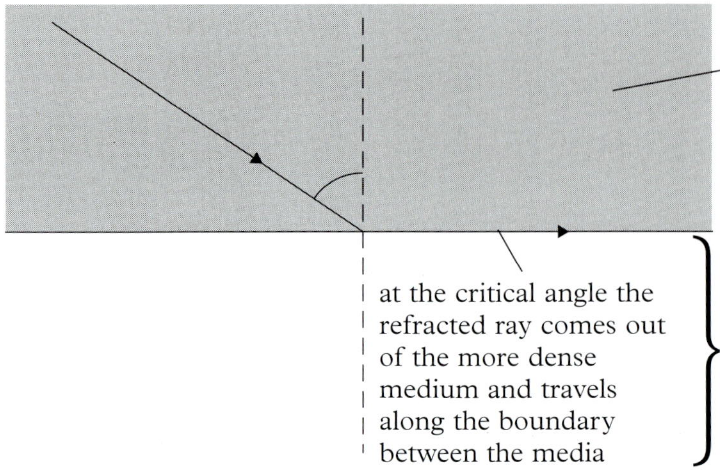
this light ray approaches the boundary at a certain angle known as the **critical angle**
more dense medium
at the critical angle the refracted ray comes out of the more dense medium and travels along the boundary between the media
less dense medium

If the angle of incidence is less than the critical angle, refraction takes place. If the angle of incidence is greater than the critical angle, total internal reflection takes place.

The critical angle is different for the boundaries between different media.

For total internal reflection to take place two things must happen:

1 The ray of light must strike the boundary between the two medium at an angle greater than the critical angle.

2 The ray must be travelling from a more dense medium such as water or glass to a less dense medium such as air or vacuum.

Total internal reflection can be very useful.

Prisms and Periscopes

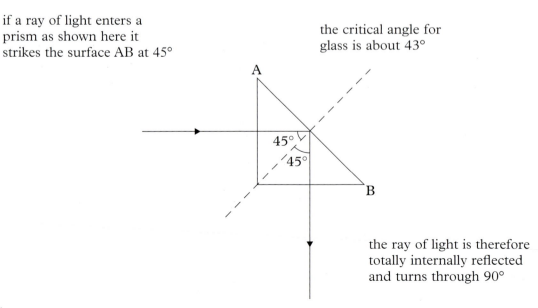

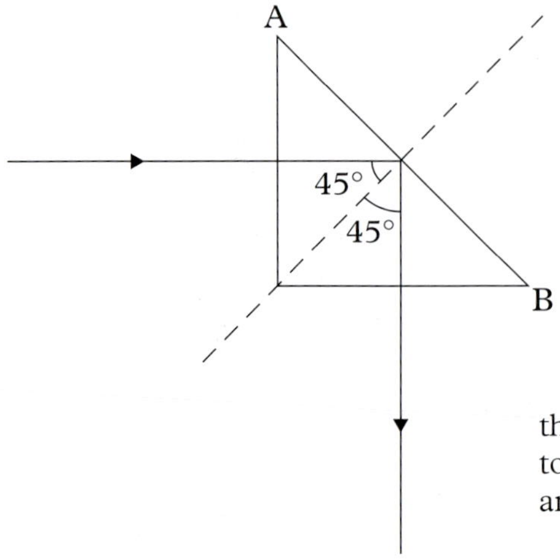
if a ray of light enters a prism as shown here it strikes the surface AB at 45°
the critical angle for glass is about 43°
A
B
45°
45°
the ray of light is therefore totally internally reflected and turns through 90°

Using two prisms it is possible to make a prismatic periscope.

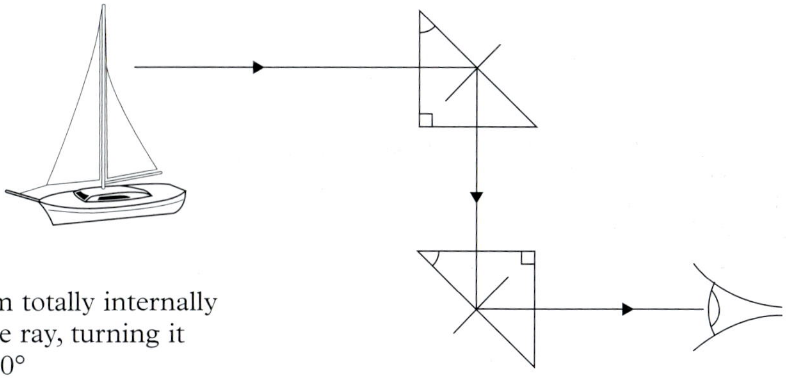

each prism totally internally reflects the ray, turning it through 90°

Prismatic periscopes are not as fragile as those that use mirrors and the images they produce are much brighter.

if a ray of light is sent into a prism as shown here it strikes the surface AB at 45° and is turned through 90°

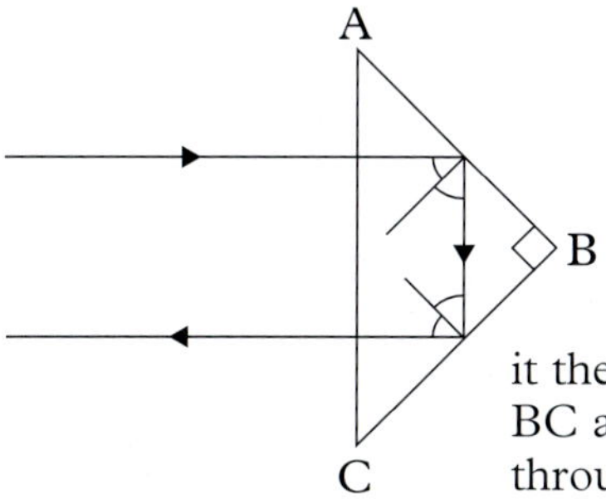

it then strikes the surface BC at 45° and is turned through a further 90°

The overall effect of these two total internal reflections is to send the ray of light back in the direction from which it came. This idea is used in a bicycle reflector.

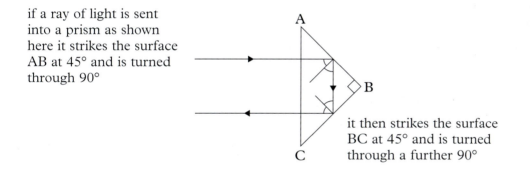

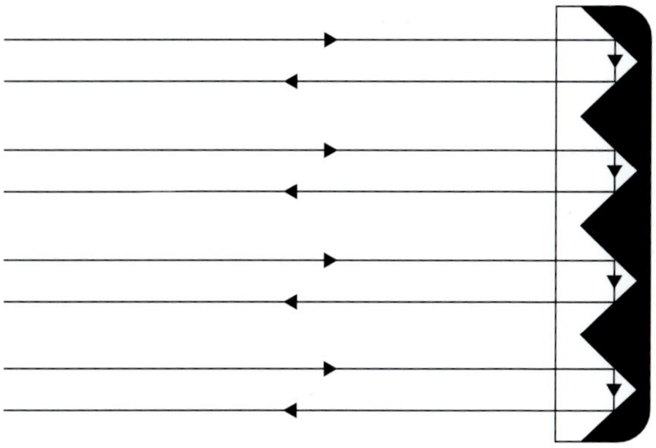

total internal reflection turns the ray through 90° twice

the reflected ray returns in the direction from which it came

A car driver can then see the reflection of his or her car headlights which alerts him or her to the cyclist.

Optical fibres

Optical fibres are thread-like pieces of glass approximately one tenth of a millimetre in diameter.

there is a thick outer coating of less dense glass
there is a narrow core of more dense glass

The inner glass is optically more dense than the outer glass.

Because the fibres are thin, rays of light travelling along the core will always strike the boundary between the two glasses at an angle greater than the critical angle. The rays will therefore undergo a series of total internal reflections before emerging at the other end. None of the light escapes through the sides of the core even when the fibre is bent. The fibre behaves like a light-tube.

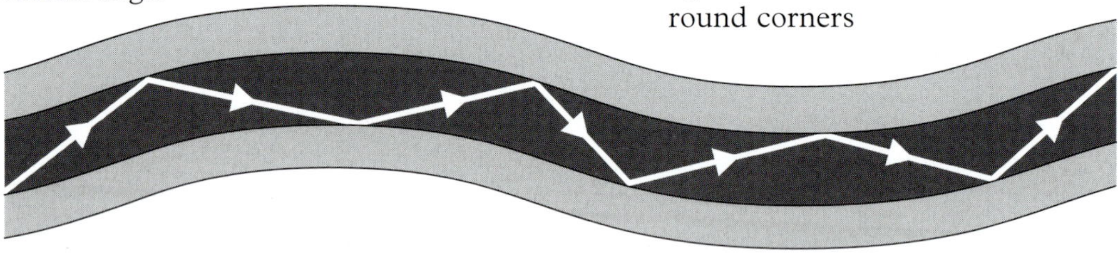
the angle of incidence is always greater than the critical angle
light is able to travel round corners

Optical fibres are used in many applications.

The endoscope
The endoscope is an instrument that allows doctors to look inside patients' bodies. It uses optical fibres to do this.

Before the invention of this instrument examination of the internal workings of the human body often meant patients undergoing an operation. Using the endoscope, doctors and surgeons can see directly the inside parts of a body such as the stomach.

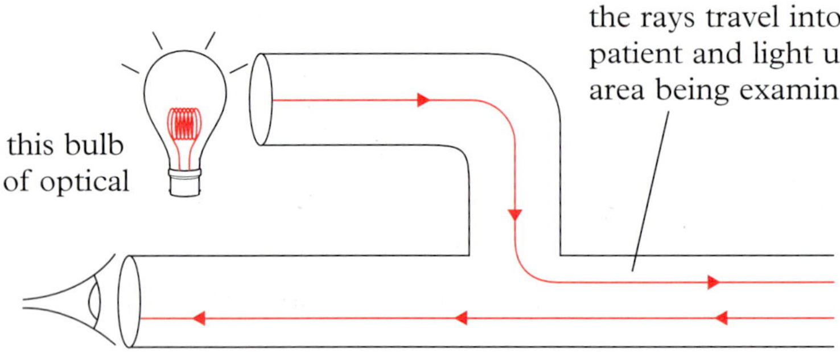
light rays from this bulb enter a bundle of optical fibres
the rays travel into the patient and light up the area being examined
the rays are then reflected back through more optical fibres and form an image for the doctor at the eyepiece lens

Communicating through optical fibres

Until recently most of our telephone conversations and cable TV programmes travelled to our homes as electrical signals passing through copper wires buried underground. Gradually these wires are being replaced by optical fibres and the electrical signals by light waves. This is because optical fibres are cheaper to install, can carry more messages at any one time and they transmit signals with very little loss of signal strength.

Questions

1. Explain the terms critical angle and total internal reflection.
2. What conditions are necessary for total internal reflection to take place?
3. a) Draw two diagrams to show how a prism can be used to turn a ray of light through
   i) 90°
   ii) 180°.
   b) Describe one use for each of these prisms.
4. Explain why the light entering one end of an optical fibre is unable to escape through the sides of the fibre.
5. Give three reasons why copper telephone wires are being replaced with optical fibres.

Summary

When a ray of light travels from one medium to a different medium, its speed changes. This change in speed may cause the ray to refract or change direction. As a ray slows down it may refract towards the normal. As it speeds up it may refract away from the normal. Because we normally expect rays of light to travel in straight lines, refraction can cause our eyes to be misled, e.g. straight objects may appear bent.

If a ray of light travelling from a more dense medium into less dense medium strikes the boundary at an angle greater than the critical angle it undergoes total internal reflection. We can use total internal reflection to change the direction of a ray of light. This can be done using prisms and optical fibres.

Key words

bundle	A large number of optical fibres grouped together.
critical angle	The angle of incidence at which a refracted ray travels along the boundary between two media instead of either crossing the boundary or being totally internally reflected.
endoscope	An optical instrument which allows doctors and surgeons to see the insides of a human body.
medium	The transparent material through which light is travelling.

Key words (continued)

optical fibre A thread of glass consisting of a dense inner core surrounded by a less dense outer coating.

refraction The change in direction of a ray of light as it crosses the boundary between two media.

total internal reflection The reflection of a ray of light from the boundary between two transparent media.

End of Chapter 11 Questions

1 Glass is optically denser than water. Water is optically denser than air. Using this information decide which of the diagrams below are correct and which are incorrect.

(a)

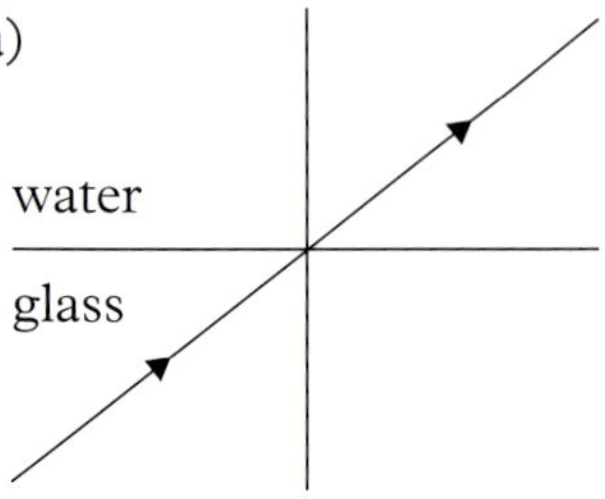

(b)

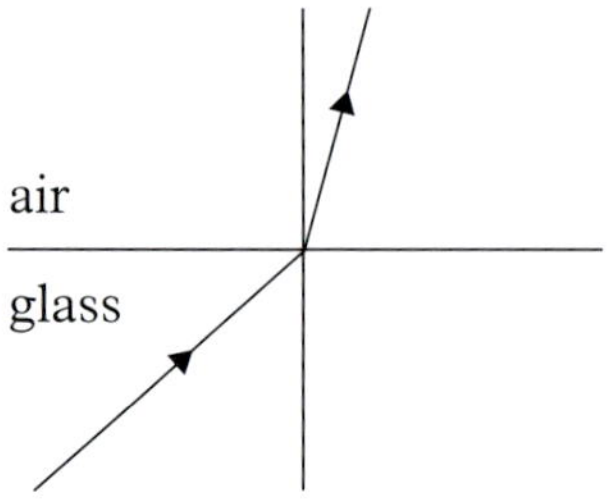

(c)

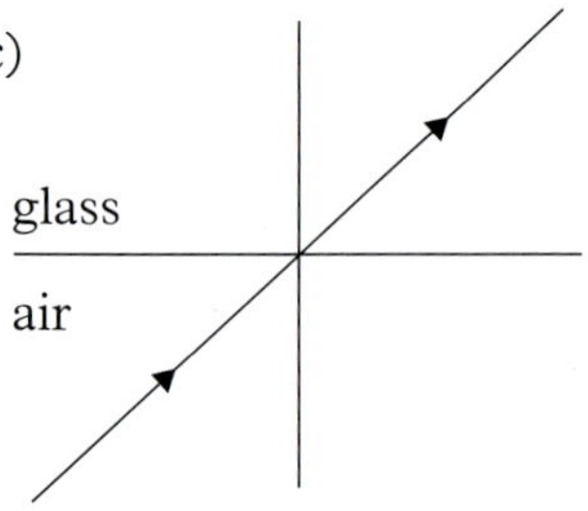

(d)

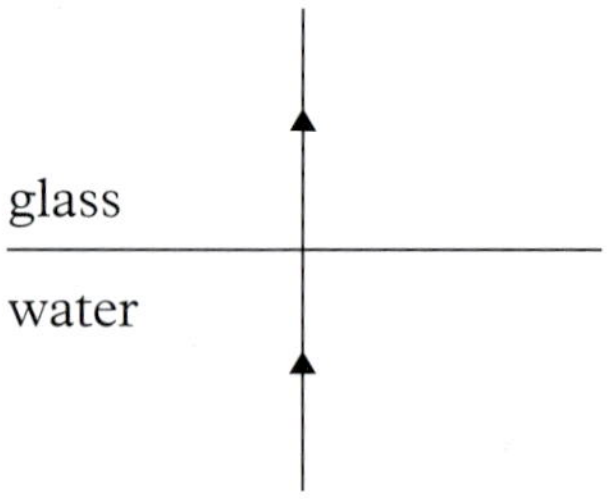

2 a) Explain why the hunter in the diagram below should not aim at the place where he sees the fish.

b) Where should he aim?

3 The critical angle for glass is 42°. Look at the diagrams below and decide which are drawn correctly and which are drawn incorrectly.

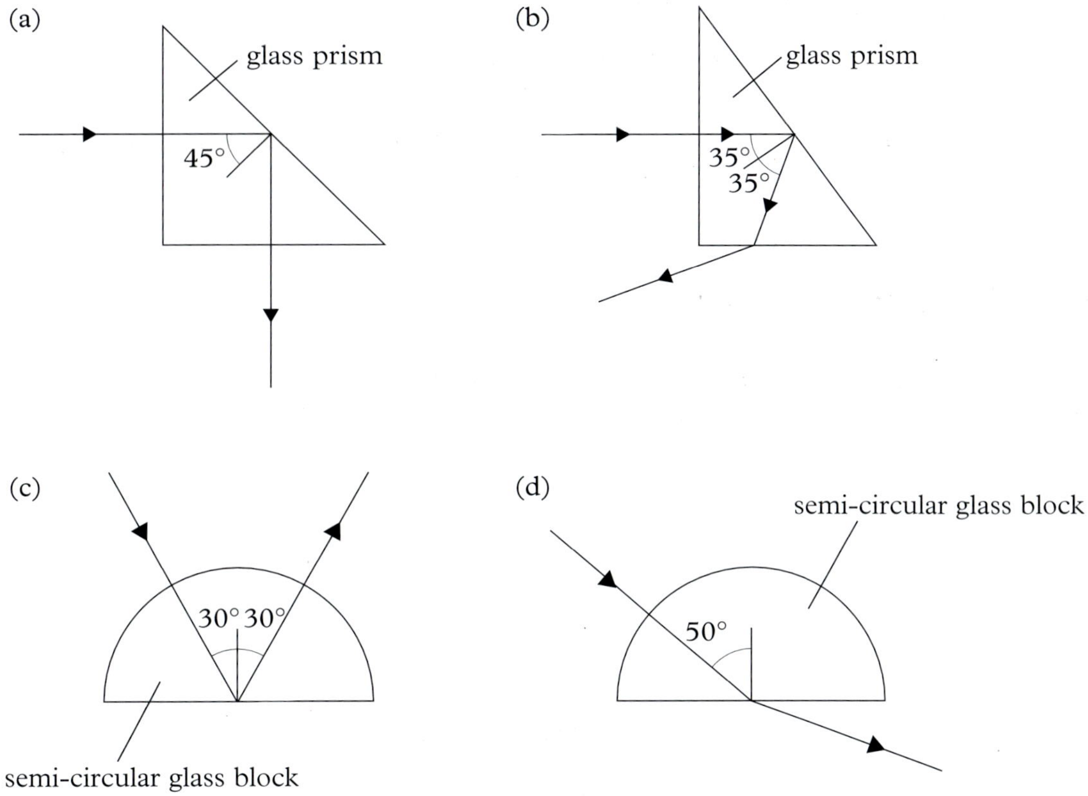

4 Explain why the core of an optical fibre must be made from a glass which is optically more dense than the outer glass. Draw a diagram of an endoscope and explain how this uses total internal reflection to allow doctors to see inside their patients.

12 Colour and the Electromagnetic Spectrum

Dispersion of white light

If we shine a ray of white light into a glass prism we will see it refract as it enters and as it leaves. These refractions cause the white light to split into a band of colours called a **spectrum**. This splitting of the light is called **dispersion**.

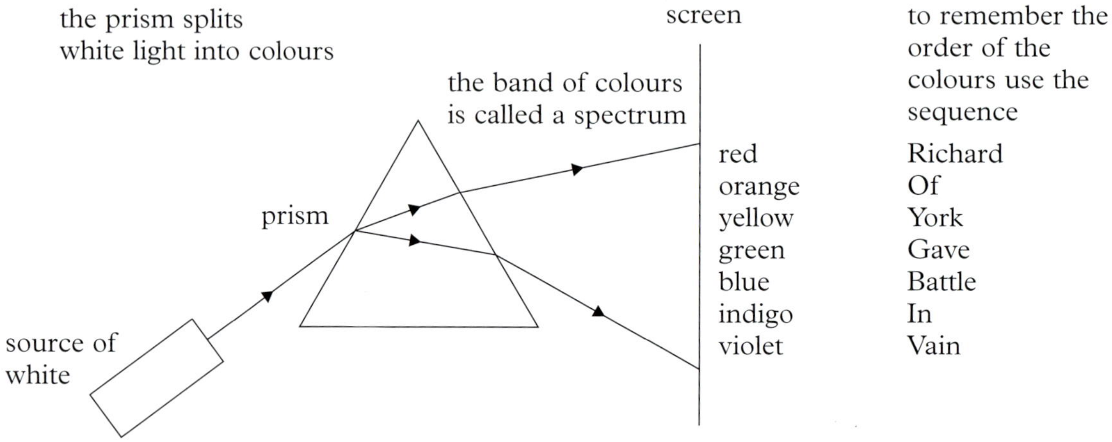
the prism splits
white light into colours
screen
the band of colours
is called a spectrum
prism
source of
white
red
orange
yellow
green
blue
indigo
violet
to remember the
order of the
colours use the
sequence
Richard
Of
York
Gave
Battle
In
Vain

Using an identical prism which has been turned upside down the colours can be made to recombine, producing white light.

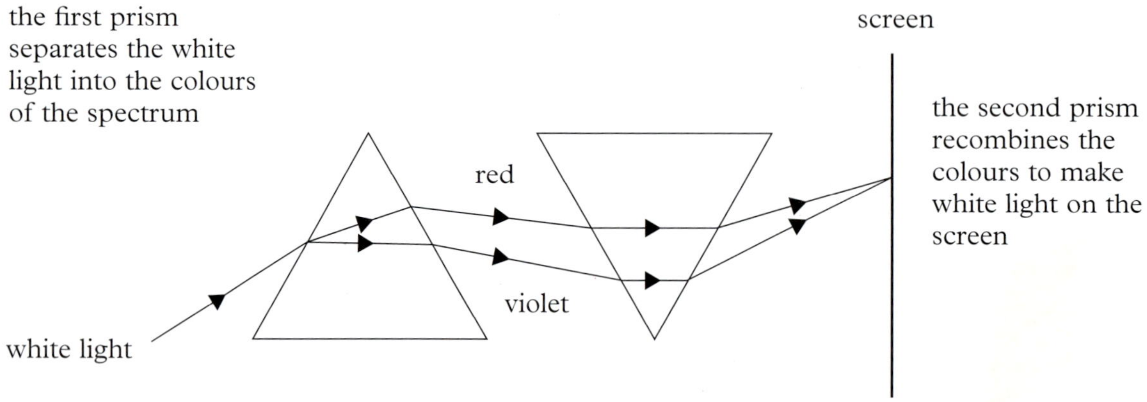
the first prism
separates the white
light into the colours
of the spectrum
screen
red
violet
white light
the second prism
recombines the
colours to make
white light on the
screen

Coloured objects

Most objects have a colour. For example grass is green, ink can be blue, black or red. They have these colours because they contain a chemical called a **dye**. The dye absorbs all the colours of light which strike it except for its own colour which it reflects.

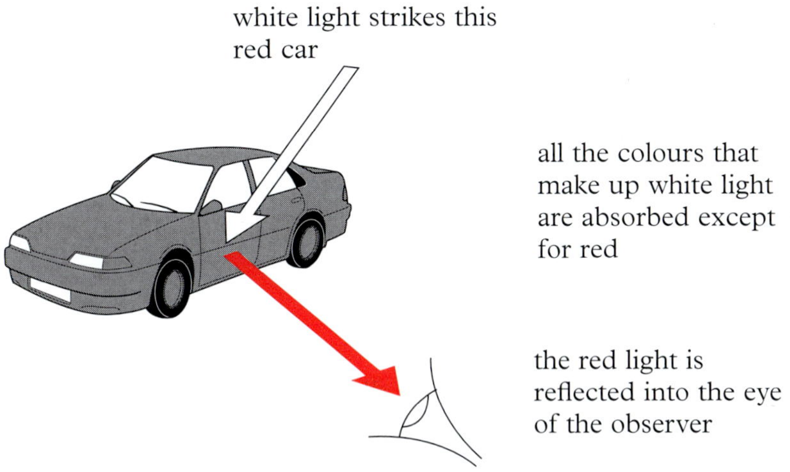
white light strikes this red car
all the colours that make up white light are absorbed except for red
the red light is reflected into the eye of the observer

This is why a red car looks red.

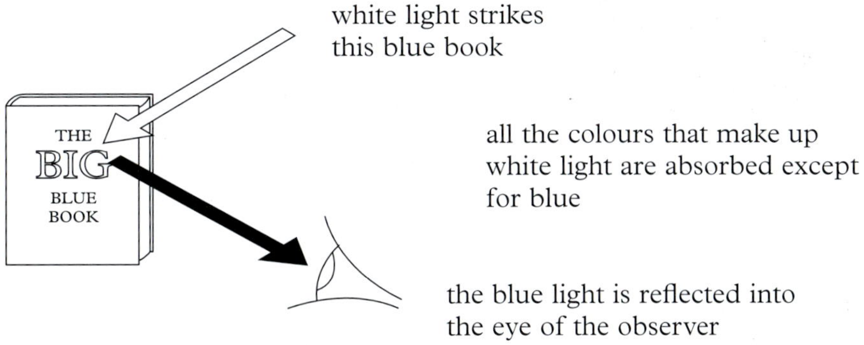
white light strikes this blue book
THE
BIG
BLUE
BOOK
all the colours that make up white light are absorbed except for blue
the blue light is reflected into the eye of the observer

This is why a blue book looks blue.

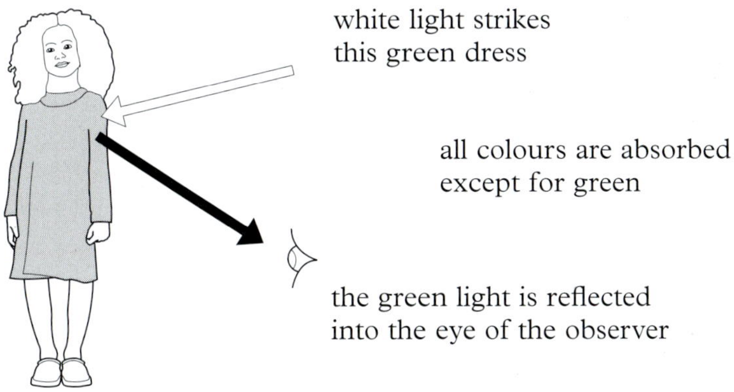
white light strikes this green dress
all colours are absorbed except for green
the green light is reflected into the eye of the observer

This is why a green dress looks green.

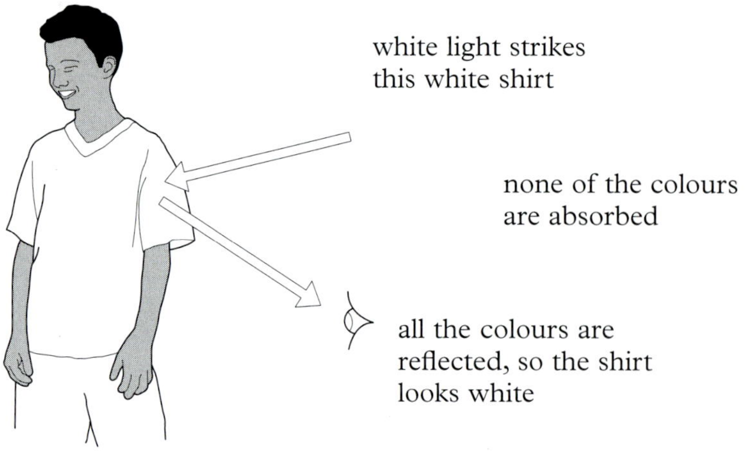
white light strikes
this white shirt
none of the colours
are absorbed
all the colours are
reflected, so the shirt
looks white

This is why a white shirt looks white.

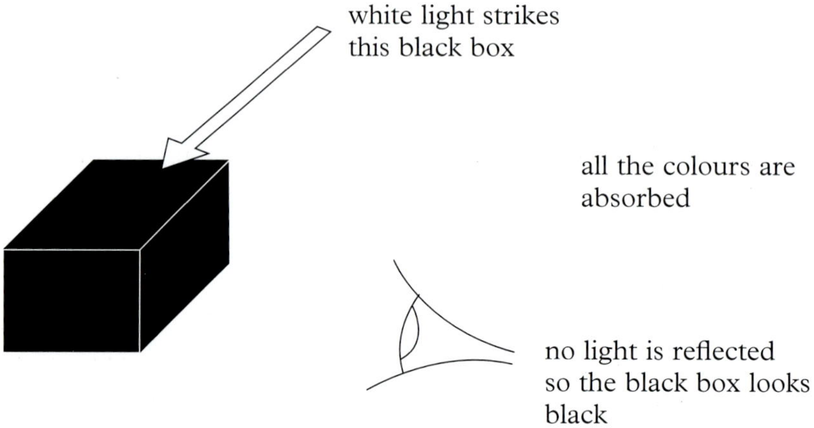
white light strikes
this black box
all the colours are
absorbed
no light is reflected
so the black box looks
black

This is why a black box looks black.

Questions

1 Write down in order all the colours of the visible spectrum beginning with red.

2 a) Draw a diagram to show how a prism can be used to produce a spectrum.
b) Give one example of a naturally occurring spectrum.

3 a) What is a dye?
b) Choose three differently coloured objects you can see and explain how you see their colours.

Adding coloured lights

If we shine coloured lights onto a white screen we can see that where the coloured lights overlap new colours are produced.

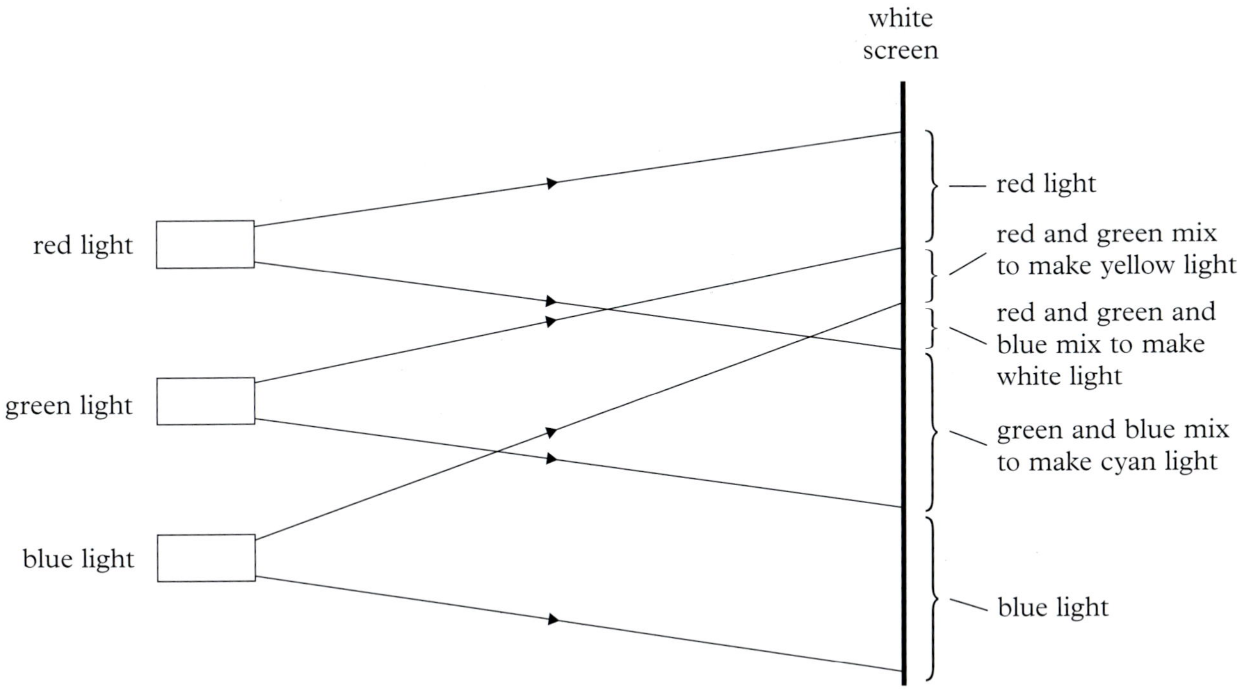

There are three colours of light which cannot be made by mixing other colours. These are red, green and blue. These are called the **primary colours**.

If all three primary colours are mixed in equal amounts they produce white light.

If equal amounts of any two of the primary colours are added together they produce the colours yellow, cyan or magenta. These are known as the **secondary colours**.

The overall effects of adding coloured lights is summarised in the colour mixing triangle.

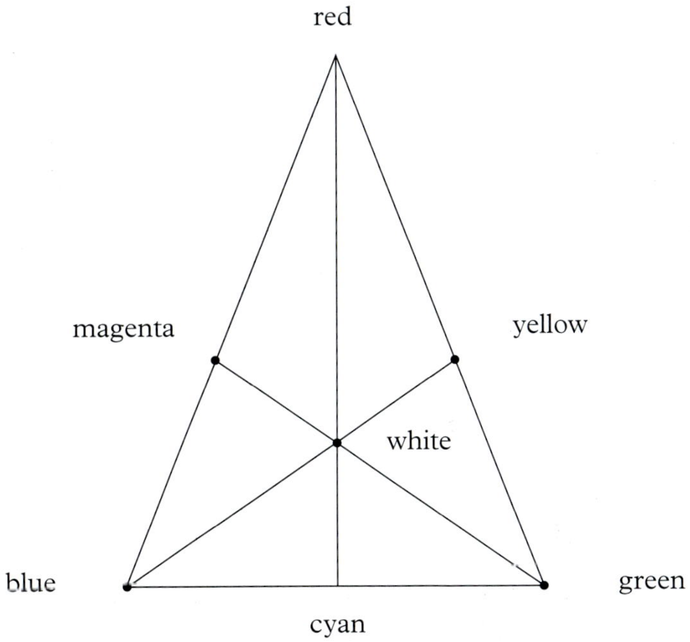

Coloured filters

A filter is a clear piece of plastic which has been coloured by a dye. Light which has the same colour as the filter can pass through it but all other colours cannot. They are absorbed.

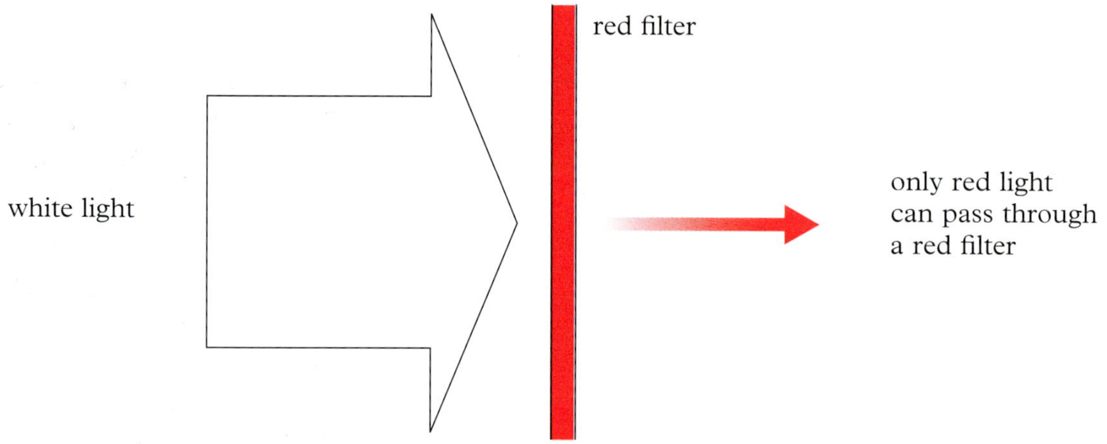
red filter
white light
only red light
can pass through
a red filter

Questions

1 Use the colour triangle to complete the table below.

Colour A	Colour B	Colour A + Colour B
Red	Green	
Red	Blue	
Blue	Green	
Yellow	Blue	
Green	Magenta	
Red	Cyan	

2 Explain what happens when white light tries to travel through the following filters:
a) a blue filter
b) a green filter
c) a red filter with a blue filter behind it.

The Electromagnetic Spectrum

Visible light is just one small part of a much larger family of waves called the **electromagnetic spectrum**. Although at first visible light seems to have little in common with the other members of the spectrum, there are in fact many similarities.

- they all travel at the same speed
- they can all travel through a vacuum
- they are all transverse waves
- they can all be reflected, refracted, diffracted and produce interference patterns

radio waves
have wavelengths
10 km–10 cm

we use radio waves to carry signals and messages over long distances

microwaves have wavelengths
1 cm–1 mm

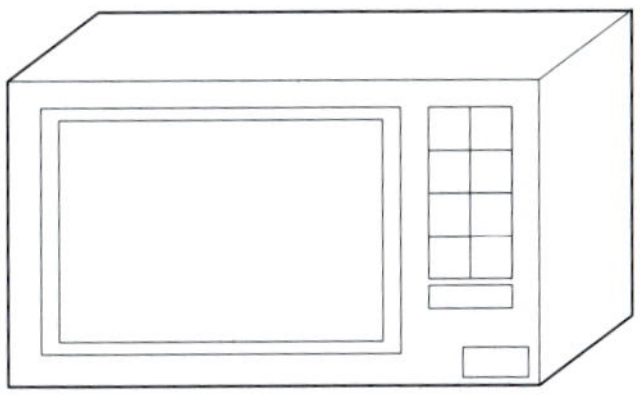

we use microwaves to cook food and carry messages from mobile phones

infra red waves
have wavelengths of
around $\frac{1}{10}$ mm

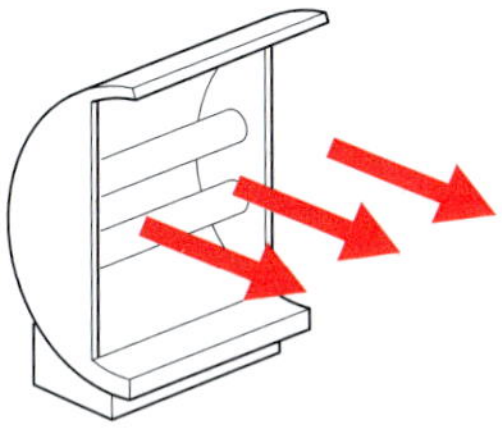

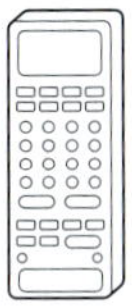

warm objects give off infra red waves which then warm up any object they meet

remote controls for TVs and stereos emit infra red waves

visible light waves have wavelengths of around
$\frac{1}{1000}$ mm

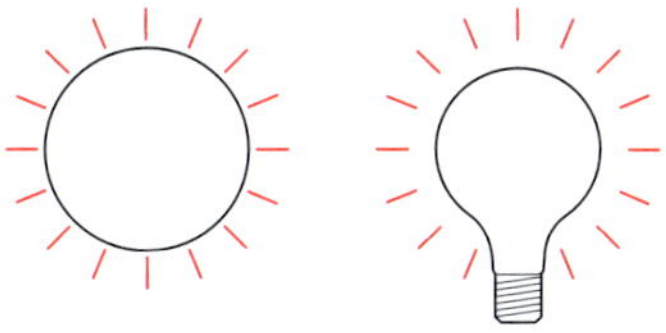

we use visible light emitted by the Sun, light bulbs and so on to see with

our eyes can detect these waves

ultraviolet waves
have wavelengths of
around $\frac{1}{10\,000}$ mm

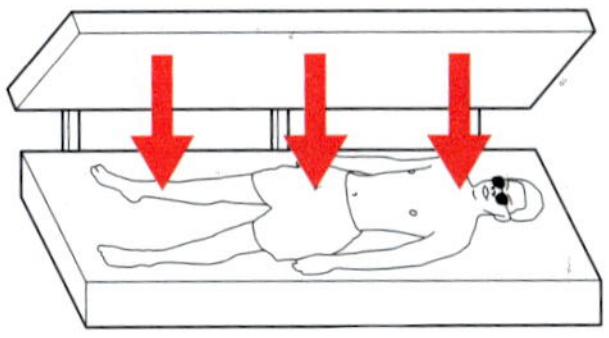

ultraviolet waves give us a suntan, but they can damage the skin and eyes

X-rays have
wavelengths of
around one-millionth
of a mm

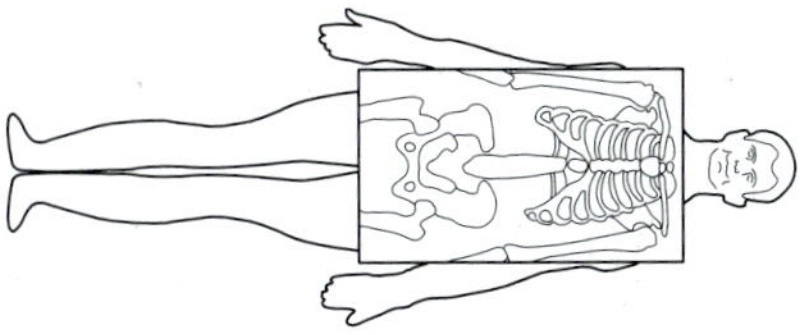

X-rays are very penetrating and are used by doctors to see the bones inside our bodies

gamma rays have
wavelengths of
around one-hundred
millionth of a mm

gamma rays are dangerous rays given out by some radioactive materials

they can cause cancer but can also be used to kill germs and sterilise surgical equipment

Questions

1 Name three properties which all the members of the electromagnetic spectrum have in common.

2 a) Name two groups of waves from the electromagnetic spectrum.
b) Describe a use for each of these waves.

Summary

White light is a mixture of different colours – the colours of the rainbow. In order, these are red, orange, yellow, green, blue, indigo and violet. When white light strikes a coloured object all these different colours are absorbed, except for light which is the same colour as the object. This light is reflected from the object.

Coloured lights can be mixed to produce new colours as shown in the colour triangle. Red, green and blue are primary colours. They cannot be made by mixing other colours.

The electromagnetic spectrum is a group of waves which have many common properties. They all travel at the same speed, they can all travel through a vacuum etc. The main members of this group, in order, are radio waves, microwaves, infra red waves, visible light, ultraviolet waves, X-rays and gamma rays. Radio waves have the longest wavelengths and gamma rays have the shortest wavelengths.

Key words

coloured filter	A piece of plastic which absorbs certain colours of light, allowing only light which is the same colour as the filter to pass through.
dispersion of white light	The splitting of white light into the colours from which it is made.
dye	A chemical which absorbs certain colours of light.
electromagnetic spectrum	A family of waves consisting of radio waves, microwaves, infra red waves, visible waves, ultraviolet waves, X-rays and gamma rays.
visible spectrum	The band of colours produced by dispersion.

End of Chapter 12 Questions

1 Use the diagram of the electromagnetic spectrum on page 137 to answer these questions.

a) Name the type of radiation which has the longest wavelength.
b) Name the type of radiation which has the shortest wavelength.
c) Name the type of radiation which is used to 'see' the bones in your body.
d) Name one type of radiation which is used to carry messages over long distances.
e) Give one use for ultraviolet radiation.
f) Give one use for infra red radiation.
g) Name two properties which all these waves have in common.

2 The white stage of a theatre is illuminated by the three, circular, coloured lights as shown in the diagram below.

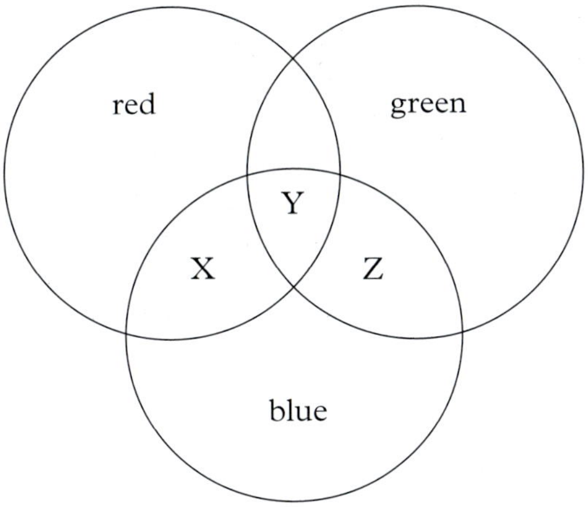

a) What colour is seen at point X?
b) What colour is seen at point Y?
c) What colour is seen at point Z?
d) Why are red, green and blue known as the primary colours?

13 Electric Charge

Sometimes when we remove some clothing we hear crackling sounds. If the room is dark we may even see small sparks. The crackling sounds and sparks are caused by the movement of **electric charges**. The photograph below shows a flash of lightning. Thunder and lightning are also caused by the movement of electric charges

Where does electric charge come from?

Everything around us is made from very small particles called **atoms.** Inside these atoms are even smaller particles called **protons, electrons** and **neutrons.** The protons have a **positive charge**, the electrons have a **negative charge** and the neutrons have **no charge**. Normally the number of protons and electrons in an atom are equal and the charges are **balanced**.

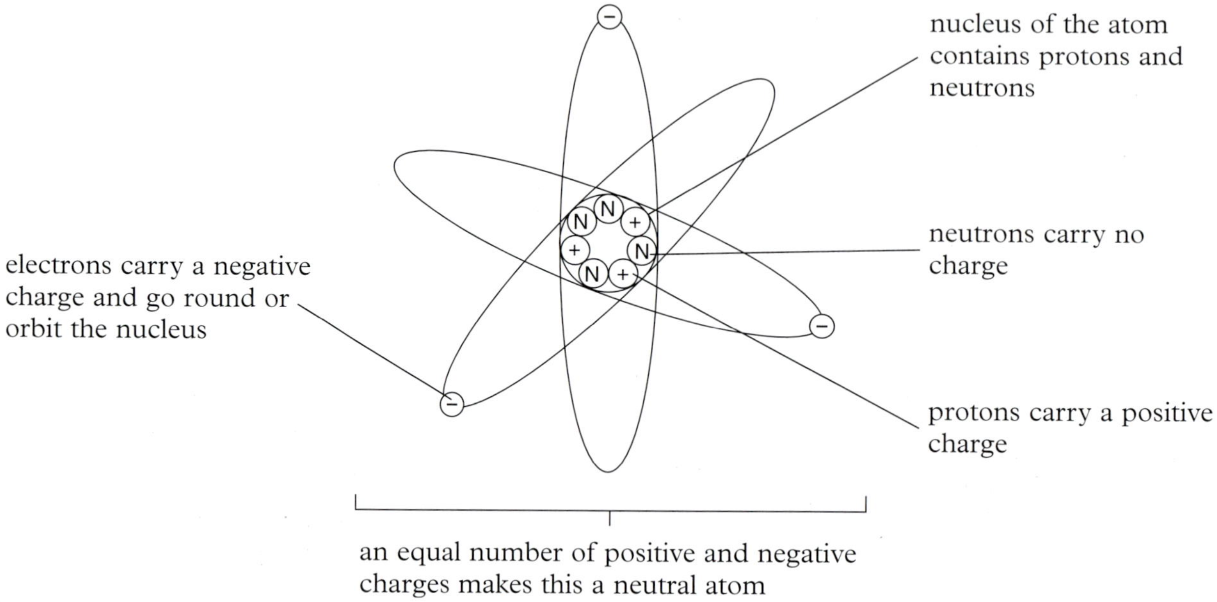
nucleus of the atom contains protons and neutrons
neutrons carry no charge
protons carry a positive charge
electrons carry a negative charge and go round or orbit the nucleus
N
+
−
an equal number of positive and negative charges makes this a neutral atom

Changing the balance of charge

It is possible to alter the balance of charge by rubbing two insulators together. An insulator is a material such as plastic or paper which does not allow charge to pass through it.

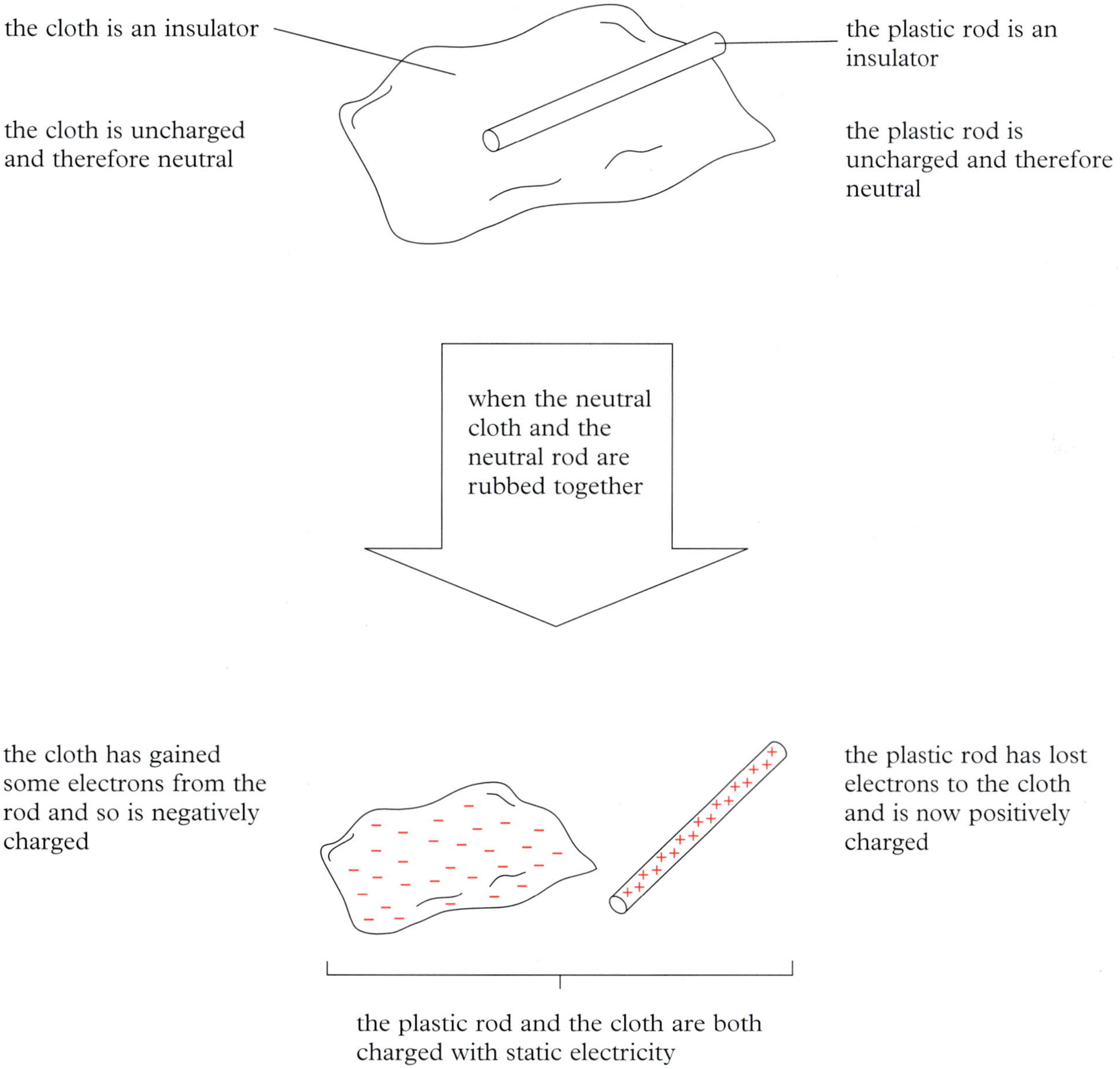
the cloth is an insulator
the cloth is uncharged and therefore neutral
the plastic rod is an insulator
the plastic rod is uncharged and therefore neutral
when the neutral cloth and the neutral rod are rubbed together
the cloth has gained some electrons from the rod and so is negatively charged
the plastic rod has lost electrons to the cloth and is now positively charged
the plastic rod and the cloth are both charged with static electricity

Questions

1 a) Name the three small particles which exist in an atom.
b) Which of these particles are charged and what charges do they carry?

2 How many positively charged and negatively charged particles does a neutral atom contain?

3 Explain, with diagrams, how an object can be charged with static electricity.

Attraction and repulsion between charged objects

If two objects with similar charges are placed close to each other they will **repel**.

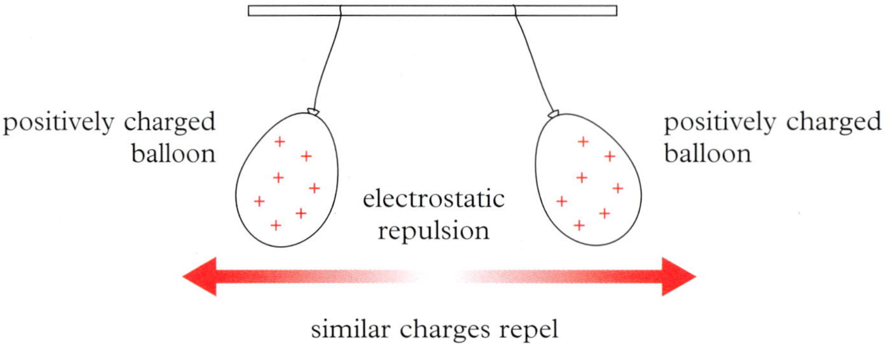
positively charged balloon
positively charged balloon
electrostatic repulsion
similar charges repel

If two objects with opposite charges are placed close to each other they will **attract.**

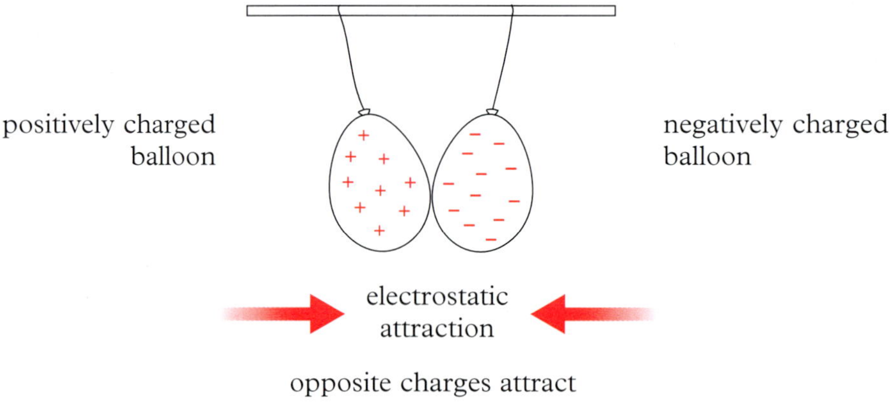
positively charged balloon
negatively charged balloon
electrostatic attraction
opposite charges attract

Static electricity

Our bodies can be charged just like the balloons. This young girl is charged with static electricity. Each strand of her hair carries the same type of charge. These charges repel, making her hair stand on end.

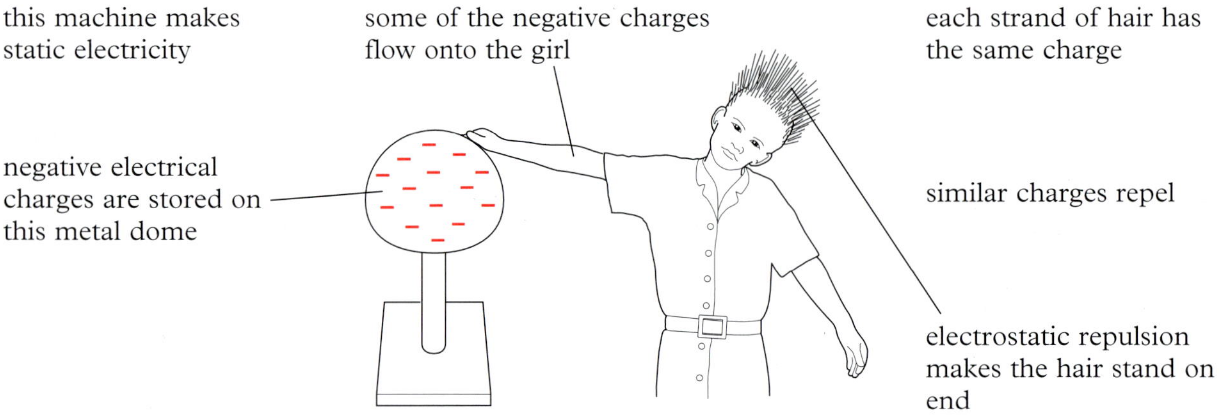
this machine makes static electricity
some of the negative charges flow onto the girl
each strand of hair has the same charge
negative electrical charges are stored on this metal dome
similar charges repel
electrostatic repulsion makes the hair stand on end

Attracting uncharged objects to charged ones

positively charged plastic comb

small pieces of uncharged paper

negatively charged electrons in the paper are attracted towards the comb

this edge becomes positively charged because electrons have been attracted towards the comb

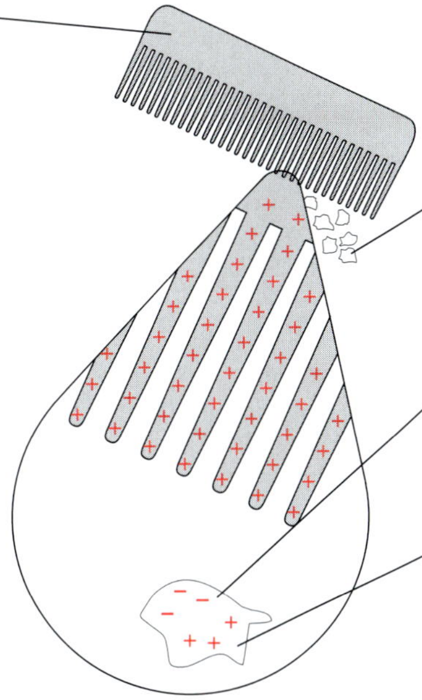

the paper is pulled towards the comb because the negative edge is closer to the comb and this makes the attractive forces stronger than the repulsive forces

the comb attracts the near edge (−) and repels the far edge (+)

In the same way that a charged comb attracts pieces of paper, a stream of water can be attracted towards a charged comb

negatively charged plastic comb

negatively charged surface

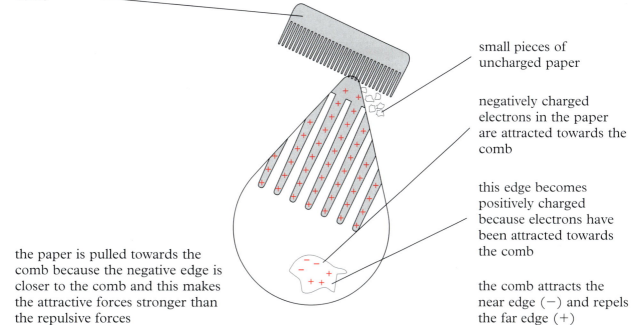

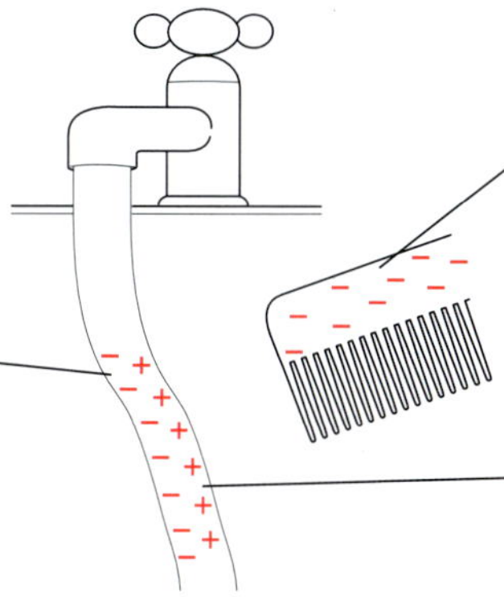

negatively charged electrons are repelled by the negatively charged comb leaving a postively charged surface

positively charged surface of water is attracted towards the negatively charged comb

Questions

1 a) Draw a diagram to show what happens if two positively charged objects are placed next to each other.
b) Draw a diagram to show what happens if a positively charged object is placed next to a negatively charged object.

2 Explain with diagrams why small pieces of paper are attracted to a positively charged comb.

Uses of static electricity

Electrostatic spraying

Painting awkwardly shaped objects like bicycle frames with a spray gun is very difficult and can waste a lot of paint. With the help of static electricity these problems can be overcome.

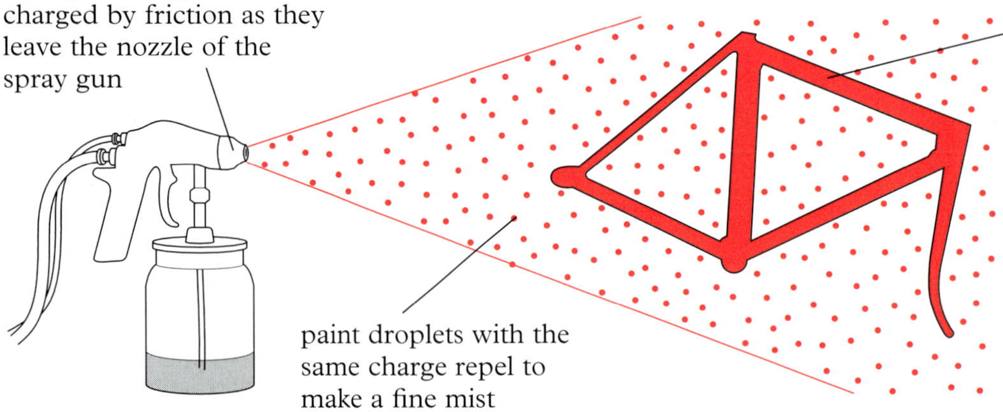
each paint droplet becomes charged by friction as they leave the nozzle of the spray gun
the bicycle frame has the opposite charge to the paint droplets
paint droplets are attracted to the opposite charge on the bicycle frame
paint droplets with the same charge repel to make a fine mist
the fine mist gives an even coat of paint

Electrostatic precipitation

In order to reduce air pollution most heavy industries such as steel works and power stations use **electrostatic preciptators**. These remove large amounts of dust from waste gases before they are released into the atmosphere.

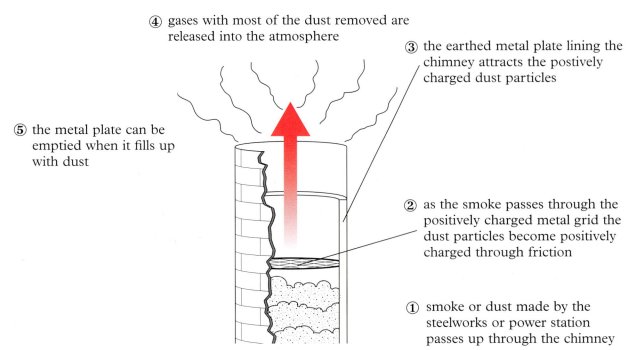

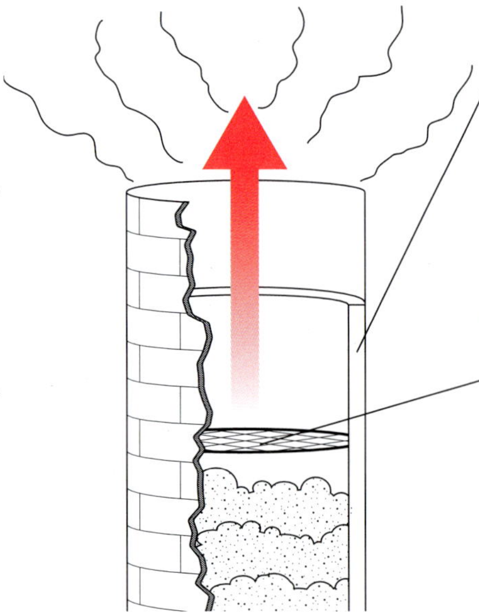
④ gases with most of the dust removed are released into the atmosphere
③ the earthed metal plate lining the chimney attracts the postively charged dust particles
⑤ the metal plate can be emptied when it fills up with dust
② as the smoke passes through the positively charged metal grid the dust particles become positively charged through friction
① smoke or dust made by the steelworks or power station passes up through the chimney

Disadvantages of static electricity

In some situations the presence of static electricity can be a nuisance or even dangerous

the plane becomes charged with static electricity as it flies through the air

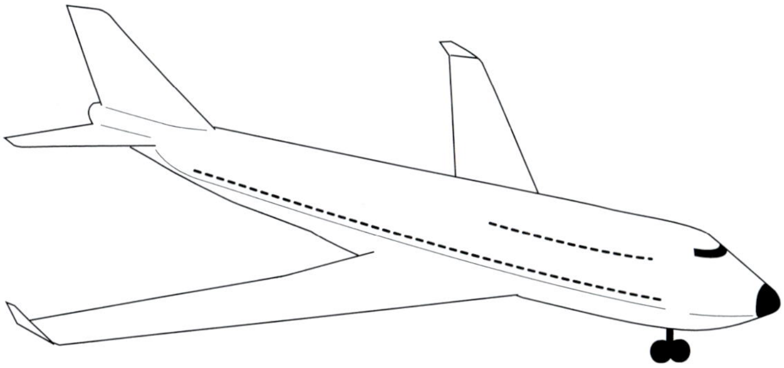

when the plane lands charges could escape as a spark and cause a fire or explosion whilst the aircraft is being refuelled

charges must be removed before the plane can be safely refuelled

charges are removed by connecting a wire between the plane and the ground

getting rid of charges by connecting a wire to the ground is called **earthing**

Other examples of the disadvantages of static electricity include:

- TV screens and computer monitors attracting lots of dust
- receiving small electric shocks when stepping out of a car or taking off some clothing
- lightning, which can cause damage to buildings and can kill any animal it strikes

Summary

When two insulators are rubbed against each other they become electrically charged. One of the insulators gains electrons and becomes negatively charged. The second loses electrons and becomes positively charged.

When two electrically charged objects are brought close together there is a force between them. If the objects have opposite charges it is an attractive force. If the objects have similar charges it is a repulsive force.

Key words

atom Small particle from which all objects are made.

earthing Providing an opportunity for charges to escape from an object so that the object becomes uncharged.

electron A small negatively charged particle found inside atoms.

insulator Any material which will not allow charge to flow through it.

neutron A small particle found inside atoms which has no charge.

proton A small, positively charged particle found inside atoms.

static electricity An object which does not have equal numbers of positive and negative charges is charged with static electricity.

End of Chapter 13 Questions

1 Copy out the sentences below then choose words from this list to complete the sentences. Each word may be used once, more than once or not at all.

static electricity positively negatively insulators electrons attract repel protons

If two different _________ are rubbed together they may become charged with _________ . One of the objects will lose _________ and become _________ charged. The other object will gain _________ and become _________ charged. If the two objects are held close to each other they will _________ .

2 a) What is an insulator?
b) Name two materials that are insulators.

3 Name two uses of static electricity.

4 Describe two situations where the presence of static electricity is a disadvantage.

5 a) What happens to a charged object which is earthed?
b) Explain why sometimes when we step out of a car we receive a small electric shock.

6 A car is being painted with an electrostatic sprayer.
a) As the paint droplets emerge from the nozzle of the spray they are charged negatively. Why does this help the spray to remain as a fine mist?
b) If the paint droplets are negatively charged what charge should be given to the car?
c) Why should the car be given this charge?

14 Simple Electrical Circuits

Although studying static electricity is interesting, it is when charges move that we can make most use of them. We can make charges move using a **cell** or **battery**. The charges that are often made to move are electrons.

Cells

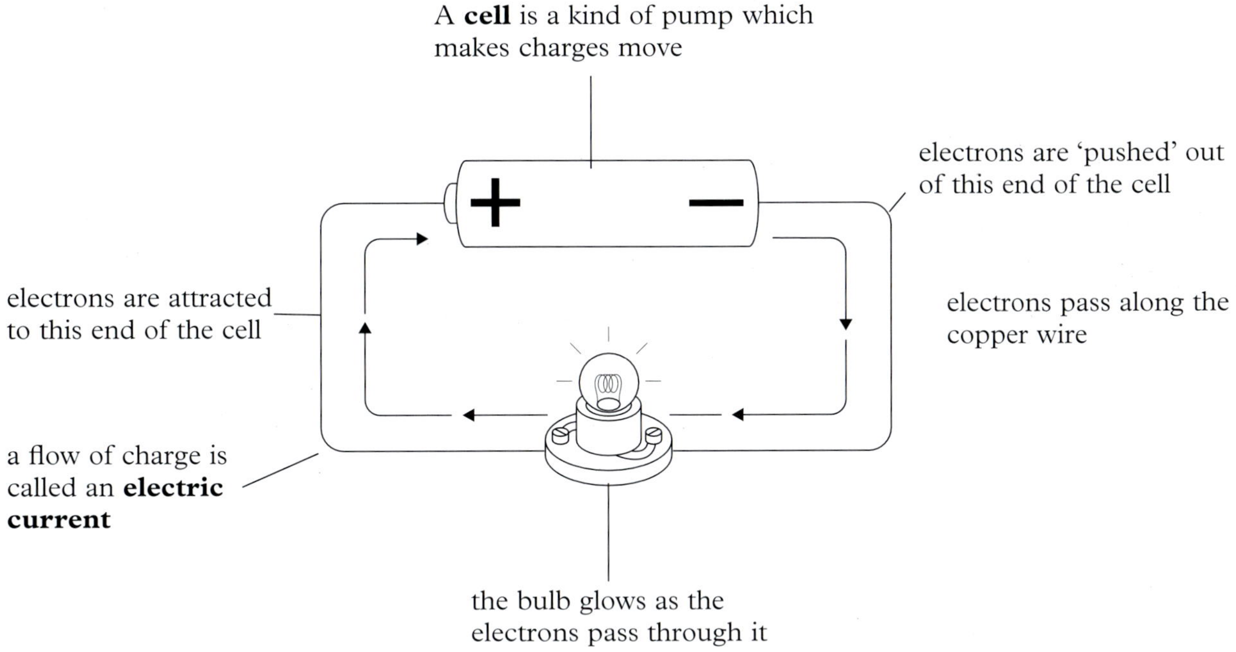
A **cell** is a kind of pump which makes charges move
+
−
electrons are 'pushed' out of this end of the cell
electrons are attracted to this end of the cell
electrons pass along the copper wire
a flow of charge is called an **electric current**
the bulb glows as the electrons pass through it

Batteries

Cells can be connected together to form a **battery**.

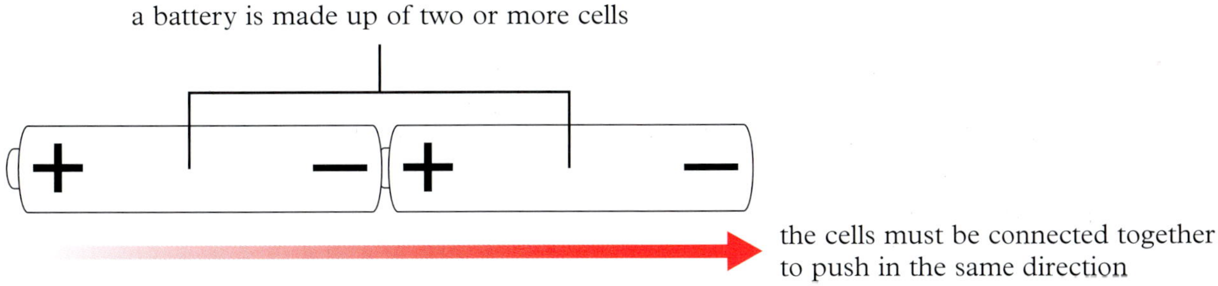
a battery is made up of two or more cells
+
−
+
−
the cells must be connected together to push in the same direction

Circuits

When you run a bath, water travels through pipes beneath the floor and comes out through the tap. Like water, charges need something to travel through. Charges travel through metal wires. Metals are good **conductors** of electricity. They allow electrons carrying negative charges to flow easily through them. The wires, cells, batteries and other electrical components are connected together to form a **circuit.**

Complete circuits

If a light bulb, a cell and some connecting wire are arranged as in the diagram below, the bulb glows, showing that current is flowing through the wires. Charge can flow from the cell around the circuit and back to the cell. This arrangement is called a **complete circuit**.

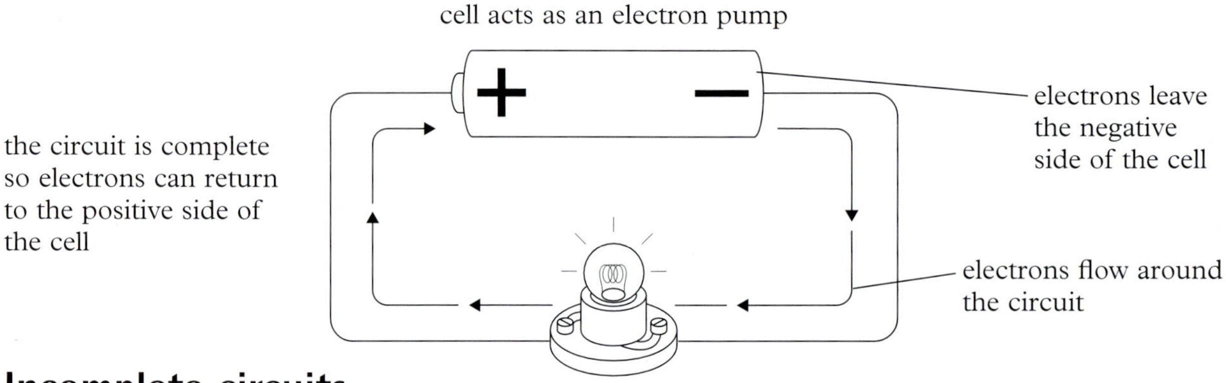

cell acts as an electron pump
+
–
electrons leave the negative side of the cell
the circuit is complete so electrons can return to the positive side of the cell
electrons flow around the circuit

Incomplete circuits

If any of the wires are removed from a complete circuit, it becomes **incomplete.** A bulb will not glow in an incomplete circuit.

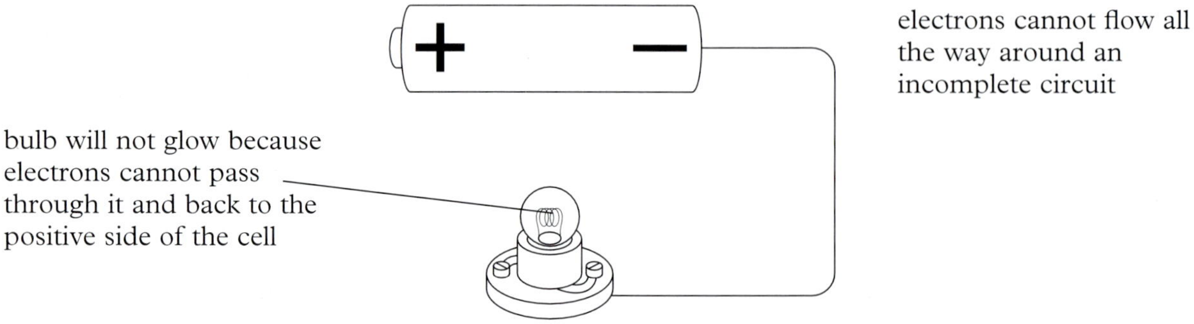

+
–
electrons cannot flow all the way around an incomplete circuit
bulb will not glow because electrons cannot pass through it and back to the positive side of the cell

Questions

1 What is an electric current?

2 a) What is a conductor?
b) Give one example of a material which is a conductor.

3 What is the difference between a cell and a battery?

4 a) What is the difference between a complete and an incomplete circuit?
b) Draw one example of each kind of circuit.

Circuit diagrams

Drawing diagrams of circuits, like those on the previous page, is not easy. It is much easier to draw **circuit diagrams** like the one below. Symbols are used to represent the different parts or components.

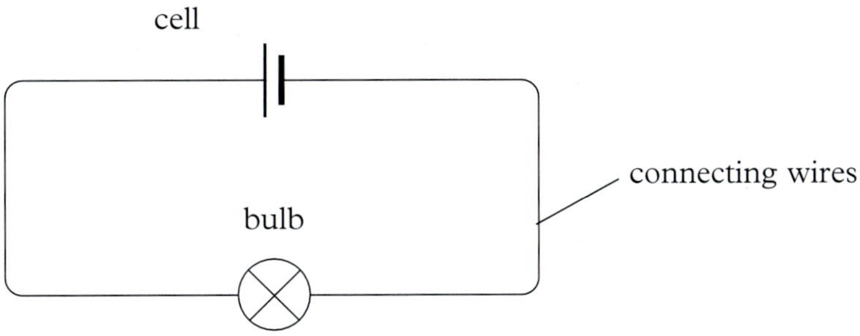

Some of the most common circuit components

This table shows the symbols for some of the most common electric circuit components

What it is	What it looks like	Symbol	What it does
cell	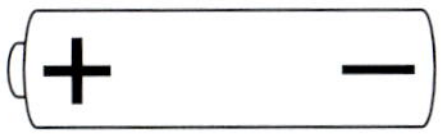	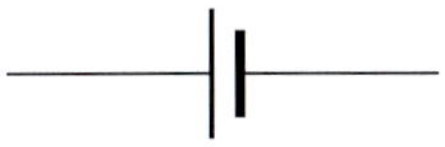	moves charge around a circuit
battery	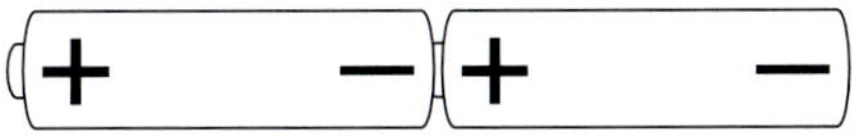		provides a larger current than a single cell
connecting wire	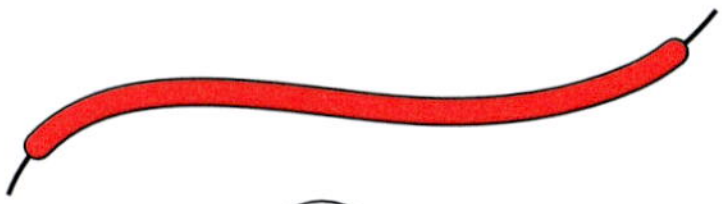		provides a path through which current can flow
lamp/bulb	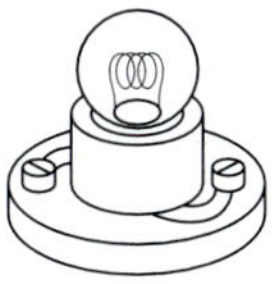	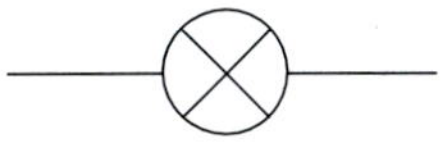	glows brightly if sufficient current flows through it
switch	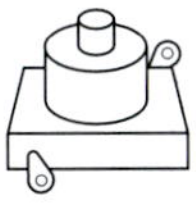		turns current in a circuit on and off
resistor	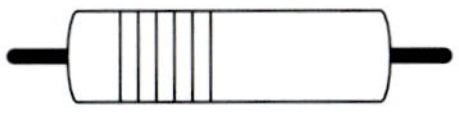	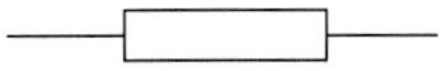	reduces the current flowing in a circuit
variable resistor			changes the size of a current by altering the value of the resistance

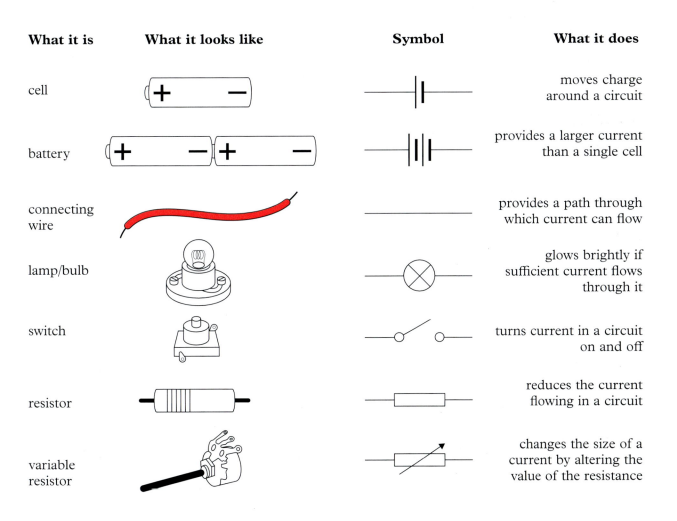

Series and parallel circuits

There are two kinds of electrical circuit.

1 A **series circuit** has no branches or junctions. The electrons have only one path to follow.

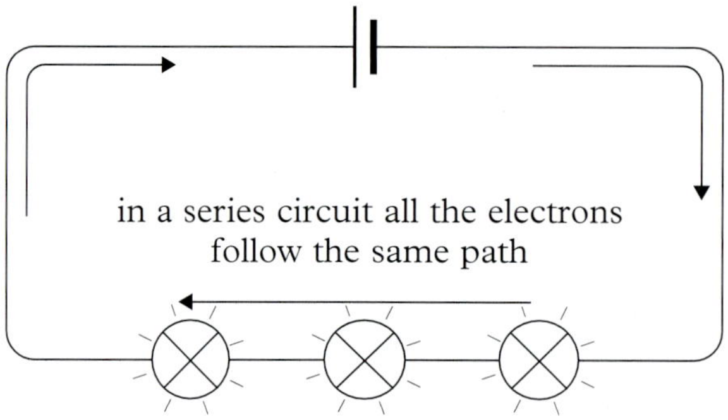
in a series circuit all the electrons
follow the same path

2 A **parallel circuit** contains branches. There are several paths the electrons can follow.

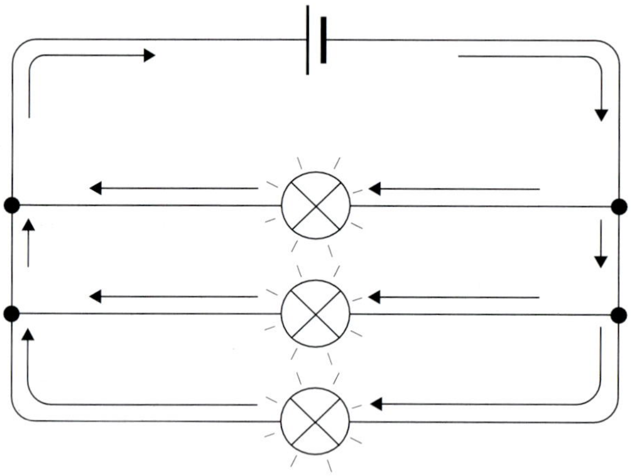
in a parallel circuit the electrons do not all follow the same path
some electrons flow through the first bulb
some electrons flow through the second bulb
some electrons flow through the third bulb
there are several routes the electrons can take

Switches

Switches turn circuits on and off by making the circuits complete or incomplete.

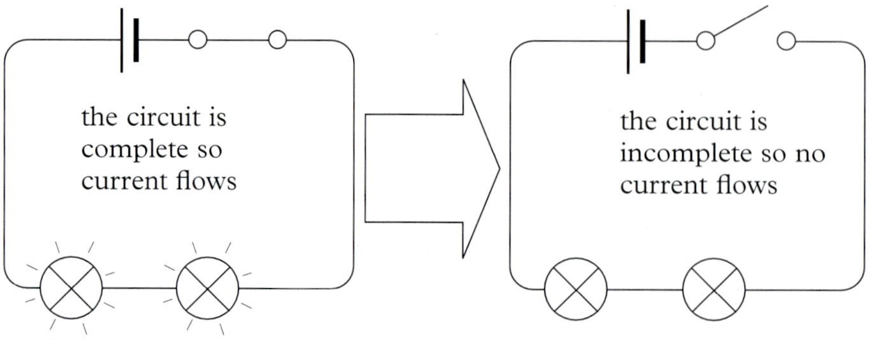
switch closed
switch open
current flows through a complete circuit
the circuit is complete so current flows
the circuit is incomplete so no current flows
current cannot flow through an incomplete circuit
all the bulbs glow
none of the bulbs glow

Switches in series circuits turn the whole circuit on or off.

Switches in parallel circuits can turn parts of the circuit on and off

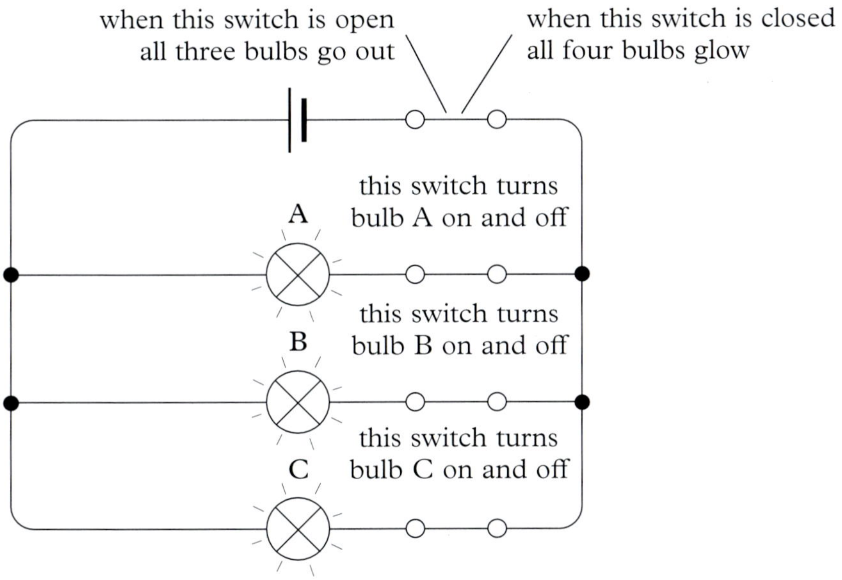
when this switch is open
all three bulbs go out
when this switch is closed
all four bulbs glow
this switch turns
bulb A on and off
A
this switch turns
bulb B on and off
B
this switch turns
bulb C on and off
C

Questions

1

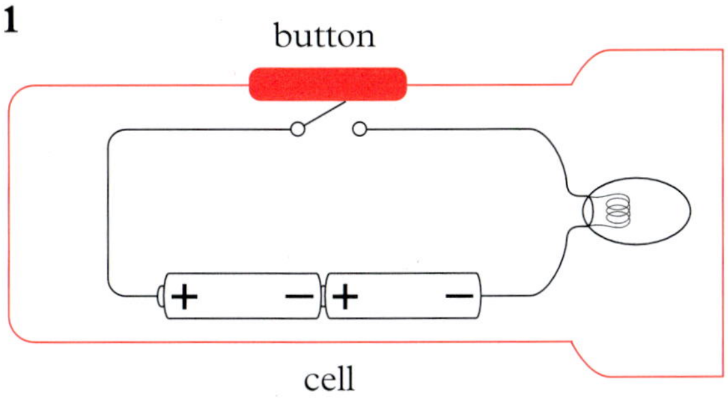
button
cell

Explain in detail what happens when the button is:
a) pressed,
b) released.

2 Look carefully at this stereo system and decide which parts of the system are working if:
a) switch A is open but all the others are closed,
b) switch B is open but all the others are closed,
c) switches B and C are open but all the others are closed,
d) switches A, B and C are open.

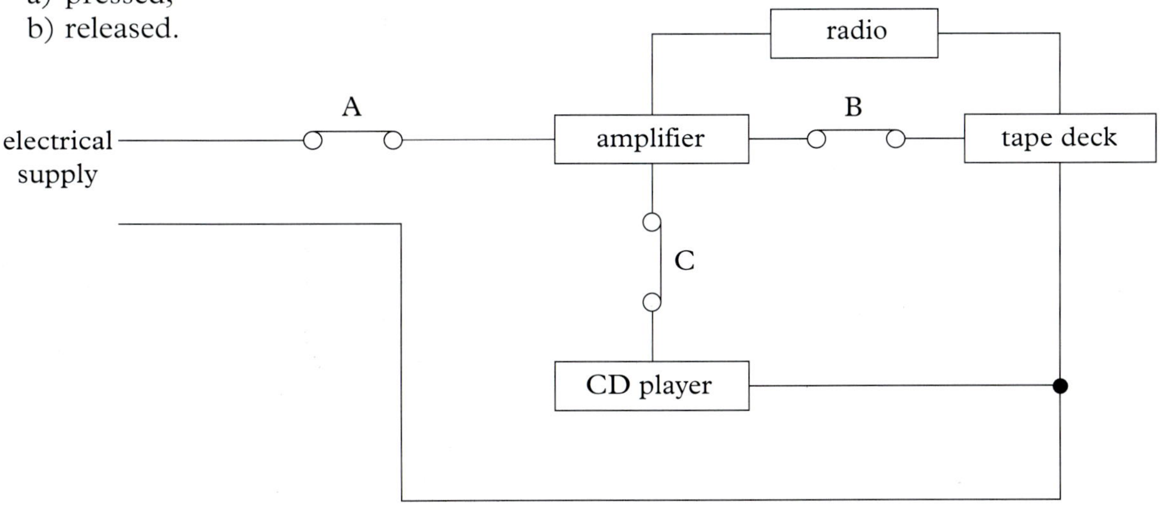
radio
electrical
supply
A
amplifier
B
tape deck
C
CD player

Electron flow and conventional current

When scientists first experimented with current electricity, they did not know which kind of charge was flowing around a circuit. They guessed that the charges were positive and flowed from the positive terminal of a cell to the negative terminal. We now know that they were wrong. To avoid too much change, scientists agreed they would continue to think of electric current as flowing from positive to negative. They called this **conventional current.** From now on all currents in this book will be conventional currents

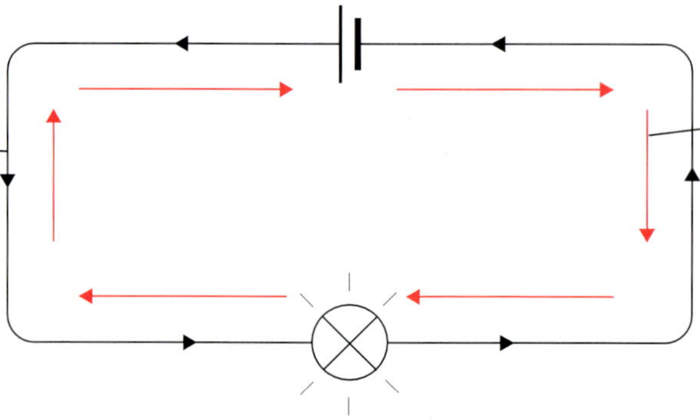
an arrow on the wire shows that conventional current flows from the positive side of the cell to the negative side of the cell
electrons flow from the negative side of the cell to the positive side of the cell

Electric charge and current

We measure electrical charge in units called **coulombs** (represented by the symbol C).

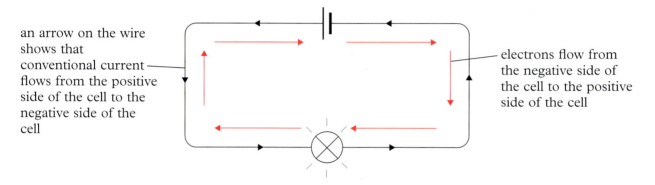

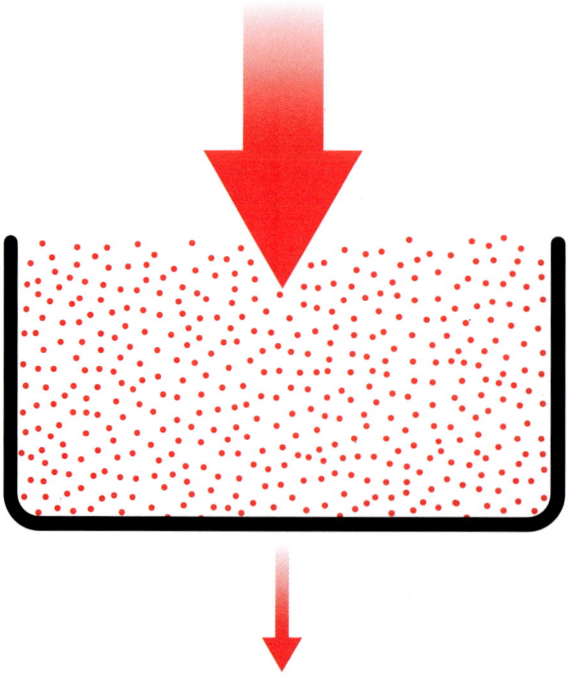
the charge carried by one electron is very very small
if six million, million, million electrons could be put into a container
there would be one coulomb of charge

Amperes

We measure electric current in units called **amperes** (represented by the symbol A).

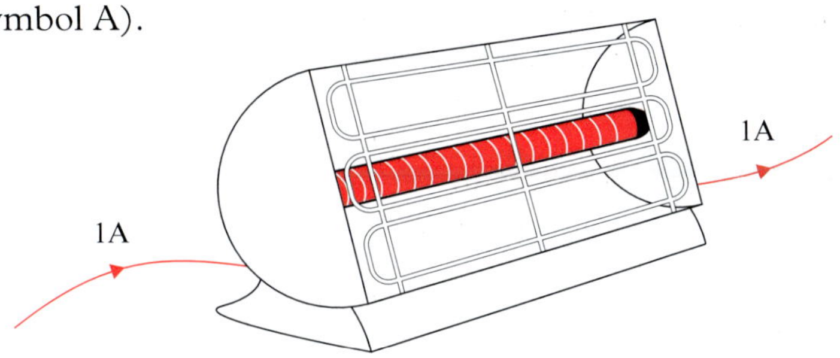

when the flow of charge is 1 C per second through this electric fire, the electric current is 1 A

If 10 C of charge flow through this heater in 5 s then the current is 2 A.

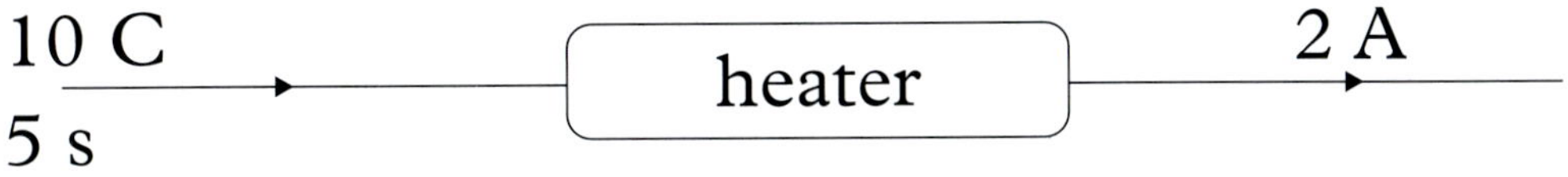

If 20 C of charge flow through this heater in 4 s the current is 5 A.

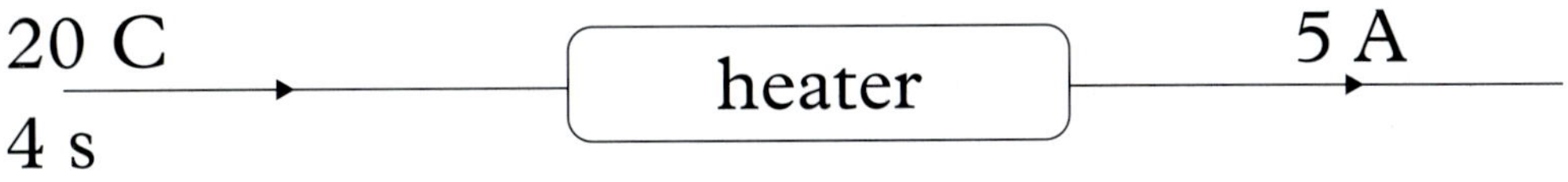

If a charge of 100 C flows through the heater in 25 s the current flowing in the circuit is 4 A.

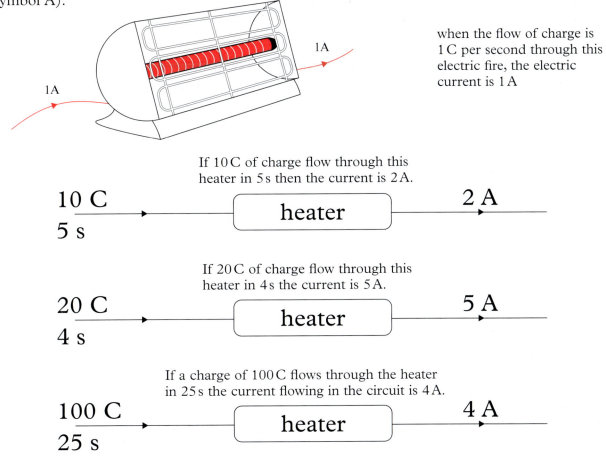

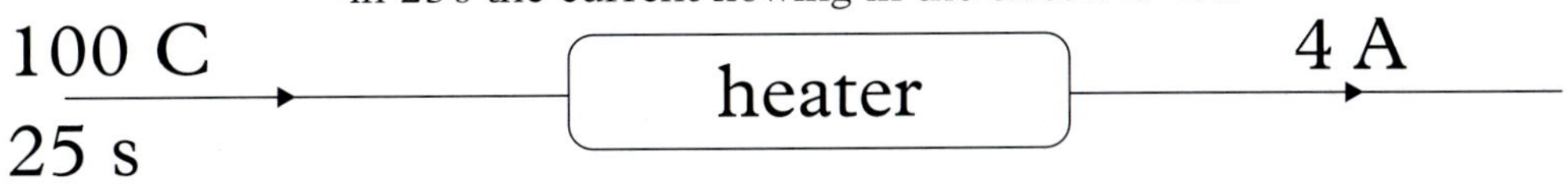

From the above we can see that the current flowing in a circuit can be calculated using the equation

$$\text{current} = \frac{\text{charge}}{\text{time}} \quad \text{or} \quad I = \frac{Q}{t}$$

Questions

1. If 40 C of charge flow through an electric fire in 8 s calculate the current flowing in the circuit.
2. If 25 C of charge flow through a bulb in 10 s calculate the size of the current in the circuit.
3. If a charging current of 3 A flows into a cell for 50 s how much charge enters the cell?
4. If a current of 2 A flows along a wire for 1 minute how much charge passes any point on that wire during this time?
5. For how long must a current of 2.5 A flow into a battery before it has received 100 C?

Measuring current

We measure the current flowing in a circuit using an **ammeter.**

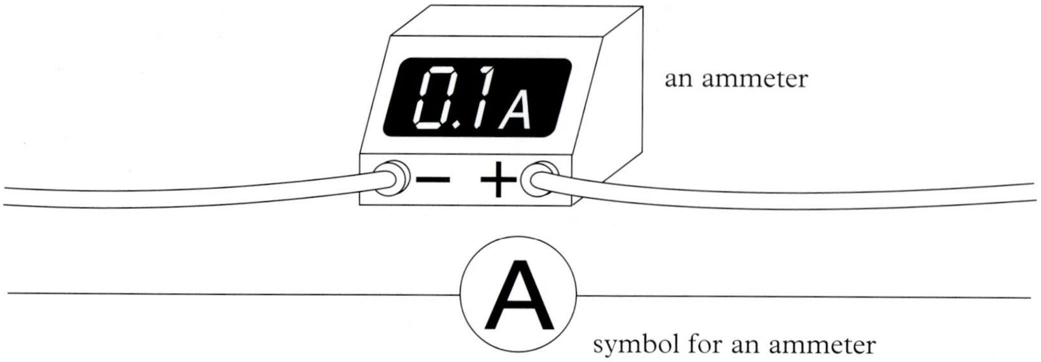

Current in series circuits

In a series circuit the current has the same value everywhere.

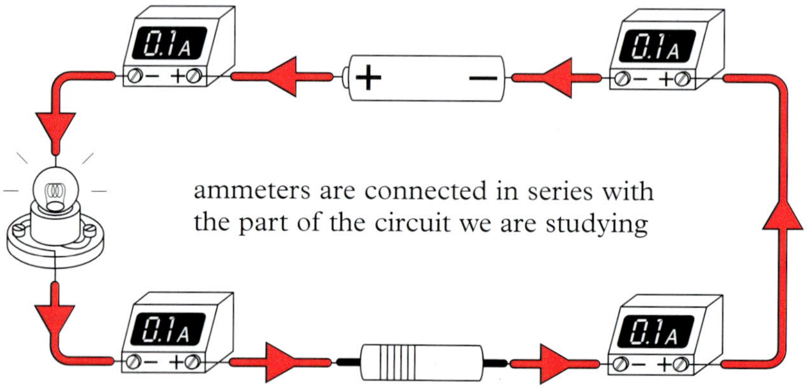

Current in parallel circuits

In a parallel circuit the current is not the same everywhere. However, the total current entering any junction is equal to the total current leaving the same junction.

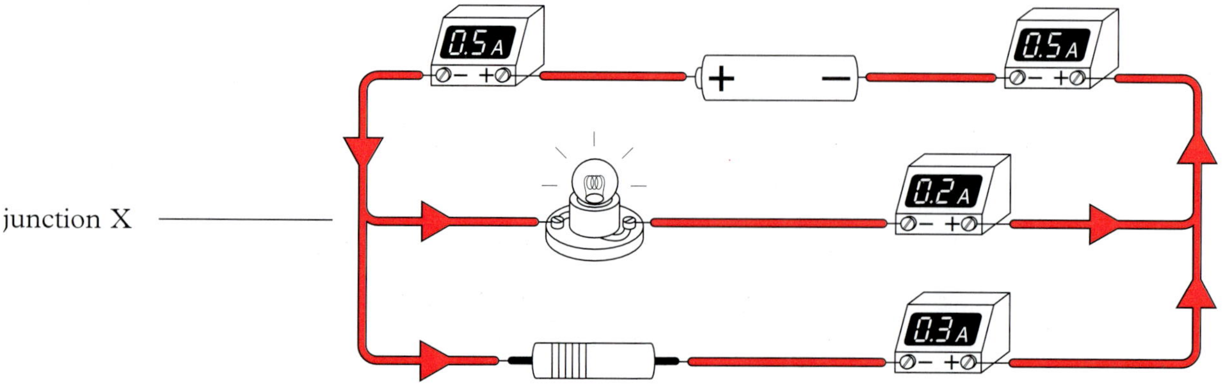

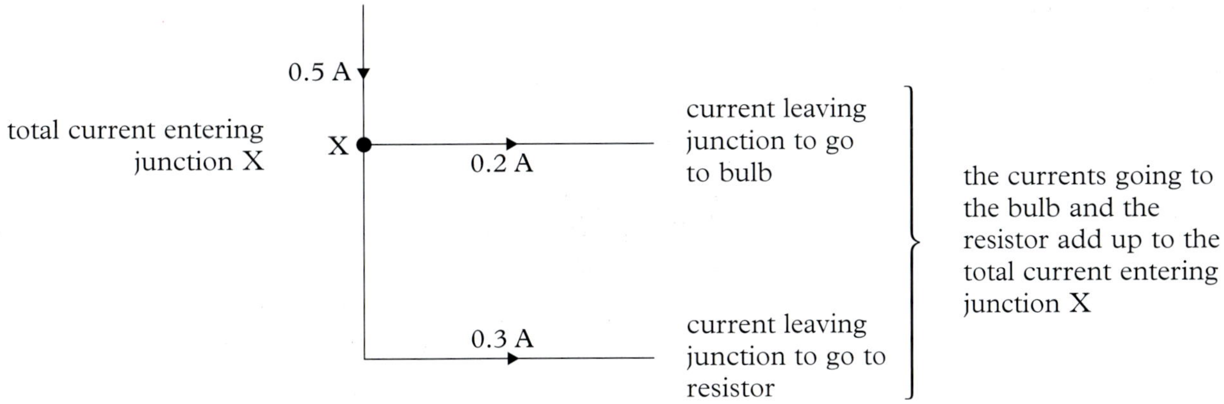

Questions

1 a) Draw a circuit diagram containing
- i) three identical bulbs in parallel,
- ii) one cell,
- iii) four ammeters.

You can decide where to position each of the ammeters.

b) If the total current leaving the cell is 0.4 A write at the side of each of your ammeters the size of the current flowing through it.

Summary

Electric currents consist of moving electrical charges. In metals the charges are carried by electrons. These electrons can be made to move around circuits by using cells or batteries. The size of an electrical current indicates the rate at which charges are flowing. Electrical current is measured in amperes (A) using a device called an ammeter.

There are two main types of electrical circuit. These are called series circuits and parallel circuits. By introducing different components into a circuit we can control currents so that they produce different effects, like making sound, light and heat.

Key words

ammeter Device for measuring current.

battery Several cells connected together.

cell Provides the energy to move charges around a circuit.

circuit Path of wires, switches, bulbs and other electrical components through which the current flows.

Key words (continued)

complete circuit Circuit which allows the current to flow out of the cell around the circuit and return to the cell.

electric current Flow of charges (usually electrons) around a circuit.

incomplete circuit Circuit which does not have a continuous path, so that the current is unable to leave and return to the cell.

parallel circuit A circuit in which there are several paths the current can follow.

series circuit A circuit in which the there is only one path for the current follow.

End of Chapter 14 Questions

1 What electrical component is used to move charges around a circuit?

2 a) What do we call several cells connected together?
b) Why is it important that the cells are connected the right way around?

3 a) What is a complete circuit?
b) What is an incomplete circuit?

4 Draw the symbol for
a) a switch
b) a cell
c) a bulb.

5 a) Draw a circuit diagram for a series circuit which contains two bulbs, one switch and one cell.
b) Why are both bulbs off when the switch is open?

6 Draw a circuit diagram for a circuit which contains one cell and two bulbs that are in parallel. Indicate with an X where in the circuit you would put a switch so that one of the bulbs could be turned on and off whilst the second bulb remained on.

7 Look at the circuit drawn below and then explain what happens to the bulb when
a) the switch is open
b) the switch is closed.

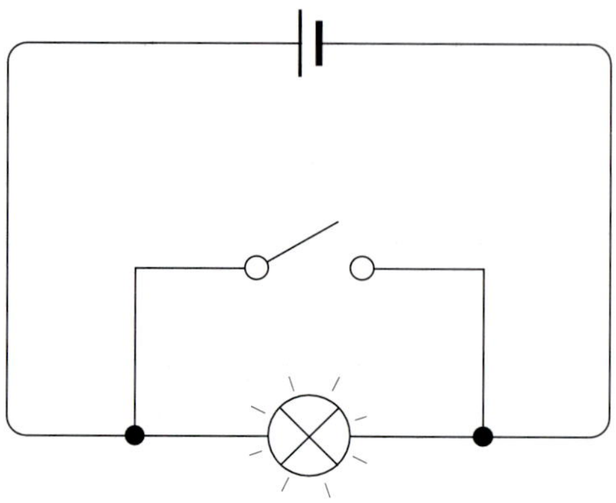

8 Calculate the values of the currents flowing in the ammeters labelled A-G in the diagram.

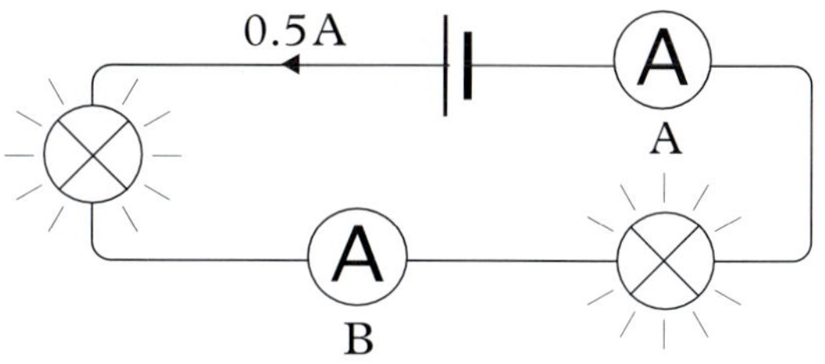

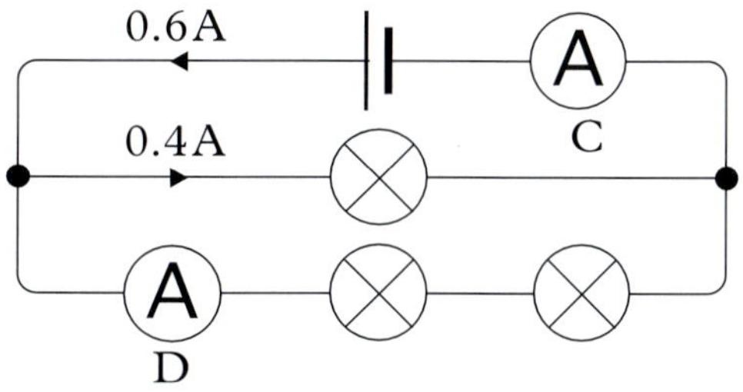

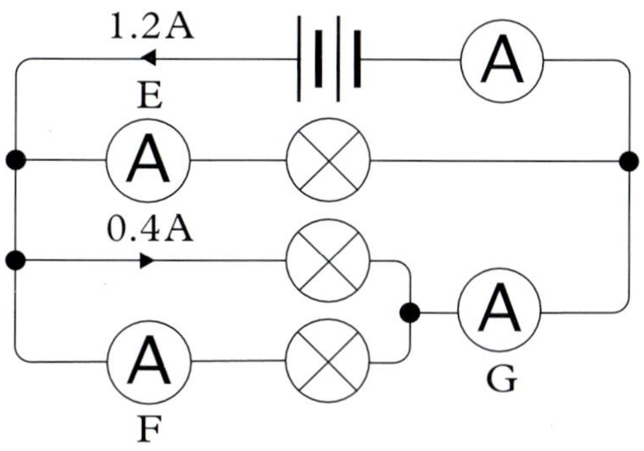

9 You are given four identical bulbs, some connecting wires and a cell.
a) How many completely different circuits can you construct using all this apparatus?
b) Draw a circuit diagram of each.
c) In which of these circuits will the brightness of the bulbs be identical?

10 Calculate the current flowing through a bulb if 120 C of charge flows through it in 2 minutes.

15 Voltage

The voltage of a cell

Charges need energy in order to move around a circuit. They may receive this energy from a cell or a battery.

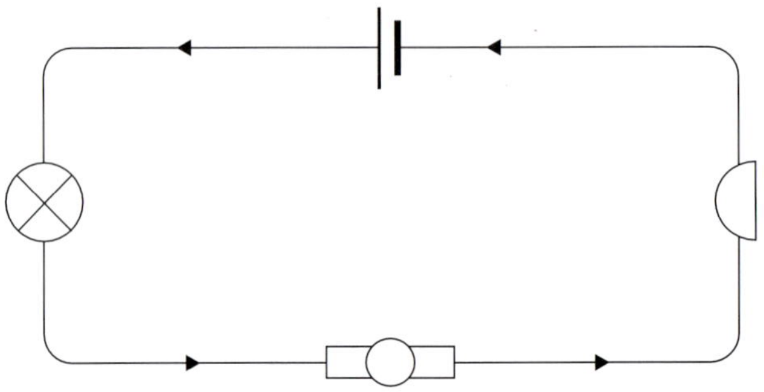
electrical energy is given to the charges by the cell
some of the electrical energy carried by the charges is changed into heat and light energy by the bulb
some of the electrical energy carried by the charges is changed into sound energy by the buzzer
some of the electrical energy carried by the charges is changed into movement energy by the motor

The amount of energy given to the charges depends upon the **voltage** of the cell. Voltage is also called **potential difference** or **p.d.** A cell which has a voltage of 1.5 V gives 1.5 J (joules) of energy to each coulomb (C) of charge which travels through it. The charges release this energy as they travel through the different components in the circuit. When the energy is released the bulb glows and becomes warm, the motor spins and the buzzer makes a sound.

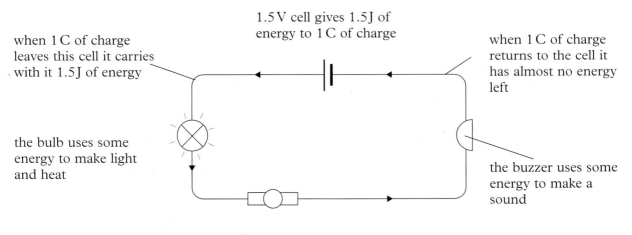

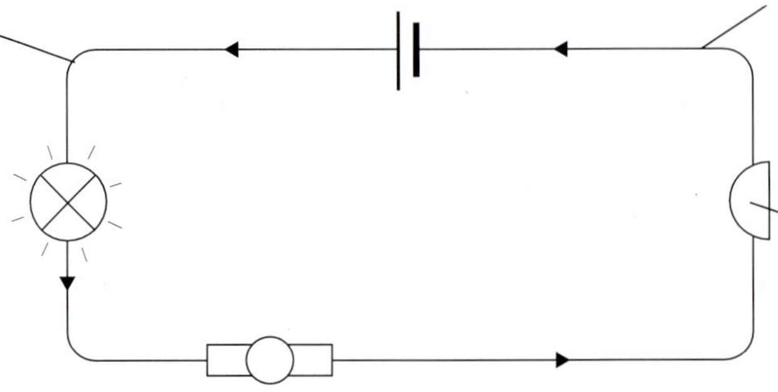
1.5 V cell gives 1.5 J of energy to 1 C of charge
when 1 C of charge leaves this cell it carries with it 1.5 J of energy
when 1 C of charge returns to the cell it has almost no energy left
the bulb uses some energy to make light and heat
the buzzer uses some energy to make a sound
the motor uses some energy to make it turn

If the 1.5 V cell is replaced with a battery which has a voltage of 4.5 V, the battery will give 4.5 J of energy to each coulomb of charge which passes through it. As they travel around the circuit the charges are able to release three times as much energy as before. This means that the bulb glows more brightly and feels a lot warmer, the motor spins more quickly and the buzzer sounds louder. The table below summarises the effect of replacing the 1.5 V cell with the 4.5 V battery.

	circuit using 1.5V cell	circuit using 4.5V battery
voltage	1.5 volts	4.5 volts
energy given to each coloumb of charge	1.5 joules	4.5 joules
what happens to: a) bulb b) motor c) buzzer	a) glows and feels warm b) spins around c) makes a sound	a) glows more brightly and feels warmer b) spins around faster c) makes a louder sound

Questions

1 a) What is the voltage of each of the batteries shown below?

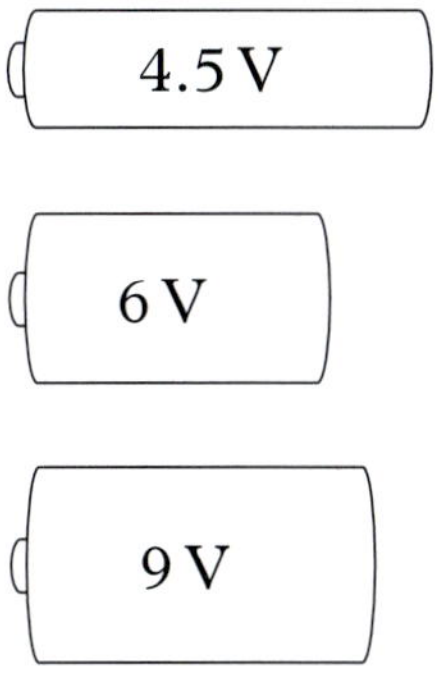

b) How much energy does each of these three batteries give to each coulomb of charge which passes through them?

c) i) What would you see if a bulb was connected in turn to each of the above batteries?
ii) Explain your answer for each battery.

d) How much energy does a coulomb of charge have when it returns to the battery?

Voltmeters and potential difference

To discover how much energy is being given up by each coulomb of charge as it travels through a component such as a bulb, we use a **voltmeter.**

When a voltmeter is connected in parallel with a component it measures the difference between the energy each coulomb of charge has as it enters and leaves the component. We call this measurement the **potential difference** (**p.d.**). In the circuit diagram on the next page the p.d. across the bulb is 2.5 V. This means that each time one coulomb of charge passes through the bulb 2.5 J of electrical energy are converted into 2.5 J of heat and light energy.

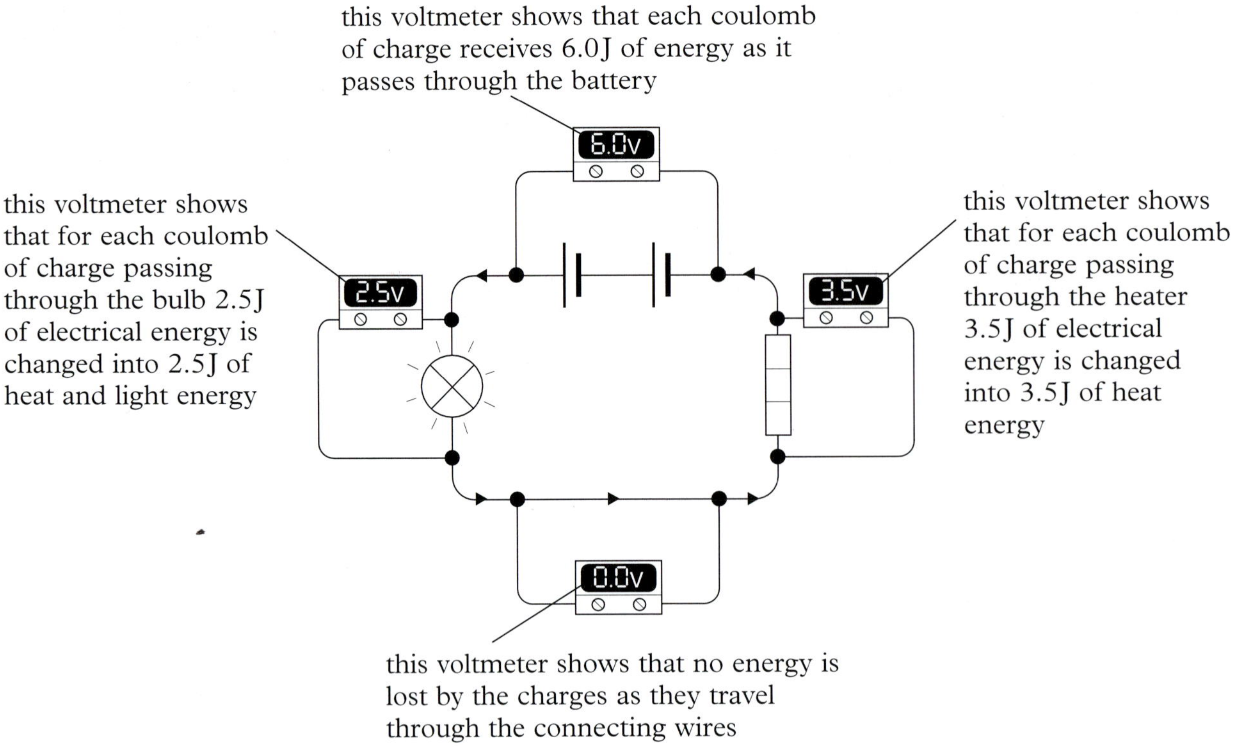
this voltmeter shows that each coulomb of charge receives 6.0J of energy as it passes through the battery
6.0v
this voltmeter shows that for each coulomb of charge passing through the bulb 2.5J of electrical energy is changed into 2.5J of heat and light energy
2.5v
this voltmeter shows that for each coulomb of charge passing through the heater 3.5J of electrical energy is changed into 3.5J of heat energy
3.5v
0.0v
this voltmeter shows that no energy is lost by the charges as they travel through the connecting wires

The energy given to each coulomb of charge by a cell or battery is all released in the external part of the circuit. Therefore the sum of the voltages across all the components in this circuit must be equal to the voltage of the battery. This may be shown as an equation:

$$V_{battery} = V_{bulb} + V_{heater}$$

In the circuit diagram above, the values would be:

$$6\,V = 2.5\,V + 3.5\,V$$

Questions

1 The diagram below shows voltmeters connected in parallel with a heater, a bulb and a buzzer. The potential difference across the heater is 4.5 V. The potential difference across the bulb is 3.0 V. The potential difference across the buzzer is 1.5 V.

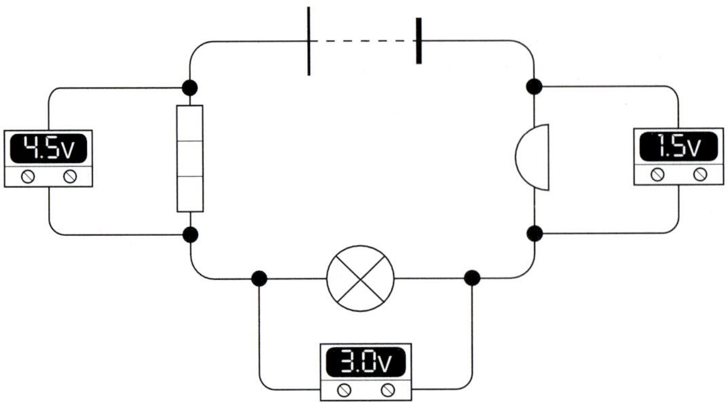
4.5v
1.5v
3.0v

Questions (continued)

a) Describe the energy changes that take place in each of these three components when 1 C of charge passes through each of them.

b) Assuming all the energy given to the charges by the battery is given to the heater, the bulb and the buzzer, calculate the voltage of the battery.

2 Explain why a voltmeter placed in a circuit in parallel with a length of connecting wire has a reading which is almost zero.

Voltages in series circuits

If all the electrical energy received from a cell or battery is changed or converted into other forms of energy by components in a series cicuit, then the sum of the p.d.s across the components is equal to the p.d. across the cell. This is shown by the formula:

$$V_c = V_1 + V_2 + V_3 \ldots\ldots$$

symbol for a voltmeter

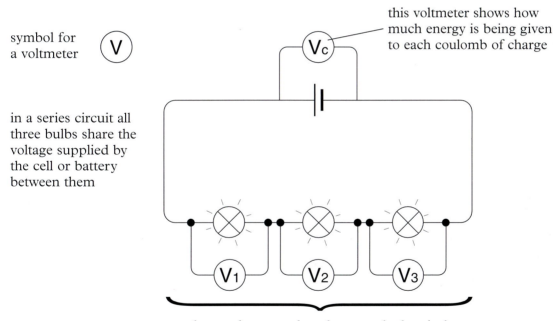

V

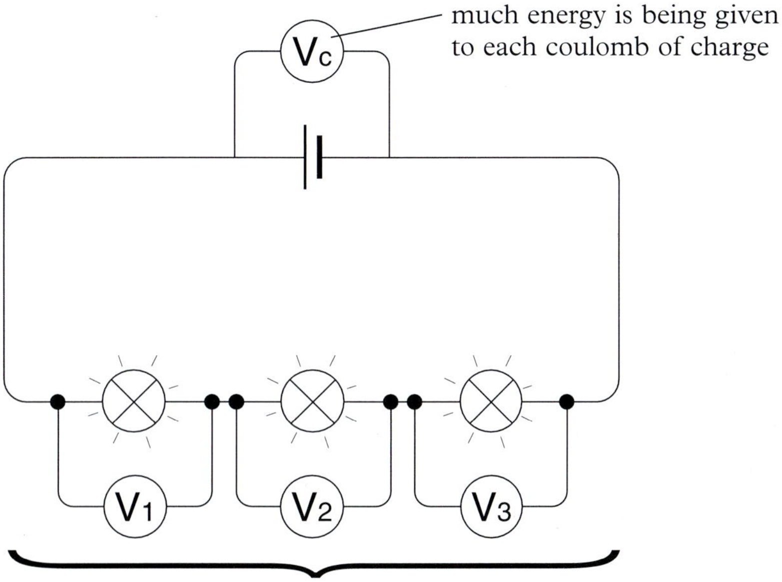
Vc
this voltmeter shows how much energy is being given to each coulomb of charge
V1
V2
V3

in a series circuit all three bulbs share the voltage supplied by the cell or battery between them

these voltmeters show how much electrical energy is being changed into heat and light energy by the bulbs

Voltages in a parallel circuit

In a parallel part of a circuit the p.d.s across all the branches of a network are the same.

The p.d.s. across the branches of a parallel circuit are shown by the formula:

$$V_c = V_1 = V_2 = V_3$$

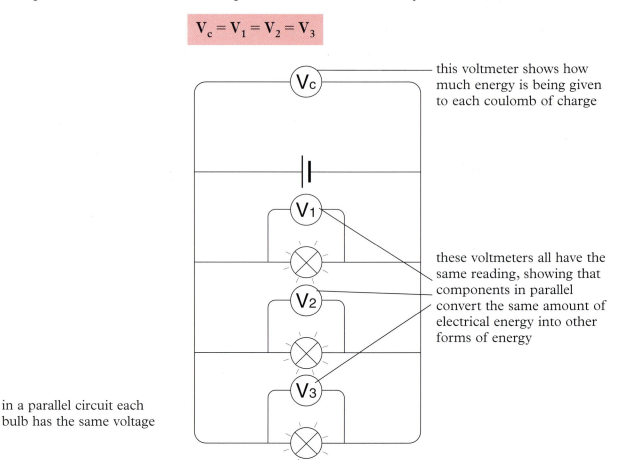

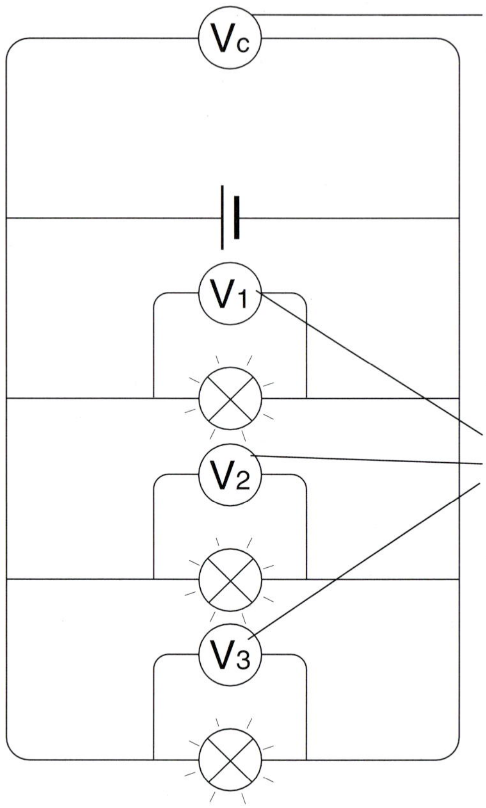
Vc
V1
V2
V3
this voltmeter shows how much energy is being given to each coulomb of charge
these voltmeters all have the same reading, showing that components in parallel convert the same amount of electrical energy into other forms of energy

in a parallel circuit each bulb has the same voltage

Key words

cell voltage Tells us how much energy is given to each coulomb of charge that passes through the cell.

potential difference (p.d.) The p.d. across a component tells us how much electrical energy is changed into other energy forms when 1 C of charge passes through.

voltmeter A piece of equipment for measuring voltages.

Summary

When charges leave a cell or battery they carry away electrical energy. This energy is changed into other forms of energy by the various components in the circuit. By using a voltmeter we can measure the amount of energy given to the charges by the cell and how much of this energy is converted into other forms by each of the components.

The voltmeter is connected in parallel with the part of the circuit we are interested in. If the p.d. across a component is 1 V this means that when 1 C of charge passes through this component 1 J of electrical energy is changed into a different form of energy.

End of Chapter 15 Questions

1 The diagram below shows a simple series circuit containing four components, a 12 V battery, a bulb, a resistor and a buzzer. The voltages across each of the components is also shown.

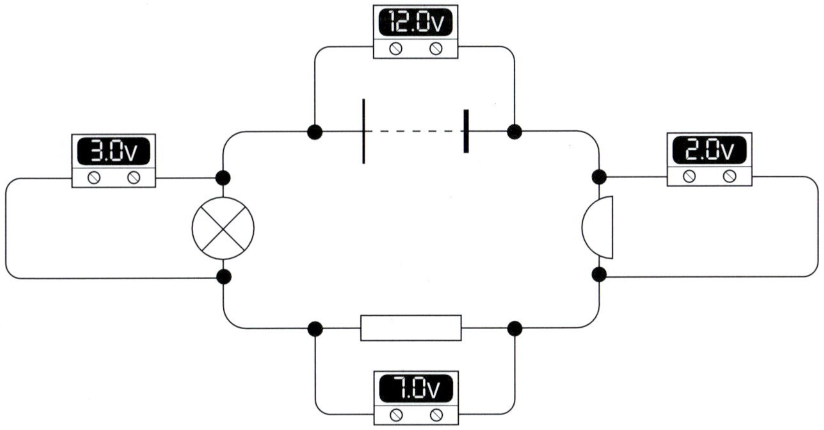

For each component write a sentence explaining what energy changes are taking place as current flows around the circuit.

2 Look at the four circuits shown below then calculate the values of the voltages:

a) W
b) X and Y
c) Z

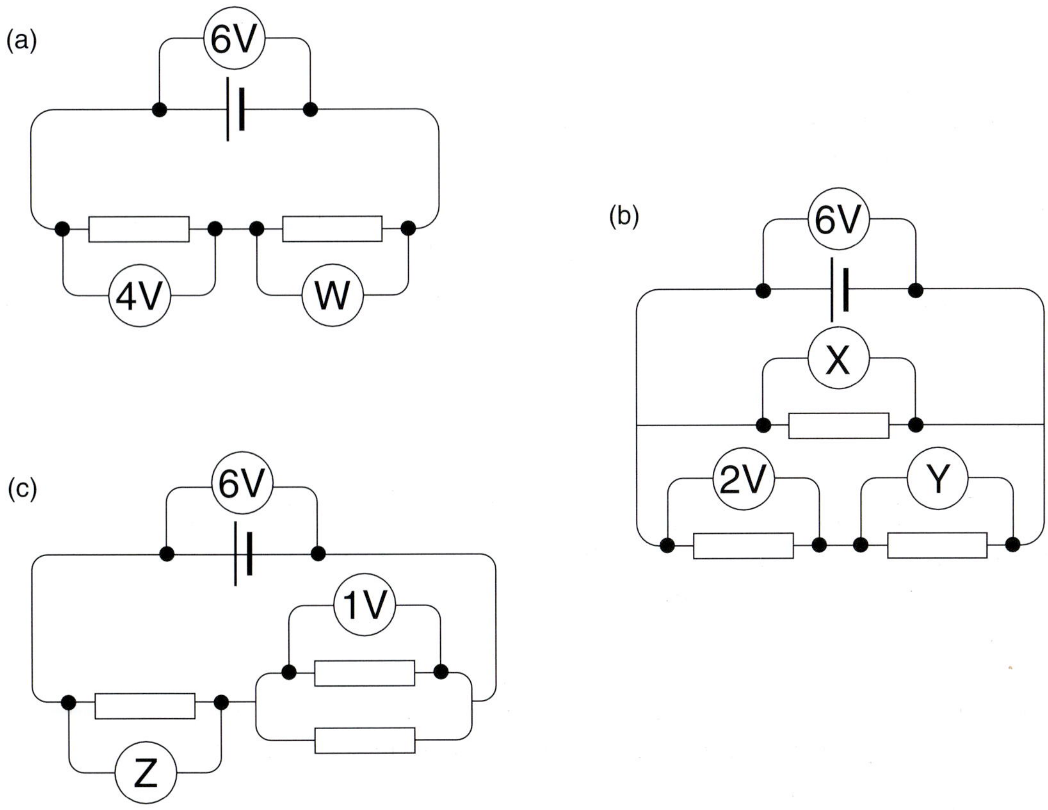

16 Resistance

When we turn up the volume of our stereo systems or alter the brightness of our TV sets we change the size of the currents that flow inside them. We do this by altering the **resistance** of the circuits.

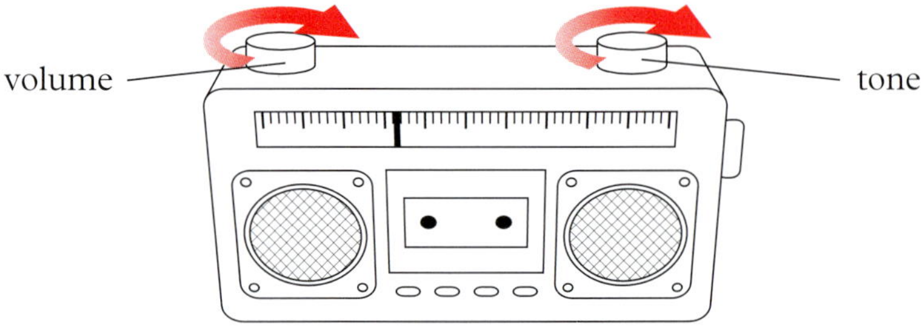
volume
tone

Electrical resistance

When electricity flows through a conductor such as a length of copper wire, electrons travel between the atoms of the metal. If the electrons can do this easily, we say that the wire has a **low resistance**. If however the electrons find it difficult then we say that the wire has a **high resistance**.

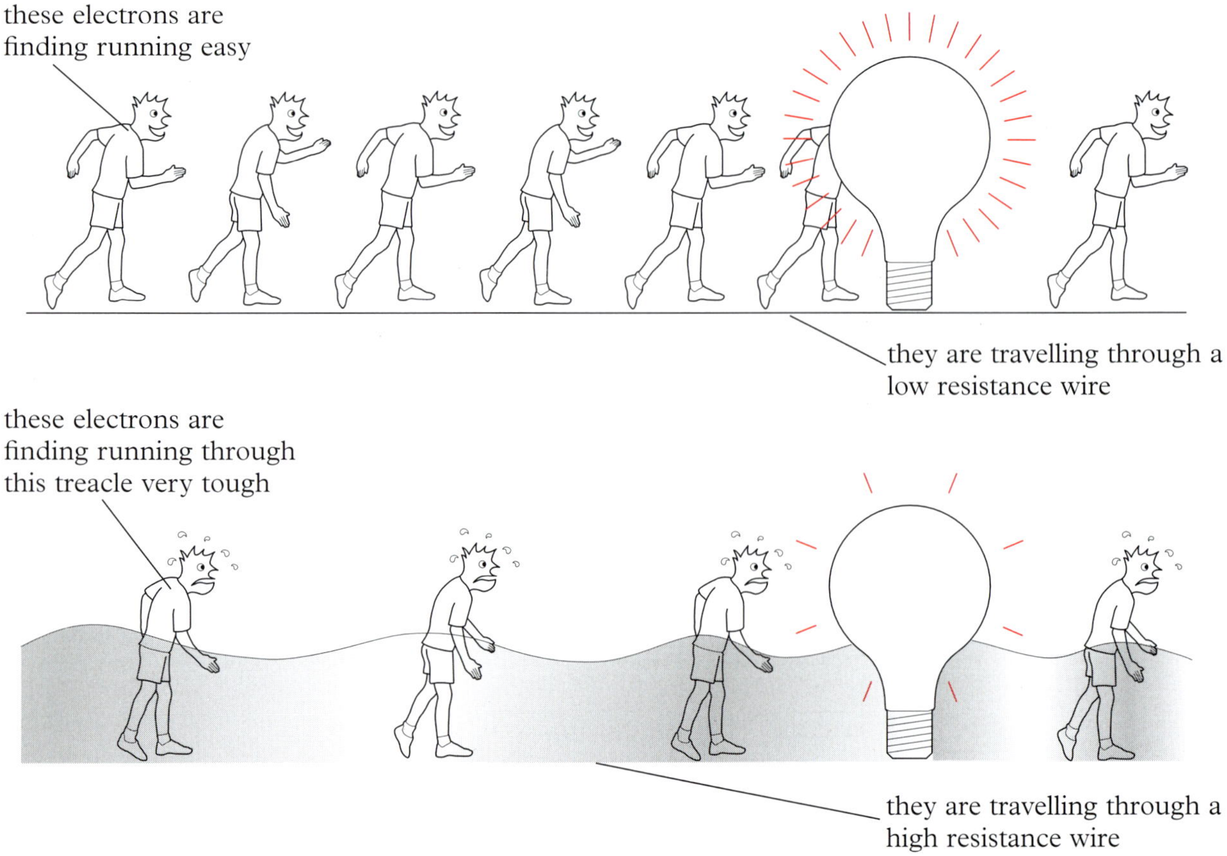
these electrons are finding running easy
they are travelling through a low resistance wire
these electrons are finding running through this treacle very tough
they are travelling through a high resistance wire

The resistance of a piece of wire at room temperature depends upon its thickness, length and the material from which it is made. The table below summarises the things that affect resistance

	low resistance	high resistance
thickness	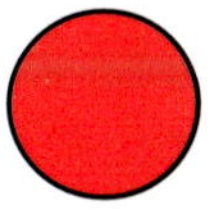 the thicker the wire the easier it is for electrons to to flow	 the thinner the wire the harder it is for electrons to flow
length	 the shorter the wire the more easily the electrons will flow through	 the longer the wire the more difficult it is for electrons to flow through it
material	 electrons flow easily through copper	 electrons cannot flow easily through a brick

Questions

1 Copy and complete the following paragraph:

'If electricity flows easily through a wire then the wire has a _________ resistance. If electricity does not flow easily through a wire then the wire has a _________ resistance.'

2 a) If the bulb in the circuit below is to glow more brightly, what should be done to the resistance of the circuit?

b) Suggest two ways in which you could increase the resistance of the circuit.

c) How would these changes affect the brightness of the bulb?

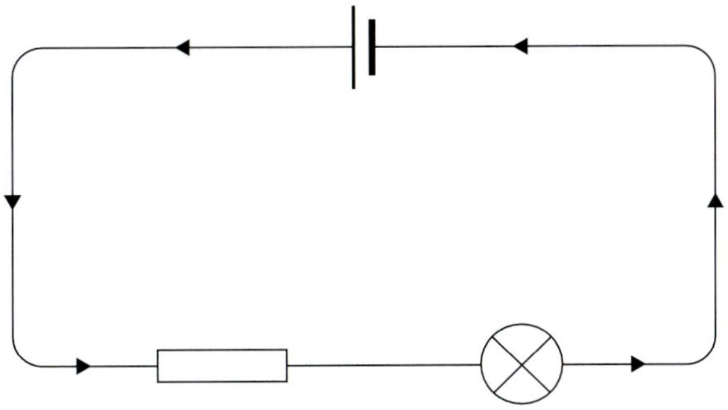

Resistors

Fixed resistors

fixed resistors are used in circuits to control current

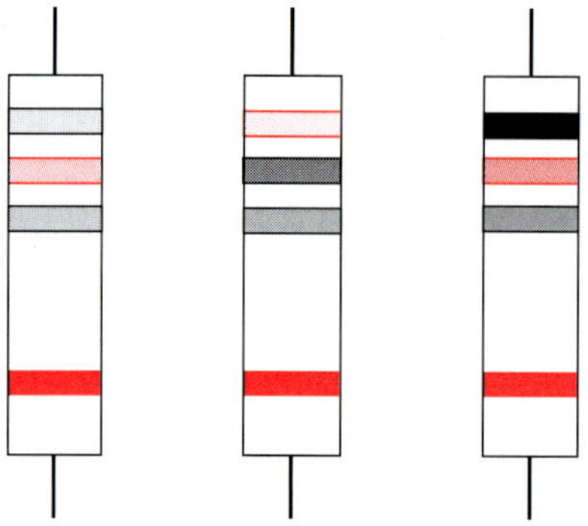

each resistor has a fixed value

coloured bands show the value of each resistor

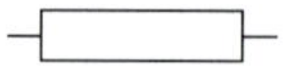

in circuit diagrams this symbol is used to show a fixed resistor

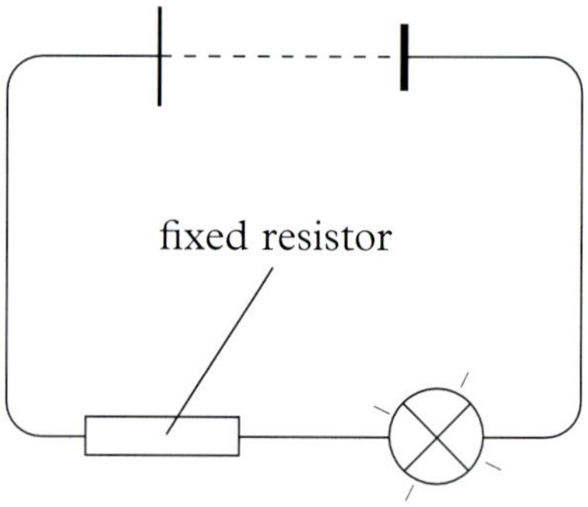

the resistor reduces the current flowing through the bulb

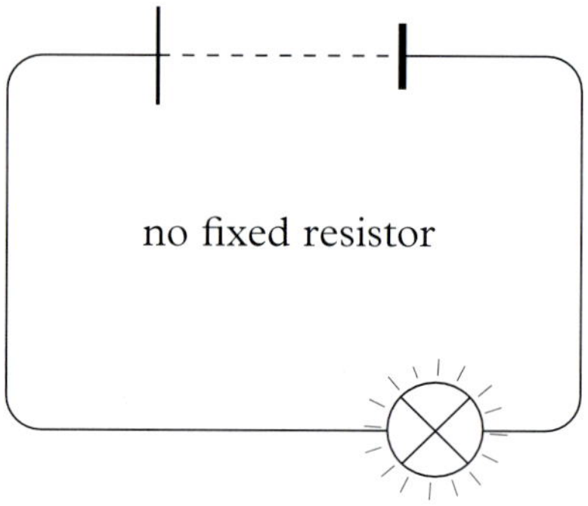

with no resistor the current may be too large and the bulb may break

Variable resistors

variable resistors are used in circuits to control current

the value of variable resistors can be altered in order to change the current

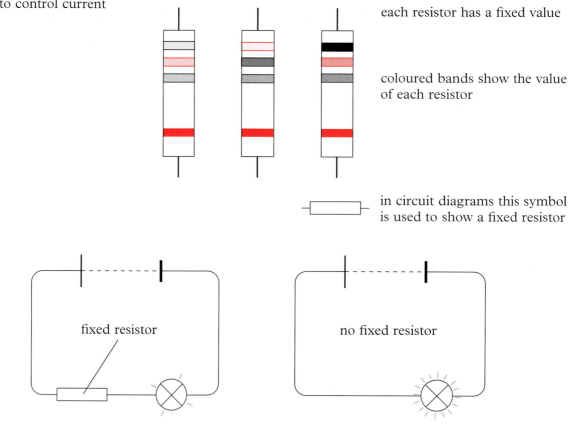

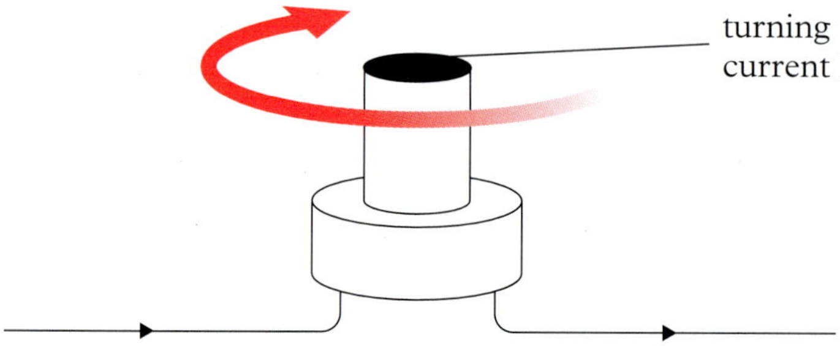

turning the knob makes the current larger or smaller

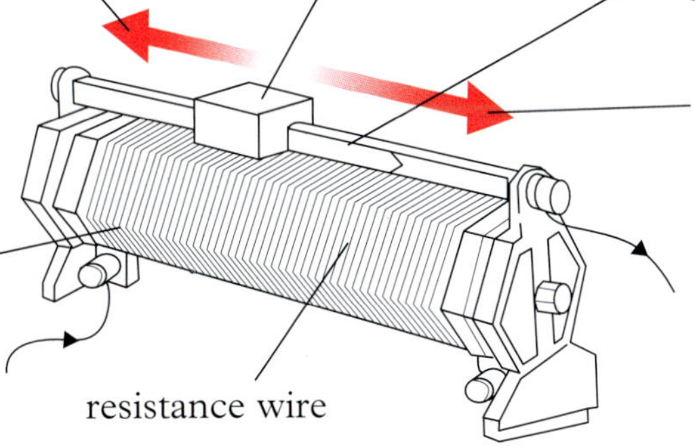
as the contact is moved to the left the current travels through a short length of resistance wire before reaching the rod with very low resistance
sliding contact
rod with very low resistance
as the contact is moved to the right the current travels through a longer length of resistance wire before reaching the rod with very low resistance
the current travels through the resistance wire from the left through the sliding contact and into the rod with very low resistance
resistance wire

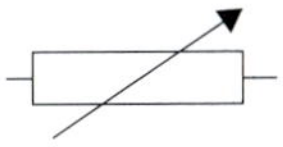
in circuit diagrams this symbol is used to show a variable resistor

Uses of variable resistors

A variable resistor is used in dimmer switches

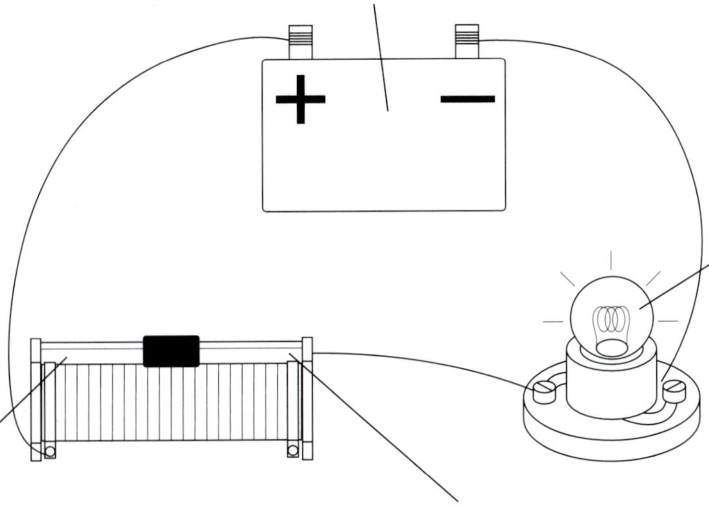
cell or battery
+
−
light bulb
as the contact is moved to the left the resistance in the circuit decreases and the light bulb becomes brighter
as the contact is moved to the right the resistance in the circuit increases and the light bulb becomes dimmer

Variable resistors have many uses. They can be used to control:

- the brightness and colour of TV screens and computer monitors,
- the loudness of music systems and radios,
- the speed of electric motors in model cars, planes, electric drills and in lots of other electrical items.

Questions

1 Explain the difference between a fixed resistor and a variable resistor.

2 Give two examples of circuits which contain variable resistors.

3 Design a simple circuit which shows how a variable resistor could be used to control the electric motor of a food mixer.

Special resistors

The resistance of some resistors can be changed by warming them or exposing them to light

The thermistor

Usually when the temperature of a piece of wire increases it is more difficult for current to flow through it, i.e. its resistance increases. But when the temperature of a **thermistor** increases, its resistance decreases.

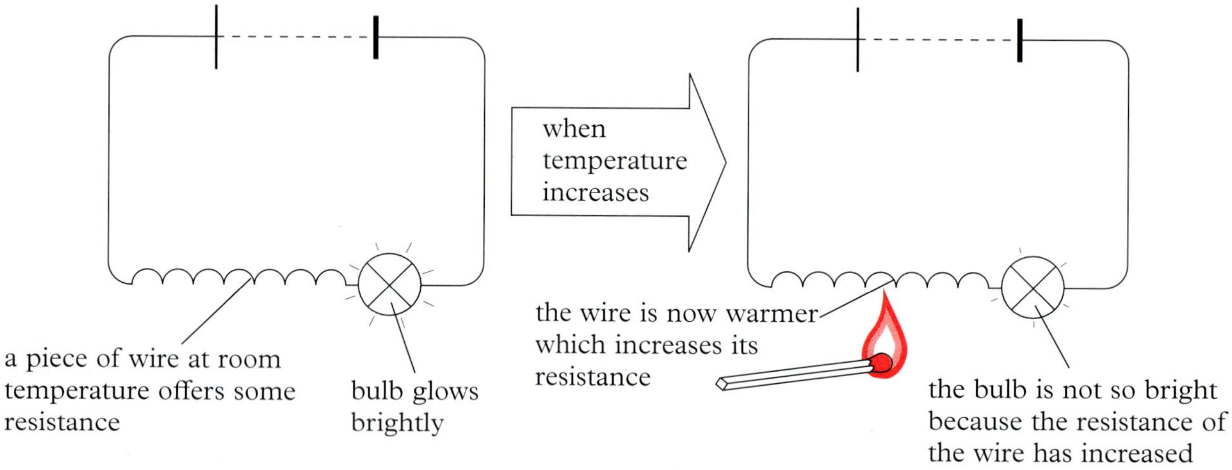
when temperature increases
a piece of wire at room temperature offers some resistance
bulb glows brightly
the wire is now warmer which increases its resistance
the bulb is not so bright because the resistance of the wire has increased

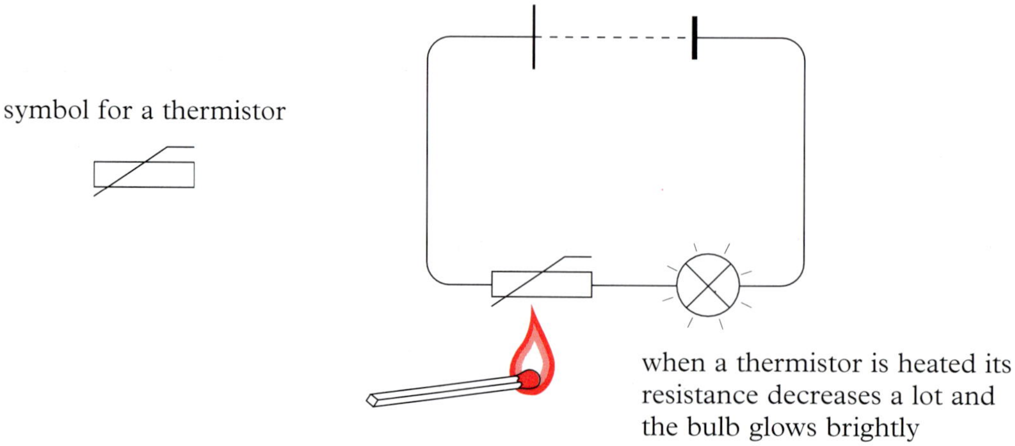
symbol for a thermistor
when a thermistor is heated its resistance decreases a lot and the bulb glows brightly

A circuit with a thermistor is very useful for monitoring temperature. It could for example be used in a fridge or freezer to warn if the temperature becomes too high.

The light dependent resistor (LDR)

A **light dependent resistor** is also a special kind of resistor. The resistance of an LDR decreases when light is shone on it.

This simple circuit shows how an LDR could be used in a light sensitive burglar alarm.

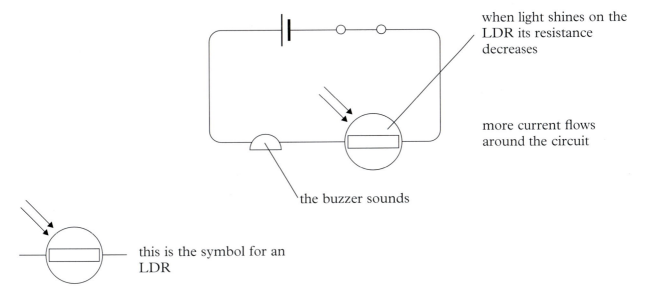

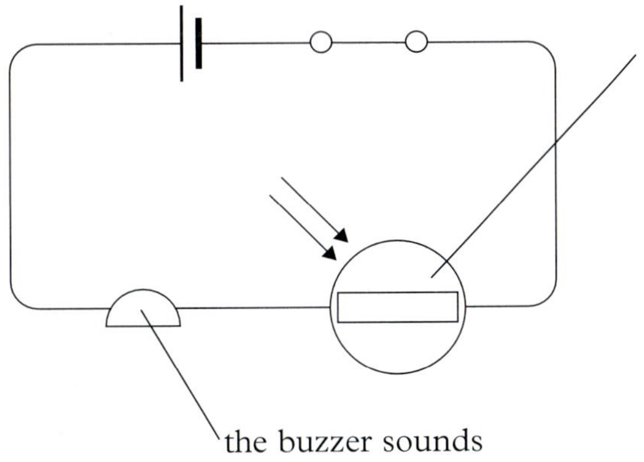
when light shines on the LDR its resistance decreases
more current flows around the circuit
the buzzer sounds

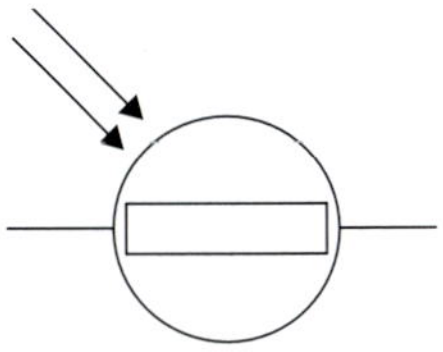
this is the symbol for an LDR

LDRs are also extremely useful in circuits which control light levels, e.g. they can be used in circuits which automatically turn on street lights when it becomes dark or in circuits which lower blinds if the sunlight becomes too bright.

Questions

1 a) What is a thermistor?
b) Give two examples of the kind of circuits which might include a thermistor.

2 a) What is an LDR?
b) Give two examples of the kind of circuits which might include an LDR.

3 What would happen to the resistance of a piece of wire if it was cooled?

Calculating resistance: Ohm's Law

If we apply a voltage across a piece of wire a current will flow. The size of the current will depend upon the size of the voltage and the resistance of the wire. The equation below shows how current, voltage and resistance are related.

$$\text{Resistance (R)} = \frac{\text{Voltage (V)}}{\text{Current (I)}}$$

This equation can be used to calculate the resistance of a wire, as in the following example.

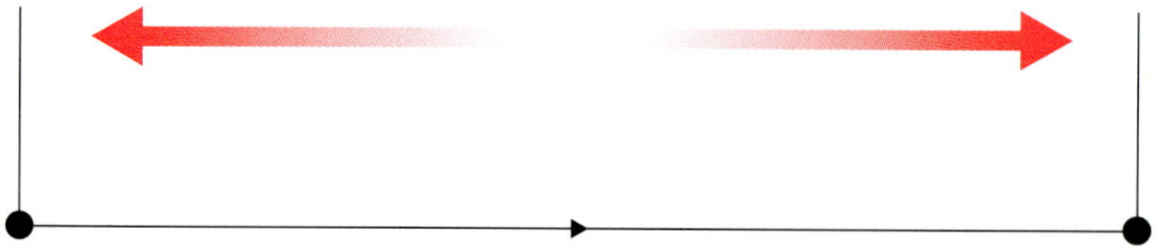
5 V is applied across the ends of the wire
2.5 A is flowing through the wire

If a current of 2.5 A flows through a wire when a voltage of 5 V is applied across its ends, what is the resistance of the wire?

$$R = \frac{V}{I}$$

$$R = \frac{5}{2.5}$$

$$\mathbf{R = 2\ \Omega}$$

We measure electrical resistance in **ohms** (Ω).

Using the equation triangle we can rewrite this equation as

$$\mathbf{V = I \times R} \quad \text{or} \quad \mathbf{I = V / R}$$

Questions

1 Calculate the resistance of a piece of wire which:
a) has a current of 2 A flowing through it when a voltage of 6 V is applied across its ends,
b) has a current of 10 A flowing through it when a voltage of 20 V is applied across its ends,
c) has a current of 0.5 A flowing through it when a voltage of 10 V is applied across its ends.

2 How large a voltage must be applied across a 100 Ω resistor in order that:
a) a current of 1 A flows?
b) a current of 5 A flows?
c) a current of 0.2 A flows?

3 Calculate the current that flows when a voltage of 12 V is applied across
a) a 6 Ω resistor,
b) a 12 Ω resistor,
c) a 120 Ω resistor.

Ohmic and Non ohmic conductors

If we set up the circuit as shown on the next page we can see how the current passing through a component varies as the voltage across it changes.

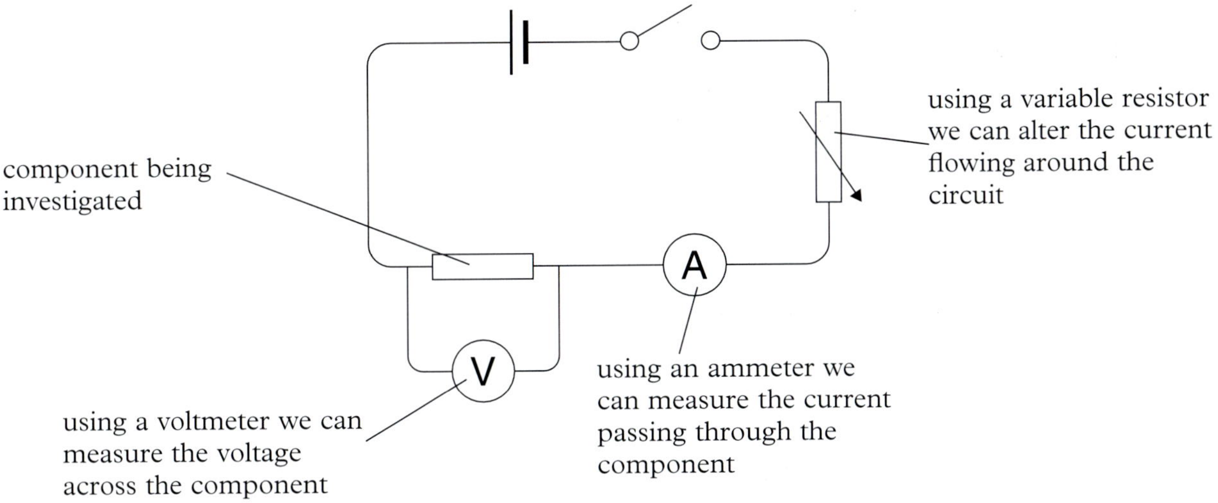
component being investigated
using a variable resistor we can alter the current flowing around the circuit
A
V
using an ammeter we can measure the current passing through the component
using a voltmeter we can measure the voltage across the component

If we present the results in the form of a graph we can clearly see the difference in behaviour of the different components.

A resistor at constant temperature

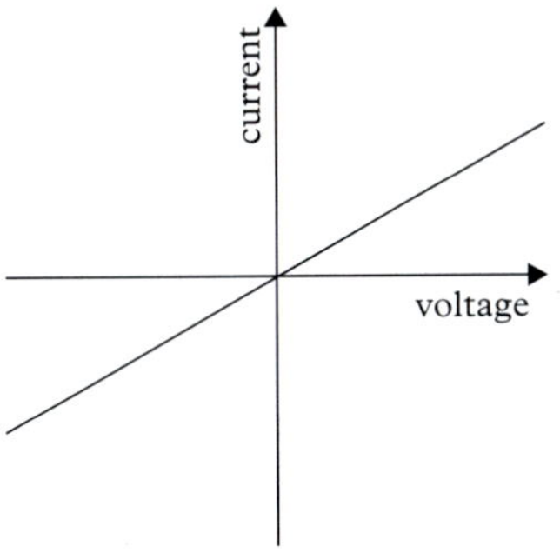
current
voltage

Conductors which produce a straight line, as in the graph for the resistor, are called **ohmic conductors**. Ohmic conductors obey Ohm's Law. This means that if the voltage across the component is doubled the current flowing through it is also doubled.

A filament light

the current/voltage graph for a filament light

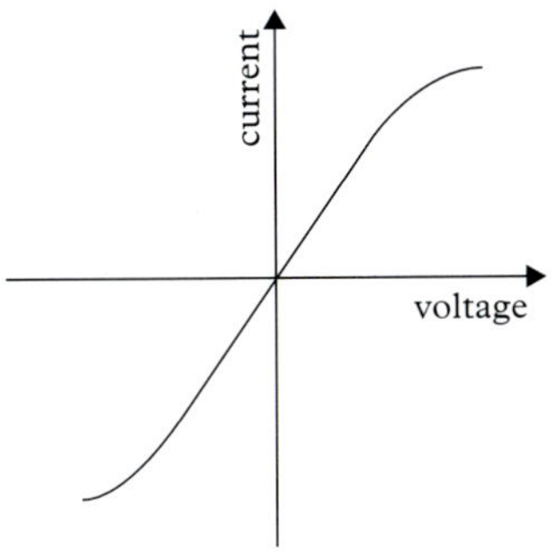
current
voltage

Components like a filament light do not produce a straight line graph. Components which do not produce a straight line graph are called **non-ohmic conductors.** Non-ohmic conductors do not obey Ohm's law. This means that if the voltage across the component is doubled the current flowing through it is not doubled.

Key words

electrical resistance	A measure of how difficult it is for current to flow through a component. Resistance is measured in ohms (Ω).
fixed resistor	A resistor of fixed resistance often used to control current or voltage.
light dependent resistor (LDR)	A resistor whose resistance decreases when light shines up on it.
non-ohmic conductor	A conductor which does not obey Ohm's Law e.g. a filament lamp.
ohmic conductor	A conductor which obeys Ohm's Law e.g. a copper wire.
Ohm's Law	The current flowing through a wire is proportional to the voltage across its ends, providing the temperature of the wire does not change.
Ohm's Law equation	Resistance (R) = Voltage (V) / Current (I)
thermistor	A resistor whose resistance usually decreases rapidly as its temperature increases.
variable resistor	A resistor whose resistance can be altered in order to vary the current or voltage.

Summary

If a current can flow easily around a circuit, the circuit has a low resistance. If it is difficult for a current to flow around a circuit, the circuit has a high resistance.

All components in a circuit have some resistance. The larger the resistance of a component the smaller the current that flows for a particular voltage. Connecting wires usually have a resistance that is so small it is taken as zero.

We measure resistance in Ohms (Ω). We can calculate the resistance of a component using the equation:

Resistance (R) = Voltage (V) / Current (I)

Conductors such as wires at a constant temperature are ohmic conductors. Filament lamps are non-ohmic conductors.

End of Chapter 16 Questions

1 The diagram below shows a simple series circuit consisting of a battery, a lamp, some connecting wires and a length of resistance wire.

Explain what will happen to the bulb if the resistance wire is replaced with:
a) an identical piece of wire which is much longer
b) an identical piece of wire which is much thicker.

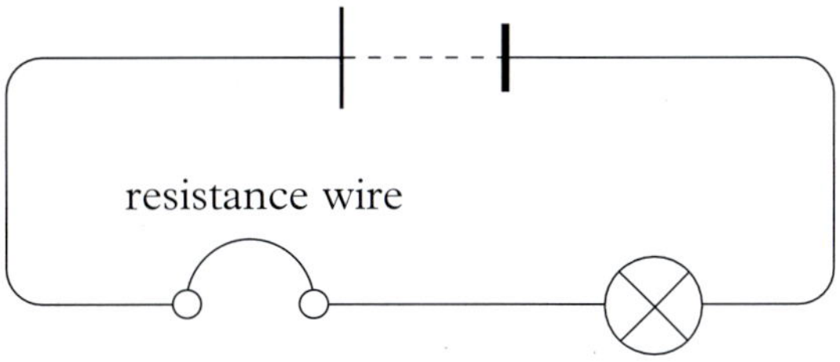

2 Look at the circuit below.

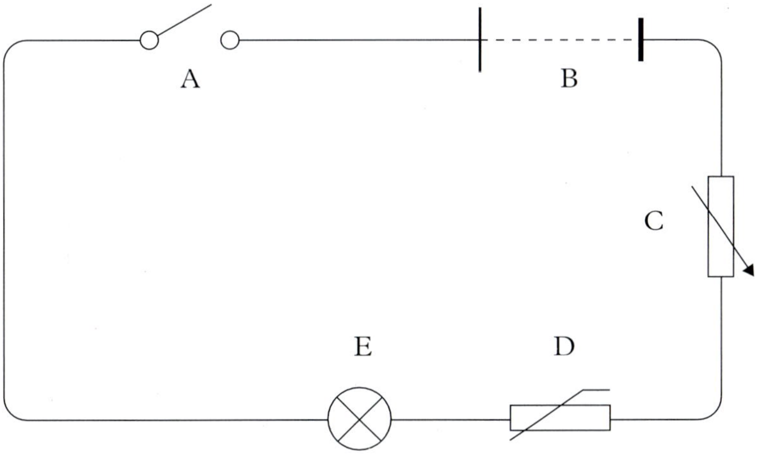

a) Identify all the components labelled A to E
b) If component A is closed and component D is warmed explain what you would see happen to component E.
c) Give one use for component D.

3 Calculate the current that will flow through a $6\,\Omega$ resistor if the following p.d.s are applied across the ends of the resistor:
a) 4 V
b) 6 V
c) 10 V

4 A pupil is given three different resistors but their labels have fallen off. Explain with diagrams how the pupil could discover which label belonged to which resistor. (Assume that standard electrical equipment like wires and cells are available).

5 A pupil builds the two circuits shown on the next page. He then increases the value of the variable resistors. He finds to his surprise that one of the bulbs becomes dimmer whilst the other becomes brighter.

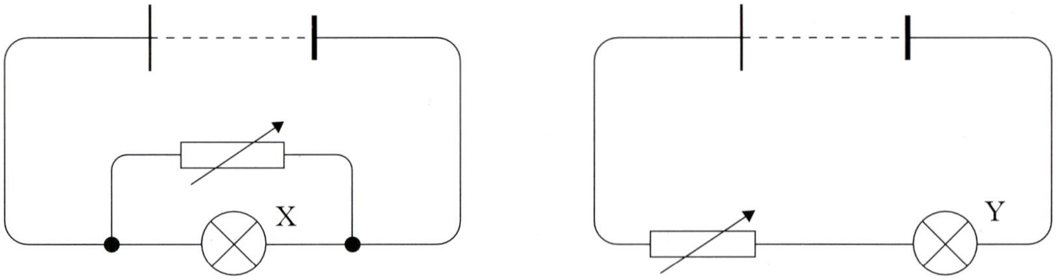

a) i) Which bulb becomes brighter?
ii) Explain why this happens.
b) Explain why the other bulb becomes dimmer.

6

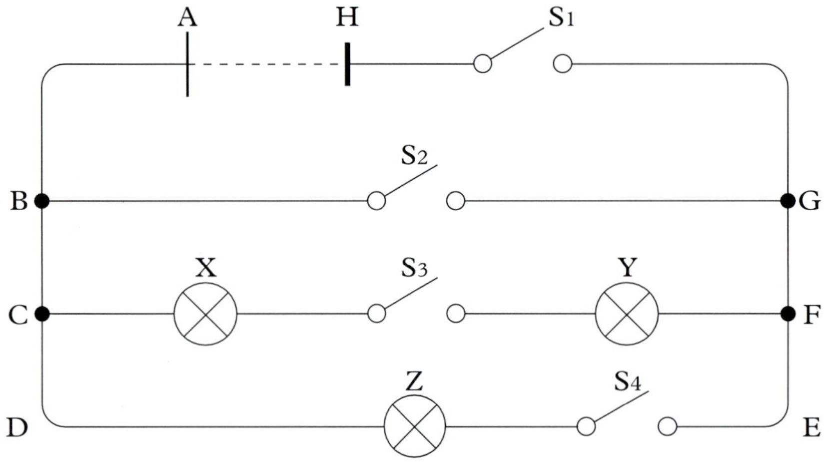

a) If all the switches in this circuit are closed which path offers the least resistance to the current?
b) What will happen to the bulbs in the circuit when all the switches are closed?
c) Why is this kind of connection called a **short circuit**?
d) Can you suggest why this short circuit will damage the cell?
e) Which bulb(s) will glow when:
  i) switches S1 and S4 are closed?
  ii) switches S1 and S3 are closed?
  iii) switches S3 and S4 are closed?

7 a)

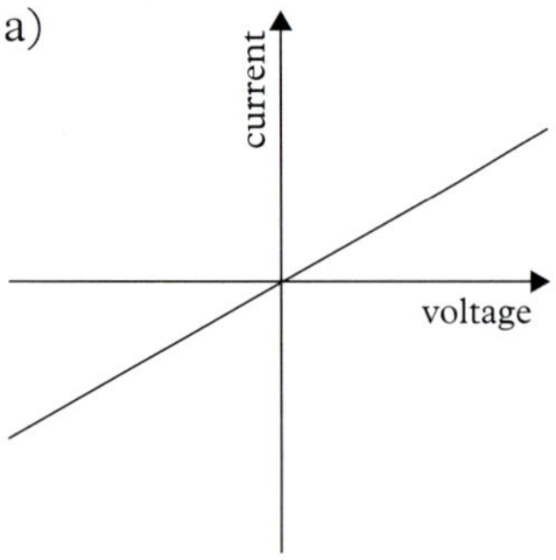

b)

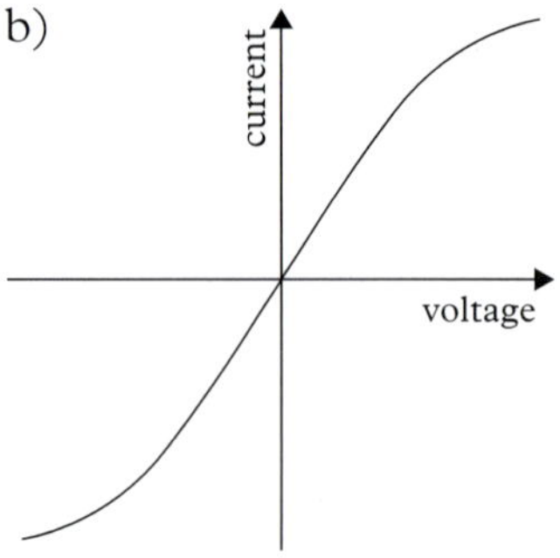

Look carefully at the current-voltage graphs above.
a) Which of these components is an ohmic conductor?
c) Give one example of a component which is:
  i) an ohmic conductor
  ii) non-ohmic conductor.

17 Magnets and Magnetism

If we spill some paperclips onto a carpet it might take a long time to pick them all up by hand. It would be much easier to pick them up using a **magnet**.

Magnetic and non-magnetic materials

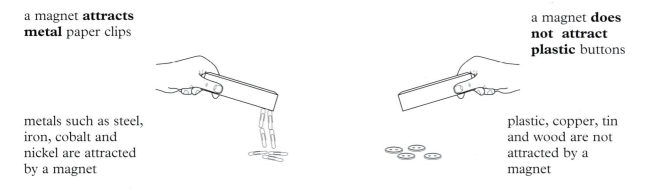

a magnet attracts some things but not others
a magnet **attracts metal** paper clips
a magnet **does not attract plastic** buttons
metals such as steel, iron, cobalt and nickel are attracted by a magnet
plastic, copper, tin and wood are not attracted by a magnet
these are called **magnetic materials**
these are called **non-magnetic materials**

The poles of a magnet

If we dip a magnet into a bowl of iron filings we notice that the iron filings are strongly attracted to certain parts of the magnet. These parts are called the **poles** of the magnet. The poles are the most magnetic parts of the magnet. Usually magnets have two poles, called the **North pole** and the **South pole**.

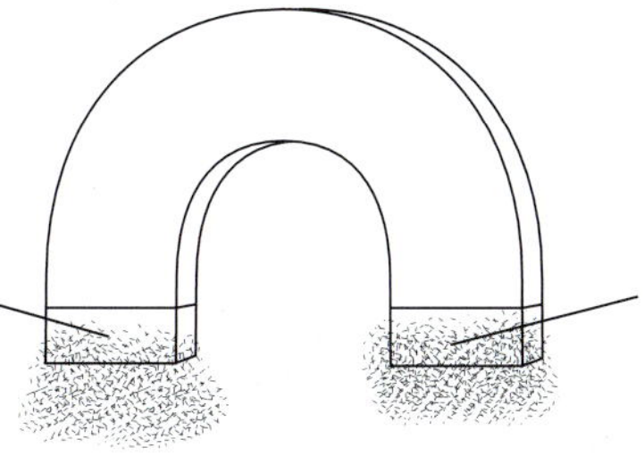
north pole
south pole
iron filings are attracted to the poles
the poles are the strongest parts of the magnet

If a bar magnet is suspended so that it is free to rotate it will come to rest with its North pole pointing northwards and its South pole pointing southwards. When this happens the magnet is behaving like a **compass**.

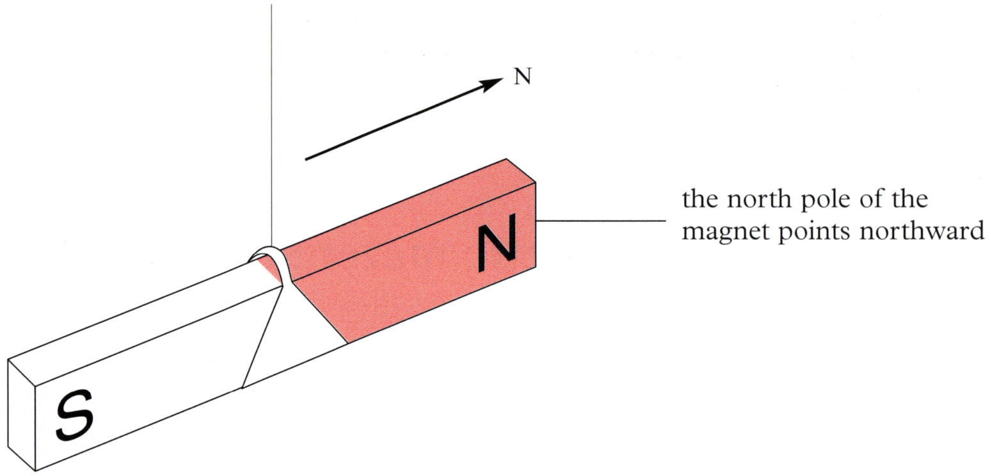
N
N
S
the north pole of the magnet points northward

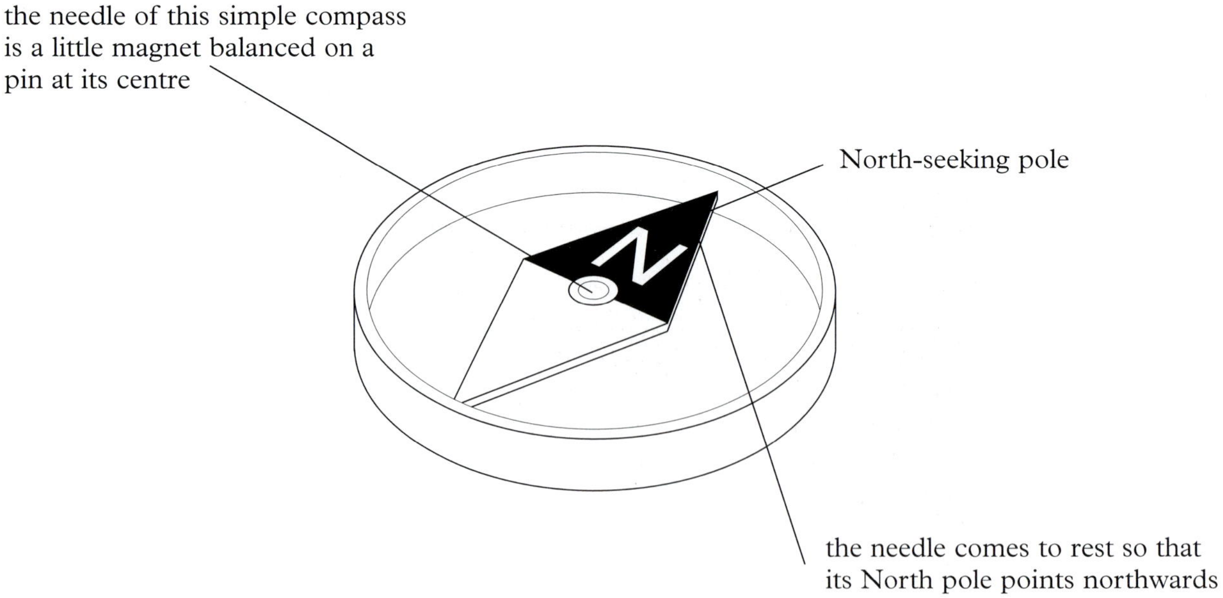
the needle of this simple compass is a little magnet balanced on a pin at its centre
North-seeking pole
N
the needle comes to rest so that its North pole points northwards

Questions

1 Name two magnetic materials.

2 Name two non-magnetic materials

3 What do we call the strongest parts of a magnet?

4 Explain with diagrams how you could use a bar magnet as a compass.

Attraction and repulsion between the poles of magnets

if the North pole of one magnet is placed close to the South pole of a second magnet they will be attracted to each other

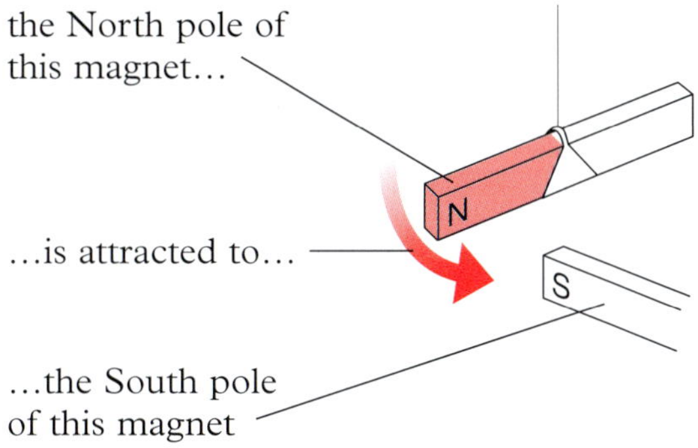
the North pole of this magnet…
…is attracted to…
…the South pole of this magnet
N
S

opposite poles attract

if two North or South poles are placed together they will push apart

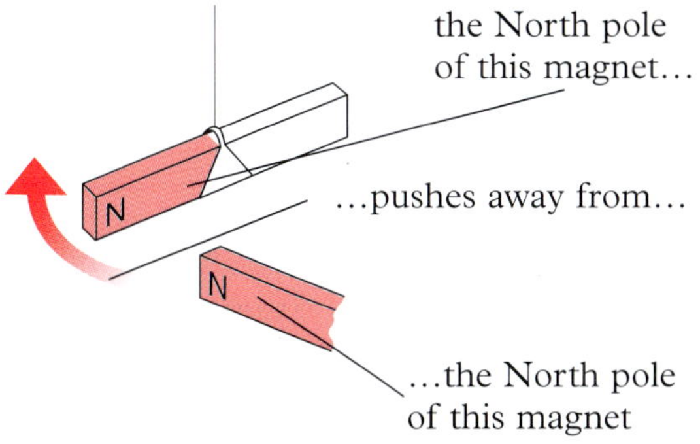
the North pole of this magnet…
…pushes away from…
…the North pole of this magnet
N
N

similar poles repel

The Maglev train shown below hovers above the rails because **similar poles repel**.

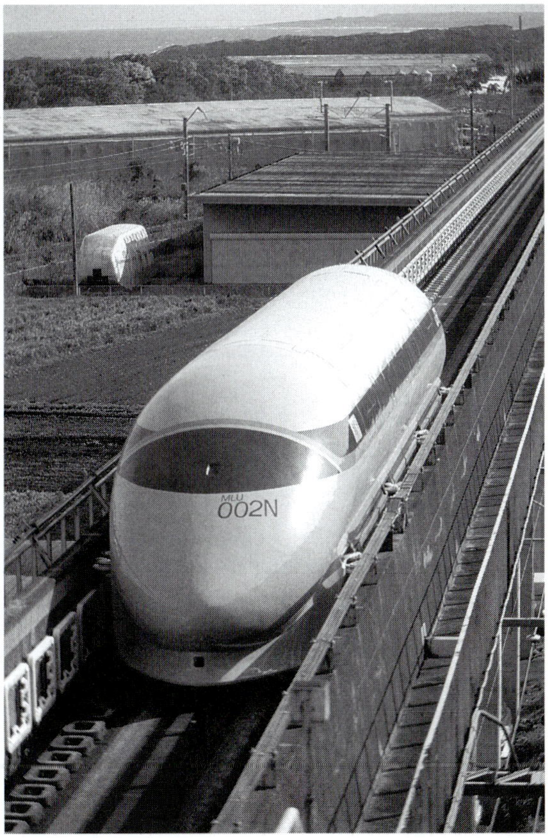
002N

Questions

1 Copy and complete these sentences:

a) If the North poles of two magnets are placed next to each other they will _________ .

b) If the South poles of two magnets are placed next to each other they will _________ .

c) If a North pole of one magnet is placed next to a South pole of a second magnet they will _________ .

Magnetic fields

Around a magnet is a **magnetic field** where we can detect magnetism. We can discover the shape of a magnetic field using iron filings or a plotting compass.

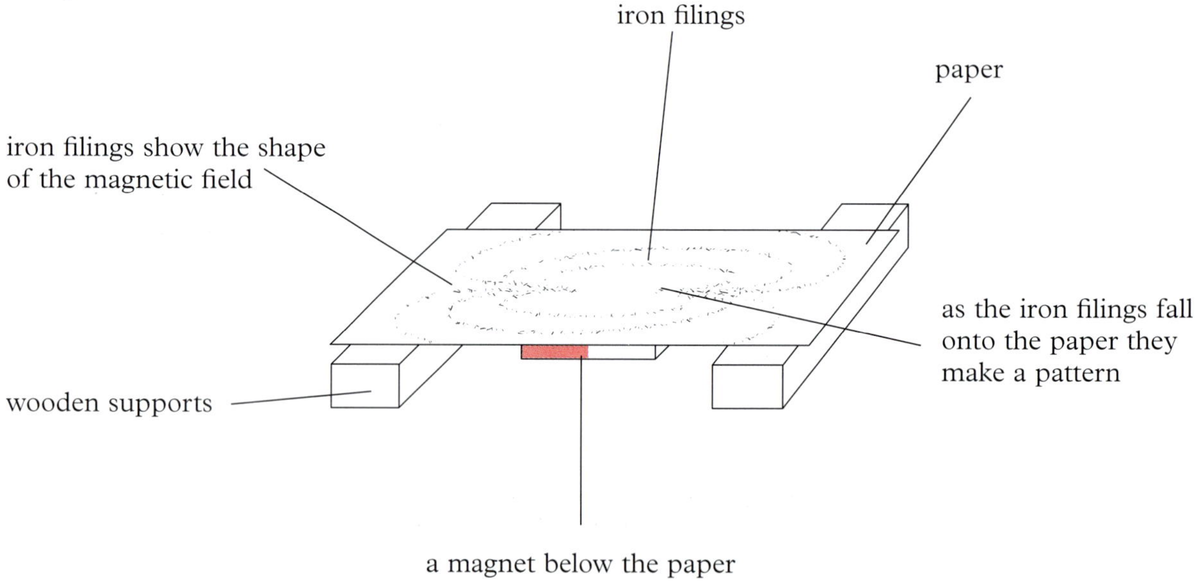
iron filings
paper
iron filings show the shape of the magnetic field
as the iron filings fall onto the paper they make a pattern
wooden supports
a magnet below the paper

The shape and direction of a magnetic field

By moving a plotting compass around the magnet we can see both the direction and shape of its magnetic field.

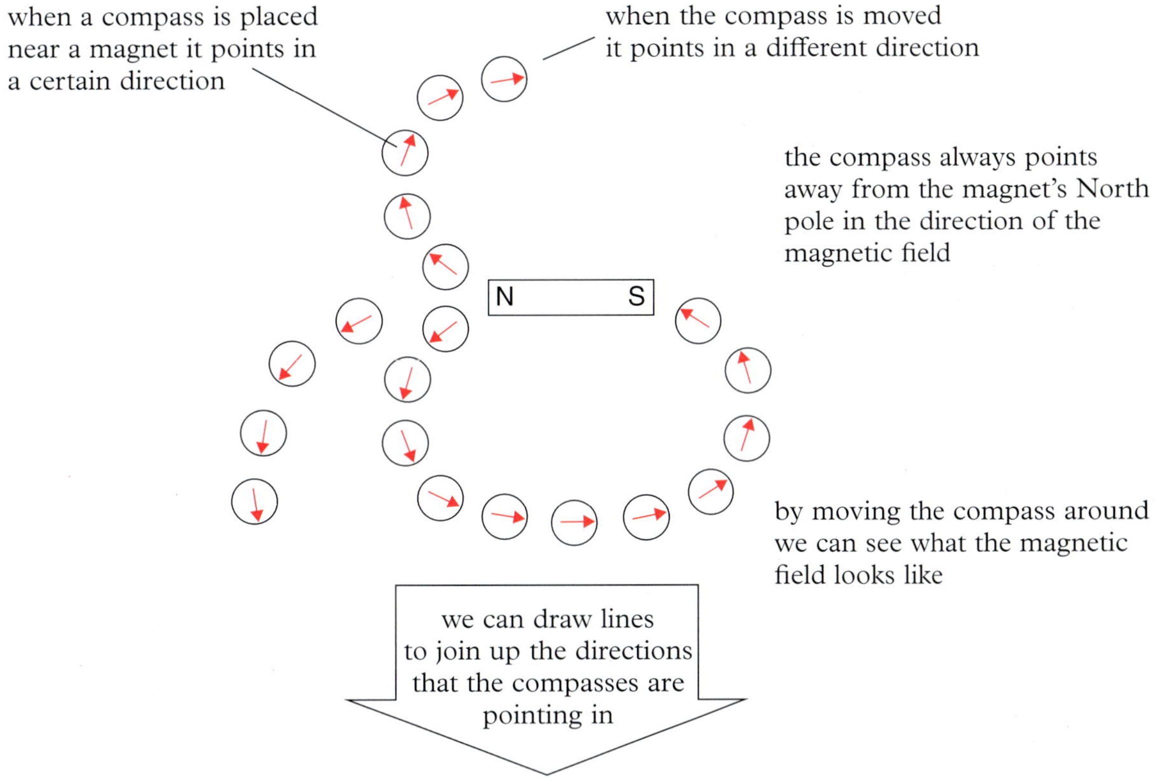
when a compass is placed near a magnet it points in a certain direction
when the compass is moved it points in a different direction
the compass always points away from the magnet's North pole in the direction of the magnetic field
N
S
by moving the compass around we can see what the magnetic field looks like
we can draw lines to join up the directions that the compasses are pointing in

the lines drawn to show a magnetic field are called magnetic lines of force

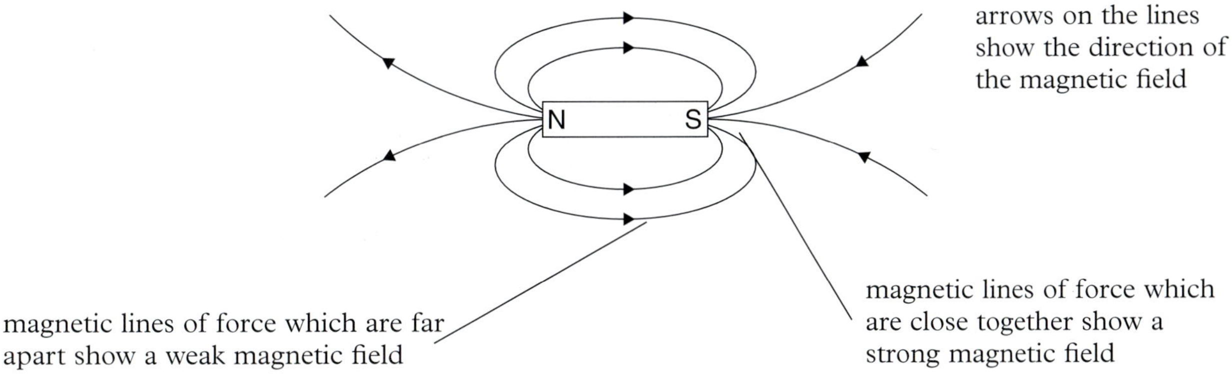
arrows on the lines show the direction of the magnetic field
N
S
magnetic lines of force which are far apart show a weak magnetic field
magnetic lines of force which are close together show a strong magnetic field

The magnetic lines of force on the diagram above show us the shape, direction and strength of the magnetic field.

Magnetic lines of force

When two bar magnets are placed near to each other their magnetic fields overlap. The diagram below shows two different overlapping magnetic fields.

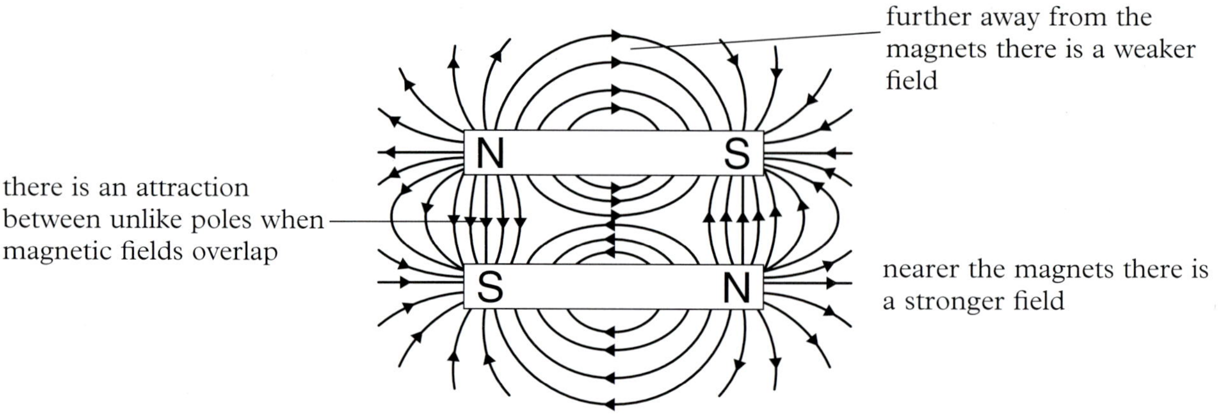
further away from the magnets there is a weaker field
there is an attraction between unlike poles when magnetic fields overlap
N
S
S
N
nearer the magnets there is a stronger field

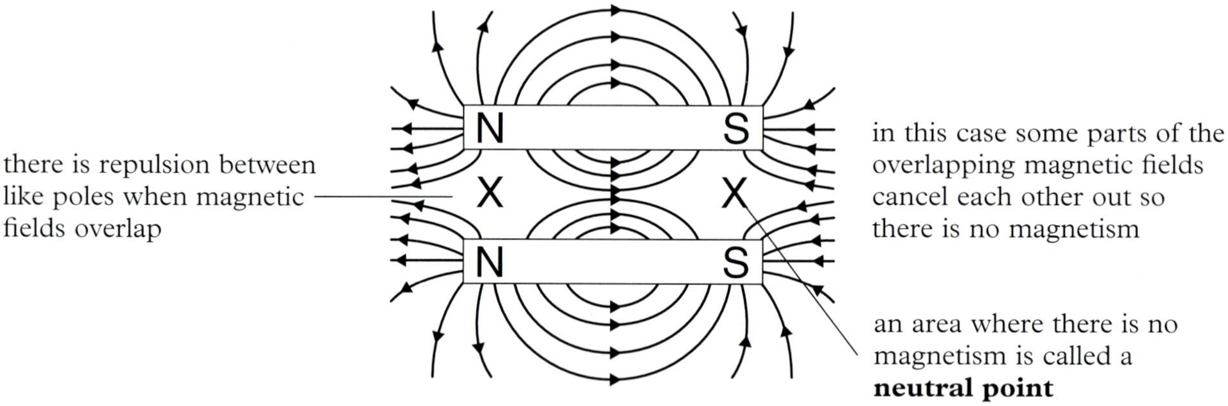
there is repulsion between like poles when magnetic fields overlap
N
S
X
X
N
S
in this case some parts of the overlapping magnetic fields cancel each other out so there is no magnetism
an area where there is no magnetism is called a **neutral point**

The magnetic lines of force on the diagram above show us the shape, direction and strength of the magnetic field.

Questions

1 a) Draw a diagram of the magnetic field around a bar magnet.
b) Label the poles of the magnet.
c) Show the direction of the magnetic field.

2 Sketch the magnetic field between two magnets which are
a) repelling each other,
b) attracting each other.

3 What is a neutral point?

Summary

Magnets are able to attract magnetic materials such as iron and steel but not non-magnetic materials such as plastic or wood. Most magnets have two poles, a North pole and a South pole. Two similar poles placed close together will repel. Two dissimilar poles placed close together will attract. If a bar magnet is suspended freely it will turn so that its North pole points north and its South pole points south. When it does this it behaves like a compass.

All magnets have a magnetic field around them. We can show the shape and strength of a magnetic field by drawing diagrams with magnetic lines of force.

Key words

magnetic field	The space around a magnet where we can detect magnetism.
magnetic material	A material which is attracted to a magnet, like iron.
non-magnetic material	A material which is not attracted to a magnet, like paper.
pole	The part of the magnet where the magnetism is strongest.

End of Chapter 17 Questions

1 Fill in the following sentence.

Magnetic lines of force are _________ where the field is strong and are _________ where the field is weak.

2 a) Describe one method for discovering the shape of the magnetic field around a bar magnet.
b) Draw a labelled diagram of the magnetic field around a bar magnet.
c) Label the area where the magnetic field is strong as **Y**.
d) Label the area where the magnetic field is weak as **Z**.

3 A pupil is given three identical bar magnets and told that one of the magnets has lost all its magnetism. How could the pupil discover which is the faulty magnet **without using any other apparatus**?

4 Explain how you could use a magnet to remove the iron filings from a mixture of iron filings and sawdust.

5 Explain why you could not use a magnet to only remove the copper filings from a mixture of copper filings and sawdust.

6 A small magnetised needle is placed on a cork which is then floated in a bowl of water.
a) In which direction will the needle point?
b) Describe what will happen if a second magnetised needle and cork are placed in the bowl.

18 Electromagnetism

If we pass an electric current through a wire, a weak magnetic field is created around it. If the current is turned off the field disappears. This connection between electricity and magnetism was first discovered in 1819 by a Danish scientist named Oersted. He called his discovery **electromagnetism**.

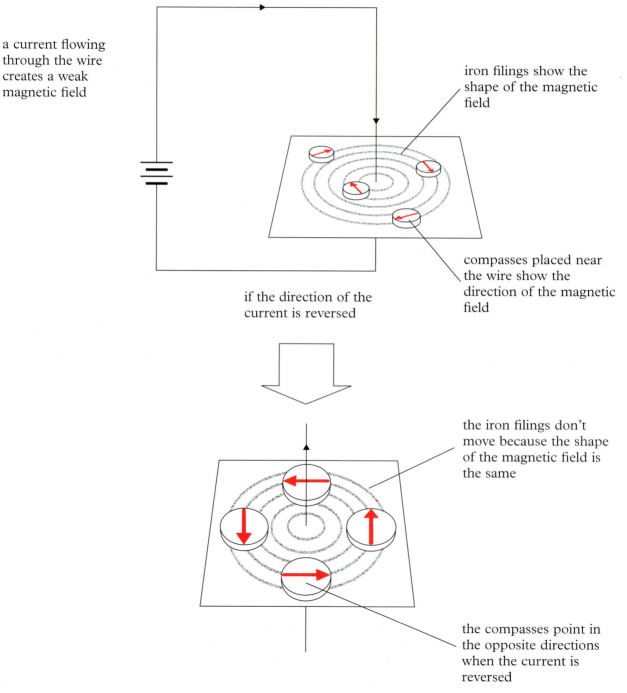

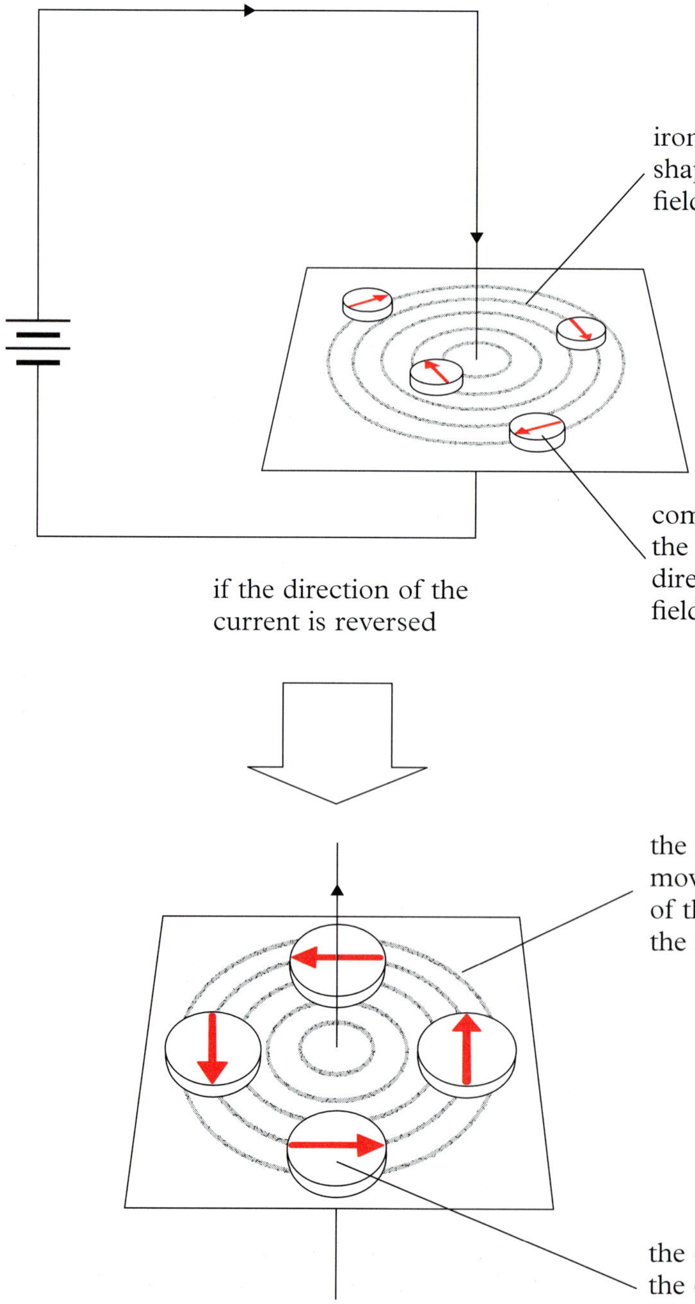
a current flowing through the wire creates a weak magnetic field
iron filings show the shape of the magnetic field
compasses placed near the wire show the direction of the magnetic field
if the direction of the current is reversed
the iron filings don't move because the shape of the magnetic field is the same
the compasses point in the opposite directions when the current is reversed

if the current is removed

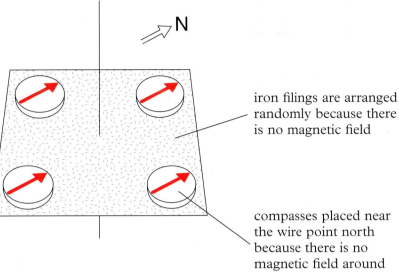

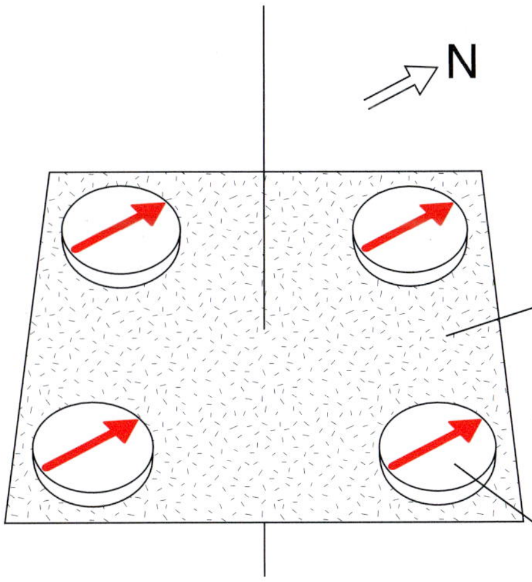
N
when no current is flowing through the wire the magnetic field disappears
iron filings are arranged randomly because there is no magnetic field
compasses placed near the wire point north because there is no magnetic field around the wire

Maxwell's Corkscrew Rule

We can predict the direction of the magnetic field using **Maxwell's Corkscrew Rule**.

Imagine you are holding a corkscrew parallel to a wire. There is a current flowing through the wire. The tip of the corkscrew is pointing in the same direction that the current is flowing in.

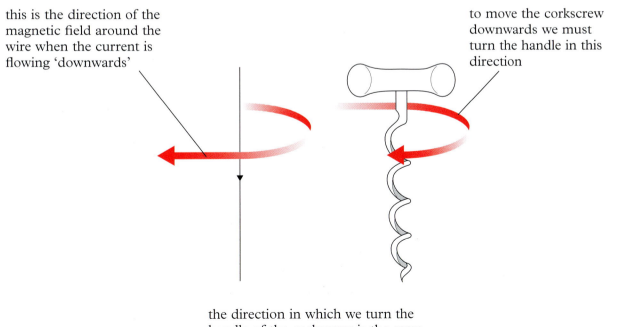

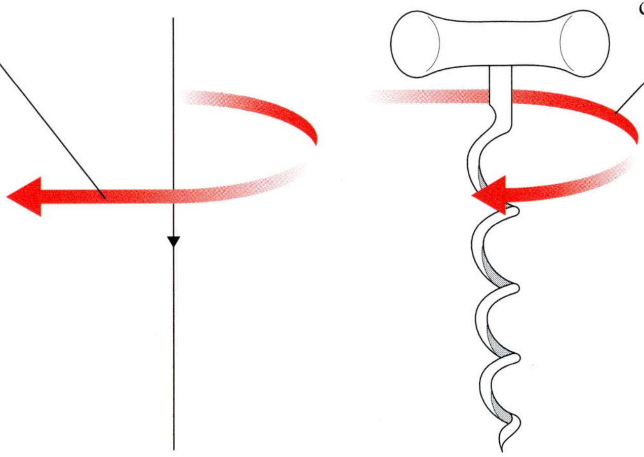
this is the direction of the magnetic field around the wire when the current is flowing 'downwards'
to move the corkscrew downwards we must turn the handle in this direction
the direction in which we turn the handle of the corkscrew is the same as the direction of the magnetic field around the wire

Questions

1 Draw a diagram to show the shape of the magnetic field around a piece of wire which has a current flowing through it.

2 a) What happens to the magnetic field if the direction of the current is changed?
b) What happens to the magnetic field if the current is turned off?

3 Describe with a diagram Maxwell's corkscrew rule.

Coils and solenoids

The magnetic field produced when a current flows through a single wire is quite weak. The strength of the field can be increased by winding the wire into several loops.

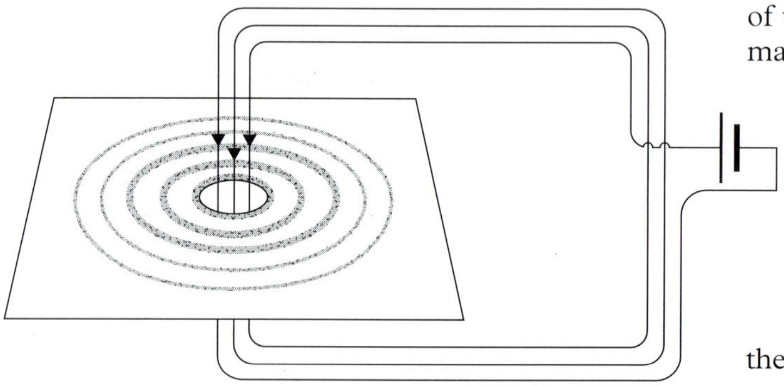
the current in each piece of wire produces a magnetic field
the magnetic fields around each wire overlap to produce a stronger magnetic field

We can increase the number of loops of wound wire.

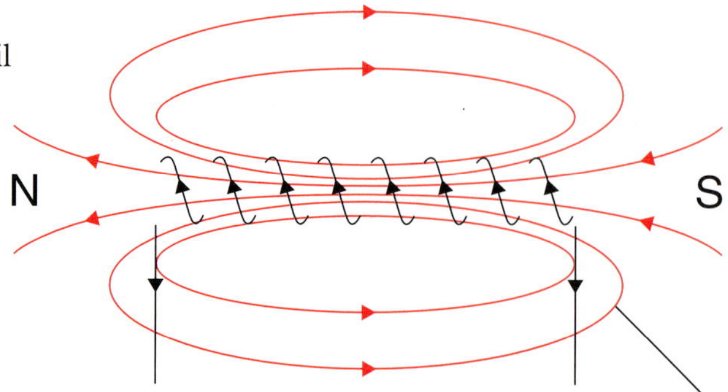
a large number of loops of wire next to each other is called a long coil or **solenoid**
a greater number of loops means that the magnetic field is even stronger
N
S
the shape of the magnetic field is similar to that of a bar magnet

The magnetic field produced by a solenoid is similar to that produced by a bar magnet but with three important differences.

1 The first difference between a solenoid and a bar magnet is that the magnetic field around a solenoid can be turned on and off

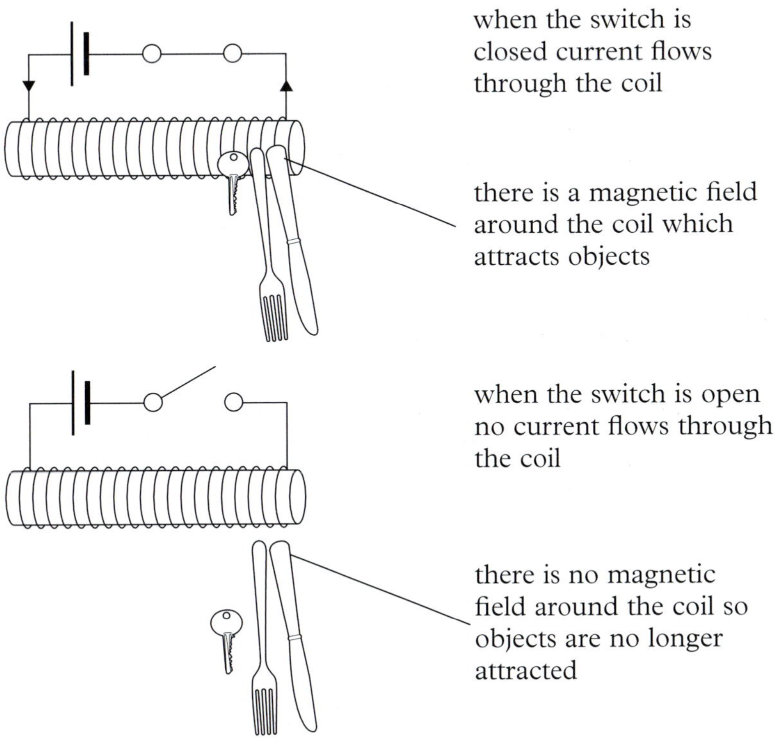
when the switch is closed current flows through the coil
there is a magnetic field around the coil which attracts objects
when the switch is open no current flows through the coil
there is no magnetic field around the coil so objects are no longer attracted

2 The second difference between a bar magnet and a solenoid is that the strength of the magnet field around a solenoid can be increased. This can be done by increasing either (i) the size of the current flowing through the solenoid or (ii) the number of turns or loops on the coil.

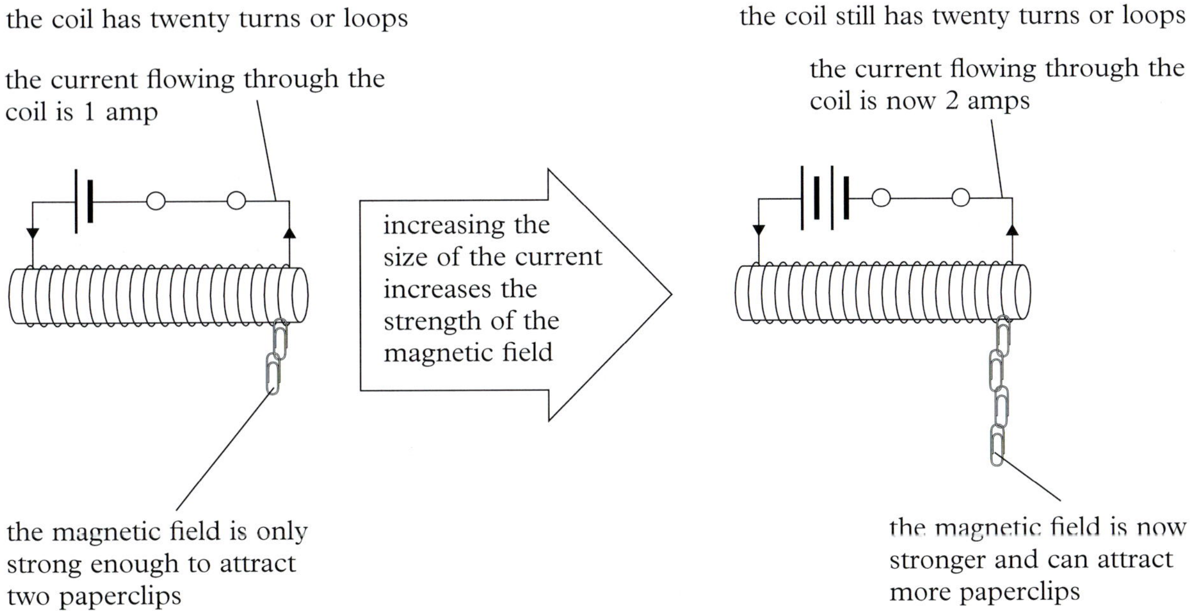
the coil has twenty turns or loops
the current flowing through the coil is 1 amp
the magnetic field is only strong enough to attract two paperclips
increasing the size of the current increases the strength of the magnetic field
the coil still has twenty turns or loops
the current flowing through the coil is now 2 amps
the magnetic field is now stronger and can attract more paperclips

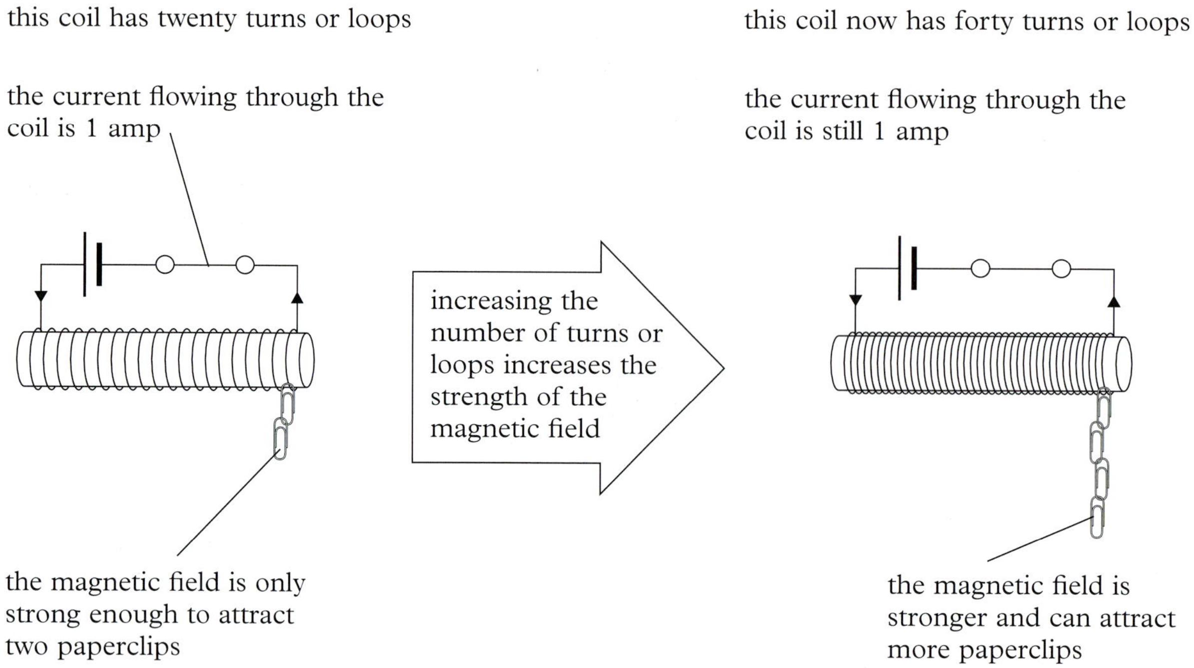
this coil has twenty turns or loops
the current flowing through the coil is 1 amp
the magnetic field is only strong enough to attract two paperclips
increasing the number of turns or loops increases the strength of the magnetic field
this coil now has forty turns or loops
the current flowing through the coil is still 1 amp
the magnetic field is stronger and can attract more paperclips

3 The third difference between a bar magnet and a solenoid is that the positions of the solenoid's north and south poles can be changed. This is done by reversing the connections to the battery so that the current flows in the opposite direction.

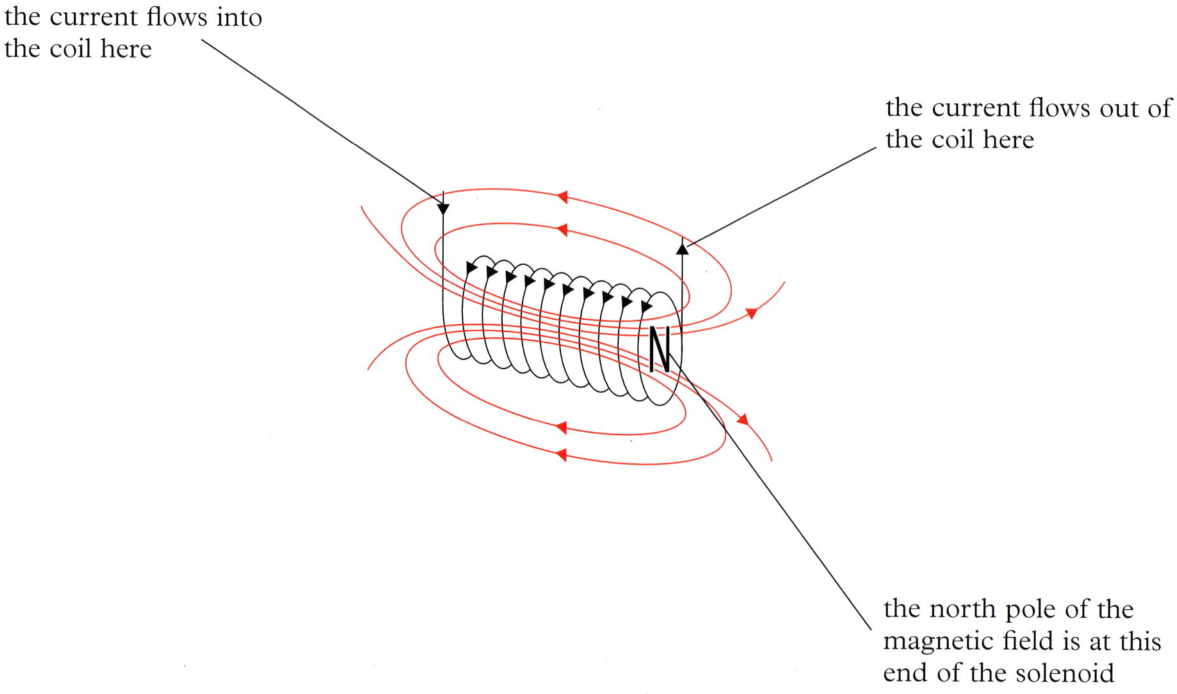
the current flows into the coil here
the current flows out of the coil here
N
the north pole of the magnetic field is at this end of the solenoid

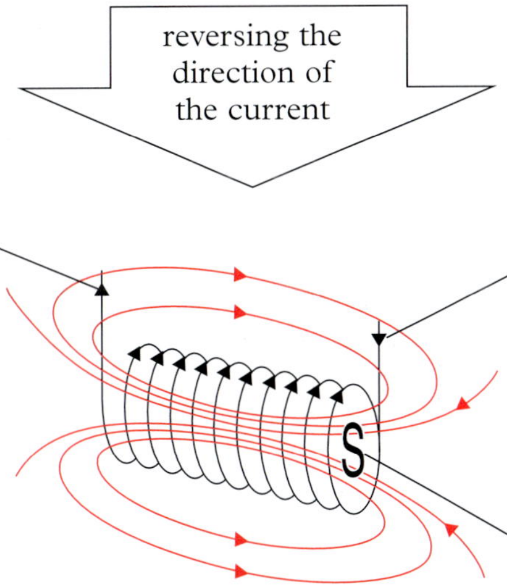
reversing the direction of the current
the current flows out of the coil here
the current flows into the coil here
S
now the south pole of the magnetic field is at this end of the solenoid

Questions

1 Draw a diagram to show the shape of the magnetic field created around a solenoid when a current is passing through it.
  a) Mark on your diagram the direction of the current in the solenoid
  b) Mark on your diagram the positions of the north and south poles of the magnetic field.

2 Suggest two ways in which you could increase the strength of the magnetic field around a solenoid.

Electromagnets

If a current is passed through a long coil which is wrapped around an iron nail, a strong magnetic field is produced. This field magnetises the iron nail. This means that the nail becomes a magnet. A coil wrapped around a magnetic material such as iron is called an **electromagnet**.

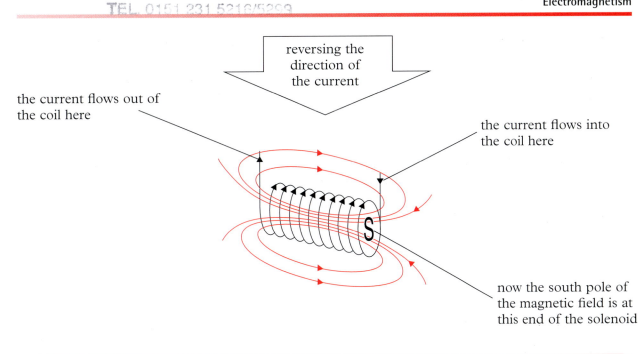

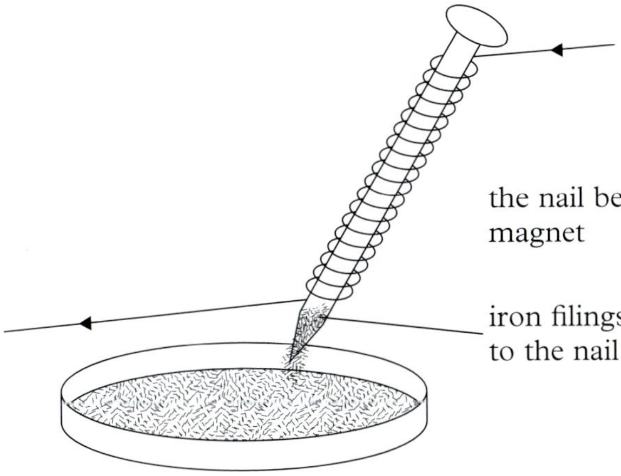
a current flows through the coil
the nail becomes a magnet
iron filings are attracted to the nail

One of the most spectacular uses of electromagnets is in the scrapyard. Here it is used to lift and move heavy pieces of iron or steel.

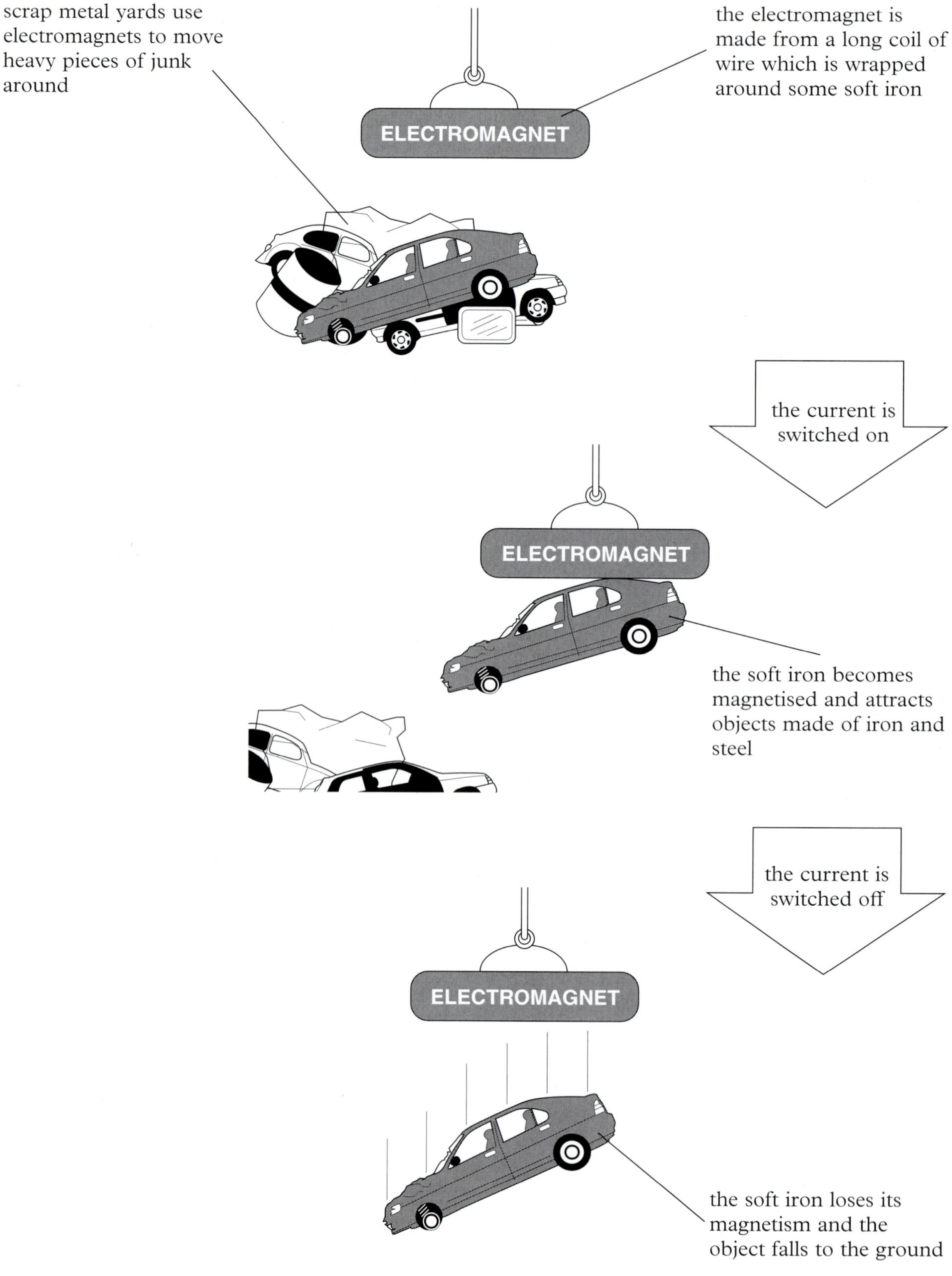
scrap metal yards use electromagnets to move heavy pieces of junk around
ELECTROMAGNET
the electromagnet is made from a long coil of wire which is wrapped around some soft iron
the current is switched on
ELECTROMAGNET
the soft iron becomes magnetised and attracts objects made of iron and steel
the current is switched off
ELECTROMAGNET
the soft iron loses its magnetism and the object falls to the ground

The material that the coil is wrapped around in an electromagnet is known as the **core**.

It is important that the core of an electromagnet is made from a **magnetically soft material** such as iron. Magnetically soft materials are easily **magnetised** and **demagnetised**. This means that they lose their magnetism as soon as the current in the coil is turned off.

If the core of the electromagnet is made from a **magnetically hard material** such as steel, the core will keep some of its magnetism when the current is turned off. Magnetically hard materials are not so easy to magnetise or demagnetise.

Questions

1 a) What is an electromagnet?
b) Draw a labelled diagram of a simple electromagnet.

2 a) What is a magnetically soft material?
b) Give one example of a magnetically soft material.

3 a) What is a magnetically hard material?
b) Give one example of a magnetically hard material.

4 Why is it important that the core of an electromagnet is made from a material which is magnetically soft?

Electric Bell

Electric bells can only work by using an electromagnet. The circuit for an electric bell is shown below.

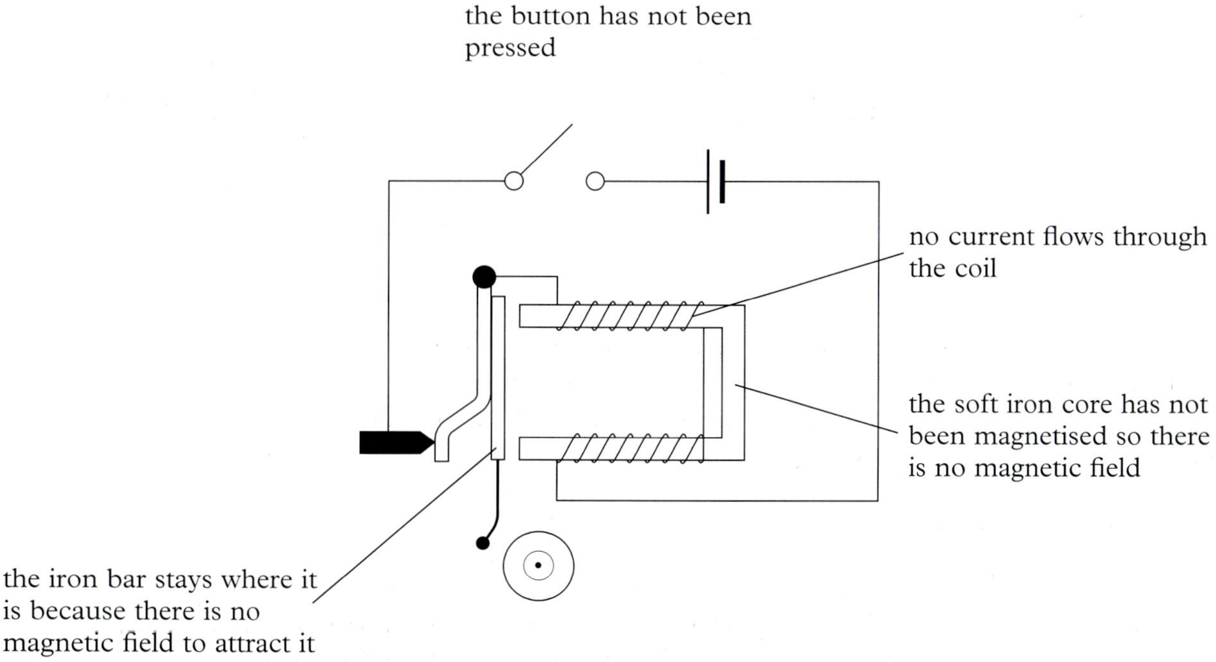

the button has not been pressed
no current flows through the coil
the soft iron core has not been magnetised so there is no magnetic field
the iron bar stays where it is because there is no magnetic field to attract it

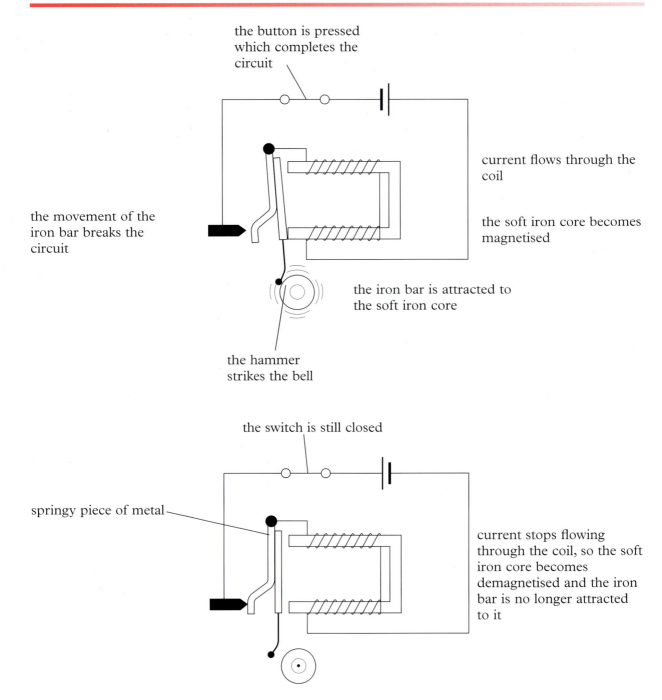

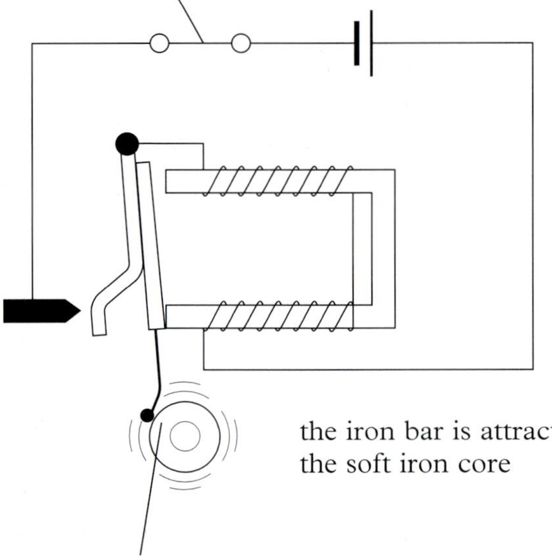
the button is pressed which completes the circuit
current flows through the coil
the soft iron core becomes magnetised
the movement of the iron bar breaks the circuit
the iron bar is attracted to the soft iron core
the hammer strikes the bell

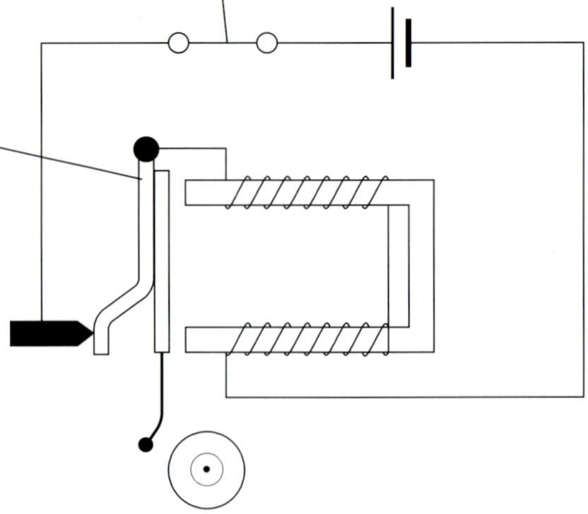
the switch is still closed
springy piece of metal
current stops flowing through the coil, so the soft iron core becomes demagnetised and the iron bar is no longer attracted to it

When the soft iron core loses its magnetism the iron bar is pulled back to its original position by the springy piece of metal. The circuit is once again complete and so the whole process begins again. The bell will continue to ring until the switch is opened.

The relay switch

Sometimes it is useful to be able to control the current flowing in one circuit by using a second circuit. This is especially true if the current flowing in the first circuit is large and therefore possibly dangerous. We can do this using a **relay switch**.

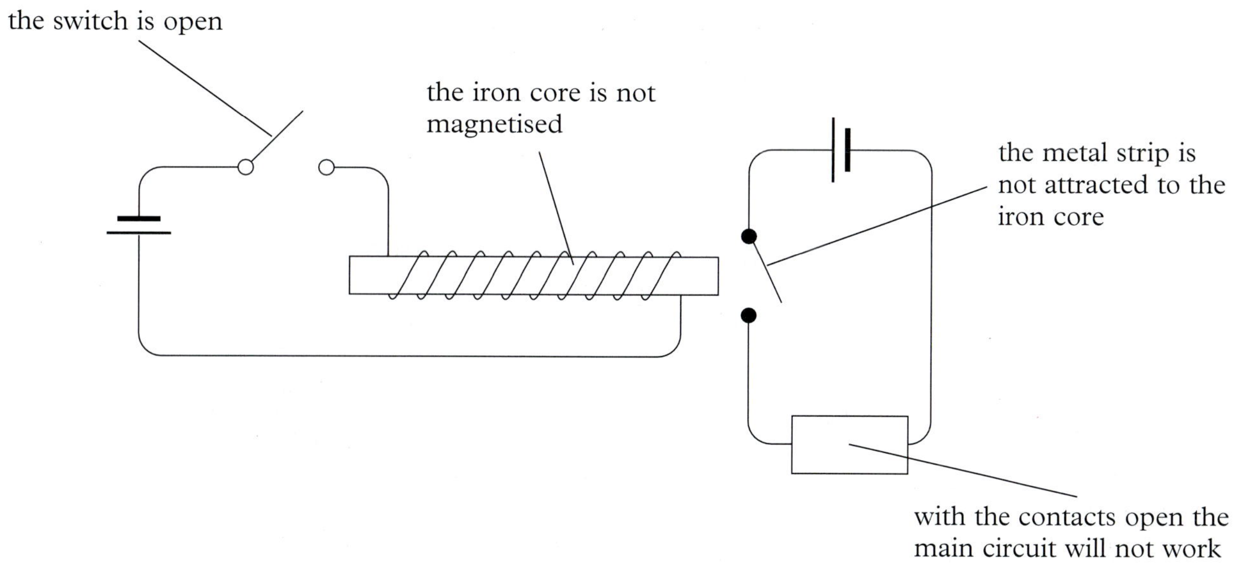
the switch is open
the iron core is not magnetised
the metal strip is not attracted to the iron core
with the contacts open the main circuit will not work

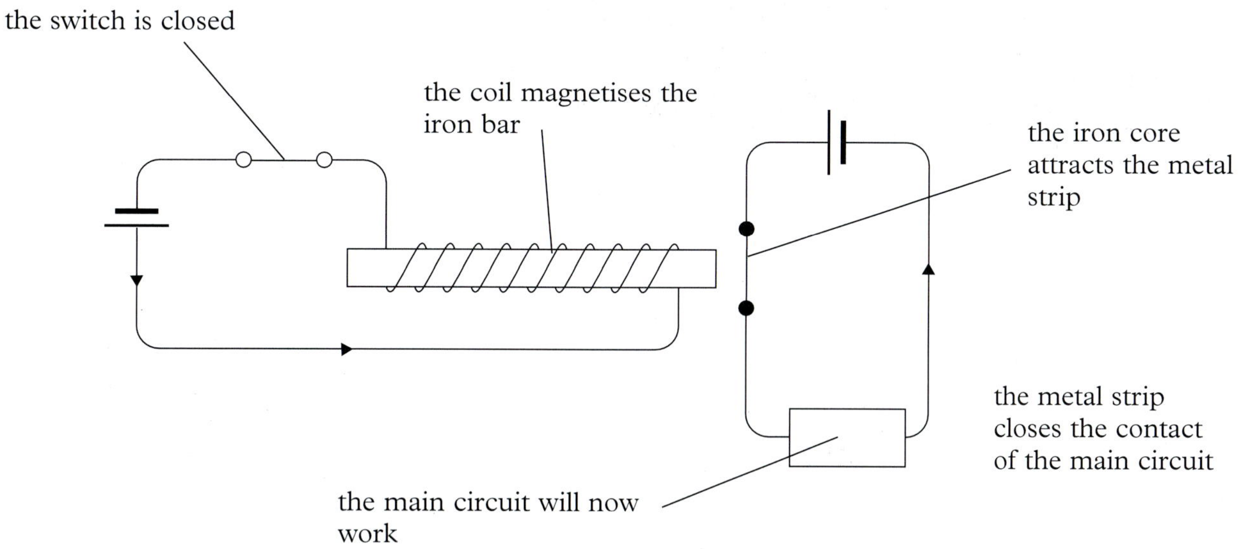
the switch is closed
the coil magnetises the iron bar
the iron core attracts the metal strip
the metal strip closes the contact of the main circuit
the main circuit will now work

Questions

1 a) Draw a circuit diagram for an electric bell.
b) Explain, in your own words, what happens when the bell push is pressed.
c) Why would the bell not work if the soft iron cores were replaced with steel cores which are magnetically hard?

2 a) What is a relay switch?
b) Describe one situation where a relay switch would be very useful.

The electric motor

A motor creates motion from an electric current. It changes electrical energy into movement or kinetic energy.

This French train 'the TGV' is one of the fastest passenger trains in the world. Its powerful electric motors enable it to travel at speeds in excess of 300 $km h^{-1}$.

Overlapping magnetic fields

It is the overlapping of two magnet fields that allows a motor to change electrical energy into kinetic energy.

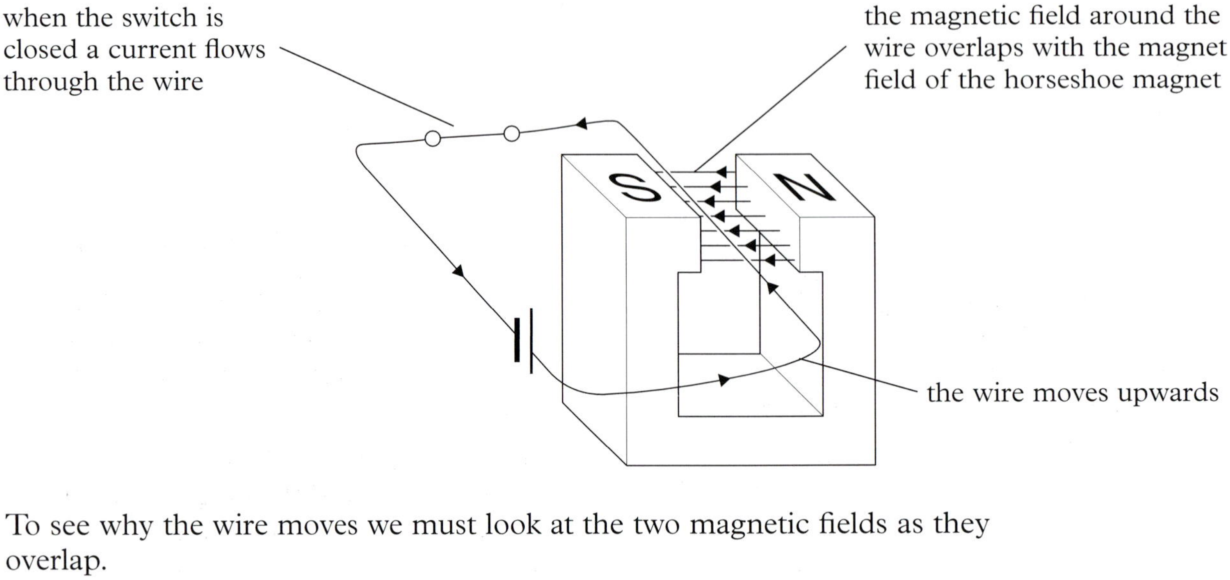
when the switch is closed a current flows through the wire
the magnetic field around the wire overlaps with the magnet field of the horseshoe magnet
S
N
the wire moves upwards

To see why the wire moves we must look at the two magnetic fields as they overlap.

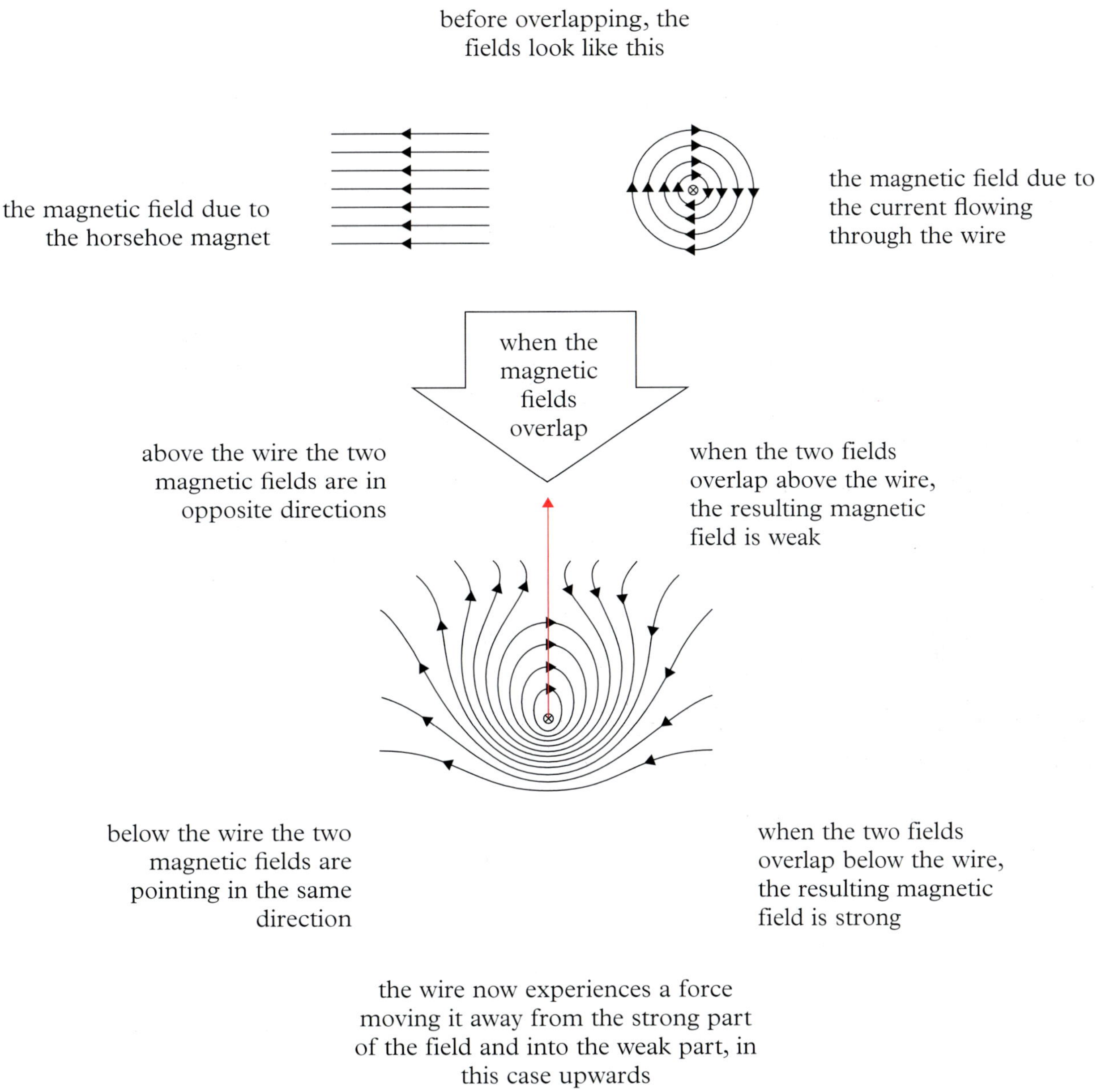
before overlapping, the fields look like this
the magnetic field due to the horsehoe magnet
the magnetic field due to the current flowing through the wire
when the magnetic fields overlap
above the wire the two magnetic fields are in opposite directions
when the two fields overlap above the wire, the resulting magnetic field is weak
below the wire the two magnetic fields are pointing in the same direction
when the two fields overlap below the wire, the resulting magnetic field is strong
the wire now experiences a force moving it away from the strong part of the field and into the weak part, in this case upwards

If the direction of the current in the wire is reversed then the wire moves in the opposite direction.

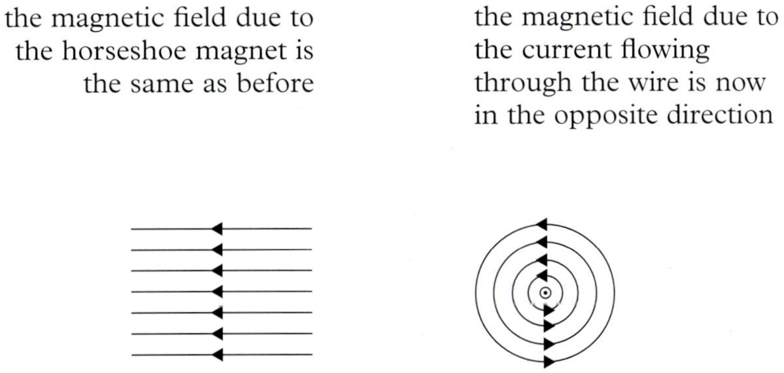
the magnetic field due to the horseshoe magnet is the same as before
the magnetic field due to the current flowing through the wire is now in the opposite direction

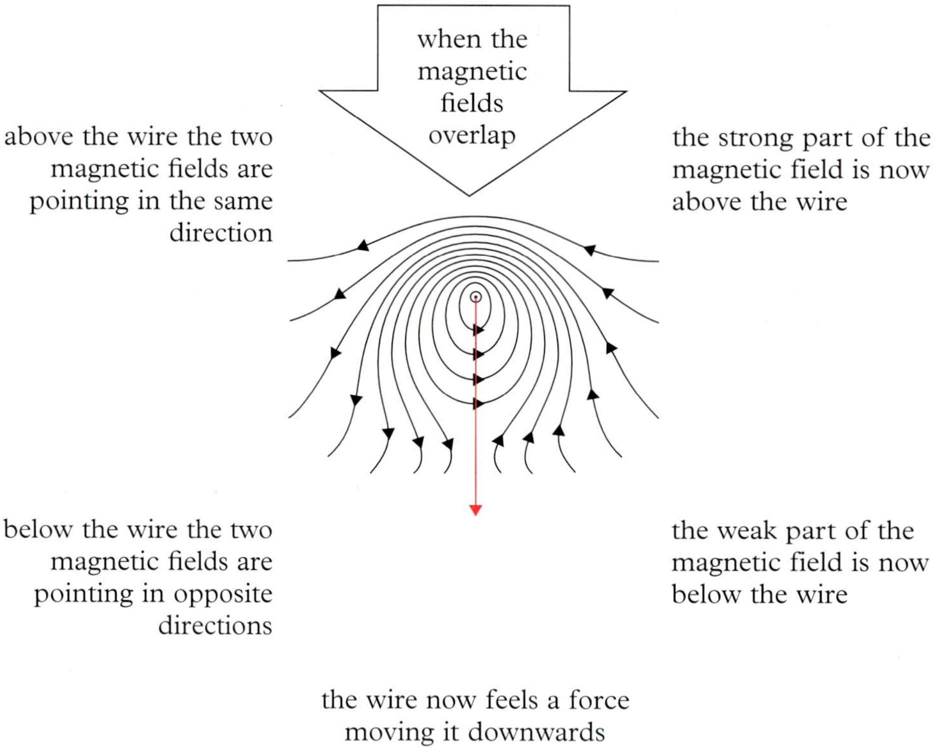
when the magnetic fields overlap
above the wire the two magnetic fields are pointing in the same direction
the strong part of the magnetic field is now above the wire
below the wire the two magnetic fields are pointing in opposite directions
the weak part of the magnetic field is now below the wire
the wire now feels a force moving it downwards

Fleming's Left Hand Rule

Fleming's left hand rule can be used to help us predict which way a wire carrying a current will move in a magnetic field.

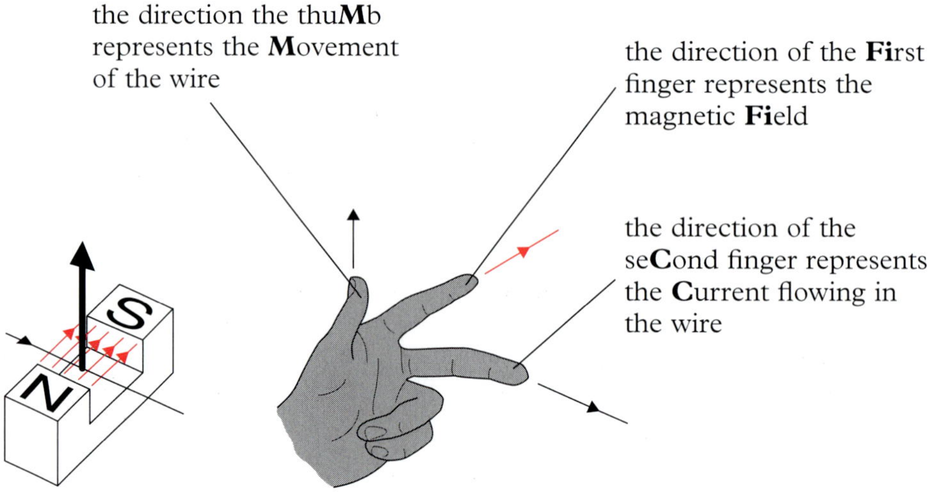
the direction the thuMb represents the Movement of the wire
the direction of the First finger represents the magnetic Field
the direction of the seCond finger represents the Current flowing in the wire
S
N

The thumb, first finger and second finger of the left hand are all at right angles to each other.

Creating a turning motion from a current

If we place a loop of wire between the poles of a magnet and then pass a current through it, one side of the loop will feel a force trying to push it up while the opposite side will feel a force trying to push it downwards. The loop will try to rotate.

the overlapping magnetic fields produce an upwards force on this side of the wire

the overlapping magnetic fields produce a downwards force on this side of the wire

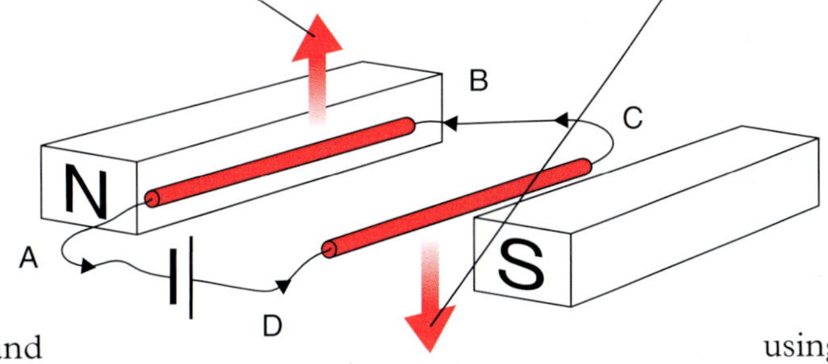

using fleming's Left Hand Rule on wire AB shows that this side of the loop will be pushed up

using fleming's Left Hand Rule on wire CD shows us that this side of the loop will be pushed down

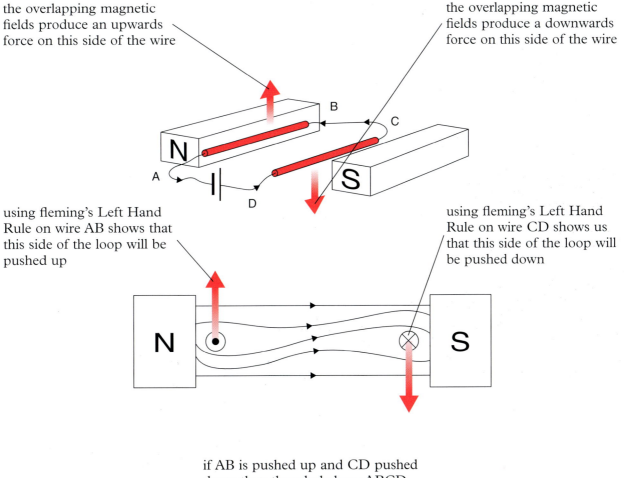

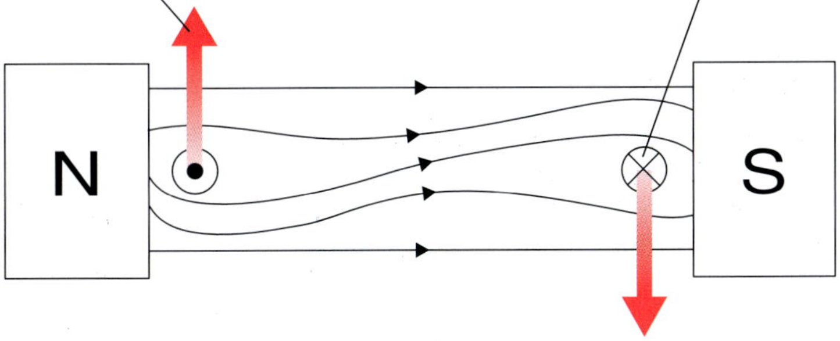

if AB is pushed up and CD pushed down then the whole loop ABCD will begin to rotate in a clockwise direction

Real motors work on this principle, although there are several differences. In real motors the single loop of wire is replaced by several coils each of which has thousands of turns of wire. The permanent magnet is often replaced by an electromagnet and a split ring is used to connect the power supply to the rotating coils. This ring allows the coils to rotate continuously.

The speed of rotation of a motor depends upon the current flowing through the coils, the number of turns on the coil and the strength of the magnetic field. Increasing any of these will increase the speed of the motor.

Moving coil loudspeaker

A loudspeaker uses the overlapping of magnetic fields to create sound.

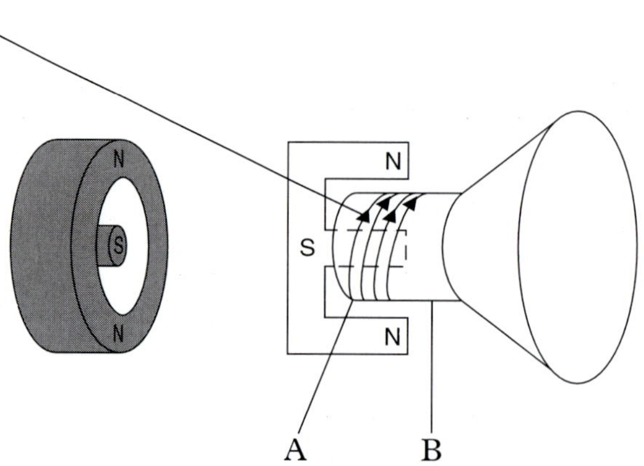
a coil of wire is wrapped around the cone of a loudspeaker which is then placed between the poles of a magnet
N
S
N
N
S
N
if the current flows from A to B through the coil, the coil and cone move to the left
if the current flows from B to A through the coil, then coil and cone move to the right
A
B

The electrical signals from a stereo system or microphone are in the form of electric currents which are continually changing direction. If these signals are passed through the wires AB, the coil and the cone will be made to move back and forth rapidly. The cone will begin to vibrate. These vibrations of the cone produce the sound waves we hear.

Questions

1 a) Draw a diagram of the magnetic field around a single piece of wire which has a current passing through it.
b) Draw a diagram of the magnetic field between the poles of a horseshoe magnet.
c) Draw a diagram of the field produced when the above two fields overlap.

2 Suggest three ways in which the speed of rotation of an electric motor might be increased.

3 Draw a labelled diagram of a loudspeaker and explain, in your own words, how it works.

Key words

electromagnet	A magnet produced by passing current around a coil.
hard material	A material that is magnetised and demagnetised with some difficulty. Ideal for permanent magnets.
magnetically soft material	A material that changes easily from being magnetised to being demagnetised. Ideal for the core of an electromagnet.
magnetise	Make into a magnet.
magnetically	A material that holds on to its magnetism once it has been magnetised.
relay switch	A switch which uses electromagnetism to turn a second circuit on and off.
solenoid	A long coil.

Summary

When a current flows through a wire it creates a magnetic field around it. This field, unlike a permanent magnet, can be turned on and off. Winding a wire into many loops creates a long coil. This is called a solenoid. A solenoid's magnetic field can be made stronger by placing an iron core through the centre of the solenoid, increasing the number of turns on the solenoid or increasing the current flowing. This combination of solenoid and iron core is called an electromagnet and is used in many devices such as an electric bell, relay switch, electric motor and loudspeaker.

If a wire carrying a current is placed inside a magnetic field it will experience a force. The strength of the force depends upon the strength of the magnetic field and the size of the current. This idea is used in the construction of the electric motor. The two sides of a coil of wire, placed between the poles of a magnet, experience forces which cause the coil to rotate.

End of Chapter 18 Questions

1 Explain the difference between a permanent magnet and an electromagnet.

2 Explain why magnetically soft materials are used for the core of an electromagnet but magnetically hard materials are used to make permanent magnets.

3 a) Name three devices which use electromagnets.
b) Using a diagram, describe how one of these devices works.

4 The diagram below shows the circuits used to start a car engine.
a) Explain how turning the key makes the starter motor work.
b) Give one reason why a relay switch like this is used when starting a car.

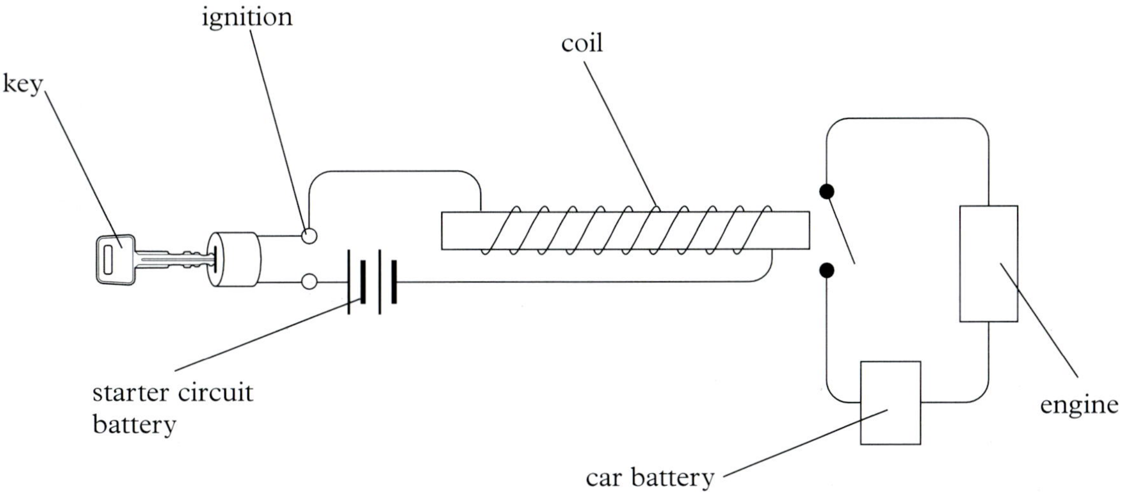

5 The diagram below shows a long wire placed between the poles of a horseshoe magnet. When current flows through the wire from A to B a force (F) acts on the wire.

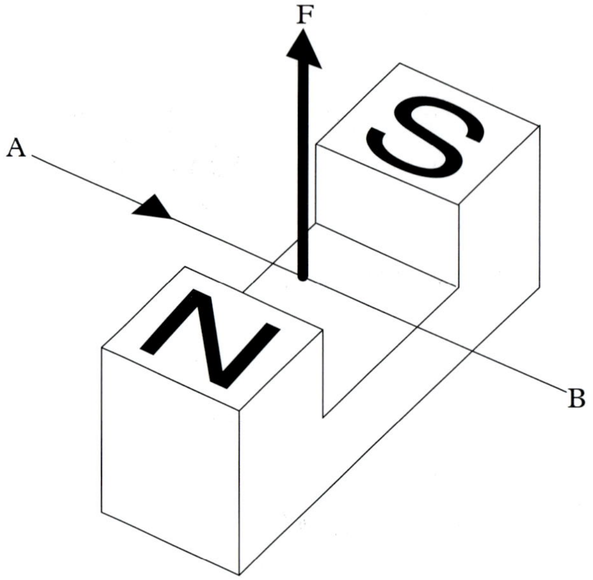

a) What happens to this force if the current in the wire is increased?
b) What happens to this force if the horseshoe magnet is replaced by a weaker magnet?
c) What happens to this force if the direction of the current is reversed?
d) What happens to this force if the wire was coiled so that several pieces of wire lay between the poles of the magnet?

19 Generating Electricity

We all use electricity. Without it life would be very different. Televisions, microwaves and fridges would not work if we were unable to generate electricity.

Generating electricity in a wire

To generate electricity we move wires through a magnetic field.

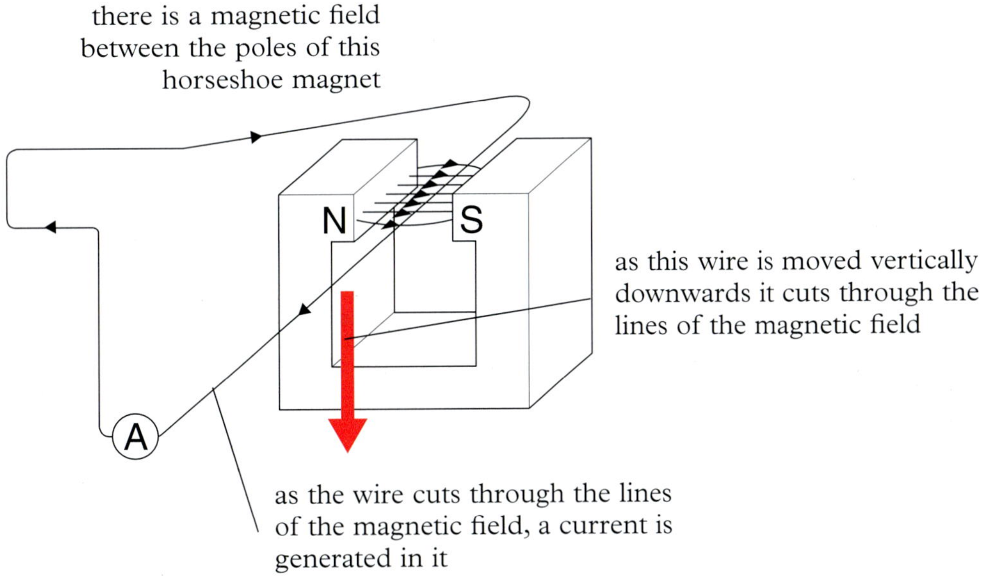
there is a magnetic field between the poles of this horseshoe magnet
N
S
A
as this wire is moved vertically downwards it cuts through the lines of the magnetic field
as the wire cuts through the lines of the magnetic field, a current is generated in it

No current is generated if a) the wire is moved horizontally between the poles of the horseshoe magnet and b) the wire is stationary. Current only flows if the wire is cutting through the lines of the magnetic field.

A larger current can be generated if:

1 the wire is moved more rapidly through the magnetic field

2 the magnetic field is made stronger

3 the wire is looped so that more pieces of wire cut through the magnetic field.

The direction of the generated current depends on a) the direction in which the wire is being moved (upwards or downwards) and b) the direction of the magnetic field.

The experiment on the next page shows that currents can also be generated in wires by moving the magnetic field and keeping the wires stationary.

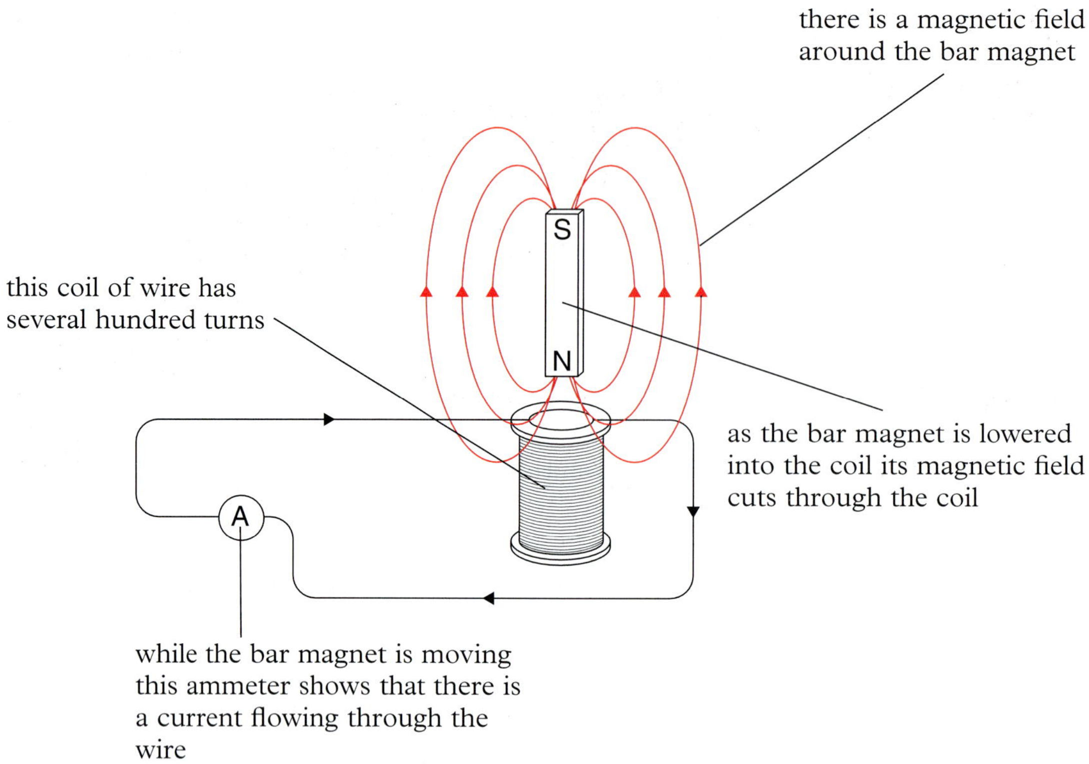
there is a magnetic field around the bar magnet
this coil of wire has several hundred turns
S
N
as the bar magnet is lowered into the coil its magnetic field cuts through the coil
A
while the bar magnet is moving this ammeter shows that there is a current flowing through the wire

This process of generating currents (and voltages) using moving wires or magnetic fields is called **electromagnetic induction**.

Questions

1

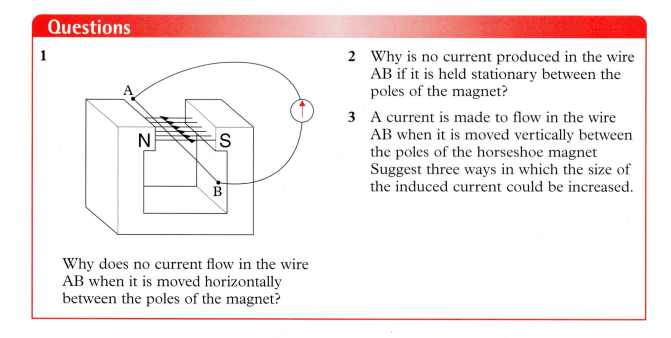

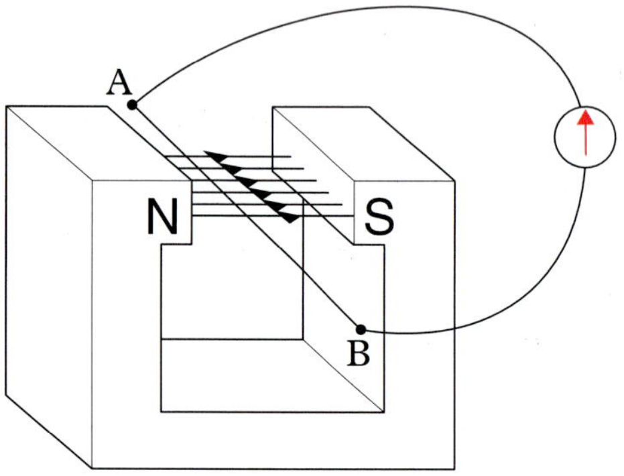
A
N
S
B

Why does no current flow in the wire AB when it is moved horizontally between the poles of the magnet?

2 Why is no current produced in the wire AB if it is held stationary between the poles of the magnet?

3 A current is made to flow in the wire AB when it is moved vertically between the poles of the horseshoe magnet Suggest three ways in which the size of the induced current could be increased.

Simple dynamo

Many bicycles use a dynamo to generate the electricity they need for their lights.

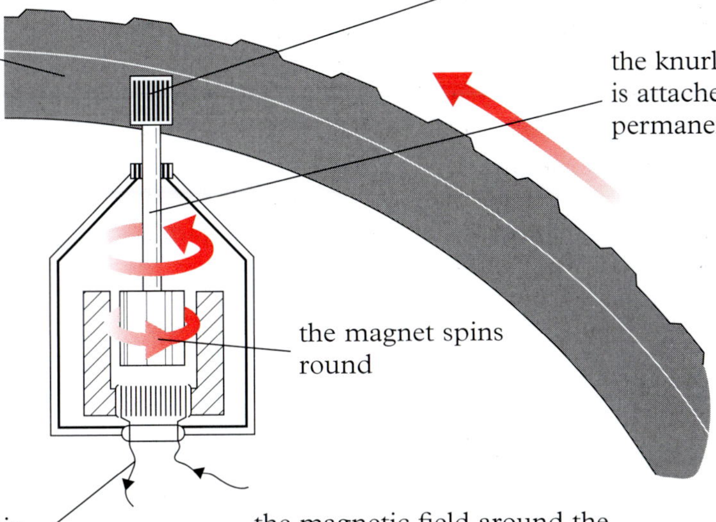
as the bicycle moves the wheel turns
the turning wheel causes the knurled knob to turn
the knurled knob is attached to a permanent magnet
the magnet spins round
a current is generated in the coil and is used to light the bulb
the magnetic field around the magnet moves with the magnet and cuts through a surrounding coil of wire

Generators

Most of the electricity we use in the home is produced by **generators** which consist of a coil of wire which is made to spin between the poles of a magnet.

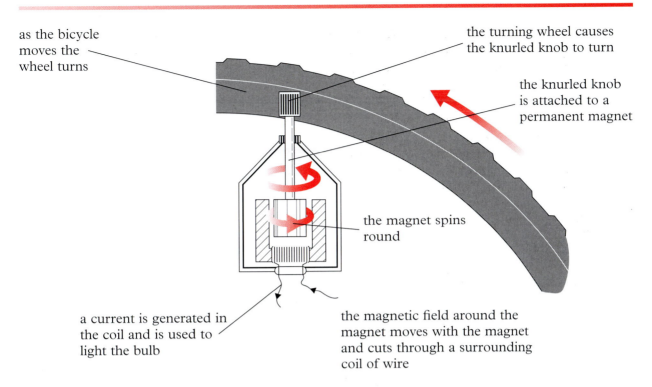

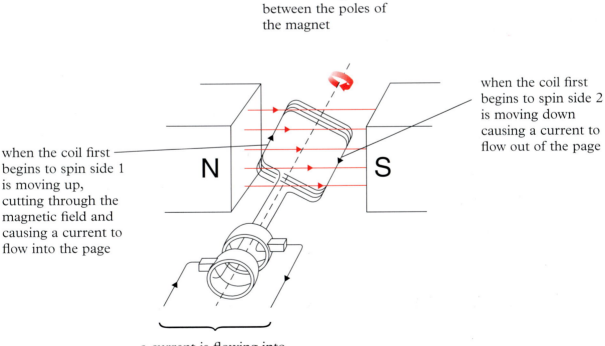

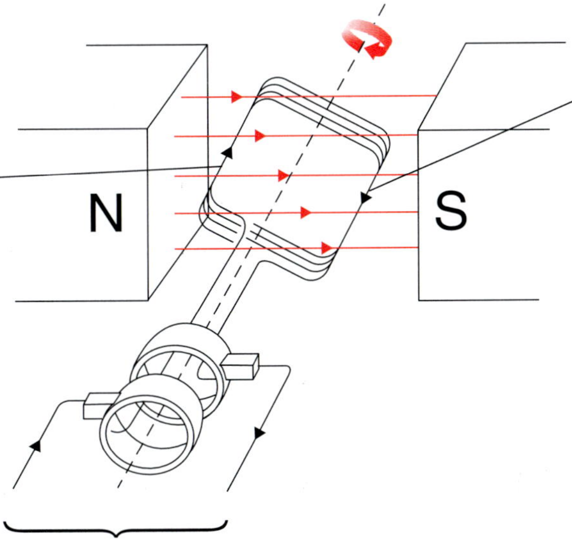
there is a magnetic field between the poles of the magnet
when the coil first begins to spin side 2 is moving down causing a current to flow out of the page
when the coil first begins to spin side 1 is moving up, cutting through the magnetic field and causing a current to flow into the page
N
S
a current is flowing into side 1 and out of side 2

As the coil rotates, its wires cut through the magnetic lines of force and a current is produced. Because the sides of the coil travel up through the field and then down again, the direction of the induced current keeps changing. This kind of current is called an **alternating current** (**a.c.**). Generators like this which produce alternating current are called **alternators**. The generators at all the power stations in Britain produce alternating current. The coils in these generators spin around 50 times every second. The frequency of the alternating current generated is therefore 50 hertz (50 Hz). Frequency is a measure of how many times something happens in one second. Frequency is measured in Hertz.

The graph below shows how the size and direction of the current changes with time.

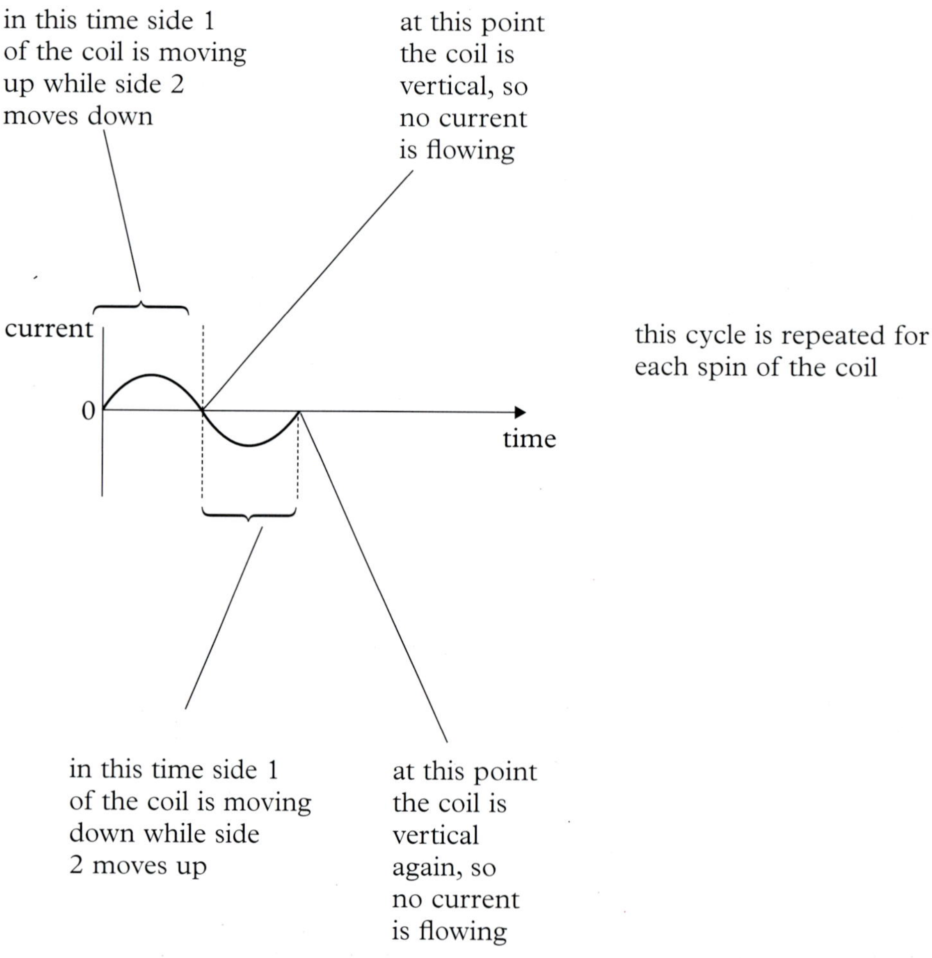
in this time side 1 of the coil is moving up while side 2 moves down
at this point the coil is vertical, so no current is flowing
current
0
time
this cycle is repeated for each spin of the coil
in this time side 1 of the coil is moving down while side 2 moves up
at this point the coil is vertical again, so no current is flowing

The size and direction of an alternating current is continually changing.

Direct current

Cells and batteries produce currents which flow only in one direction. This kind of current is called a **direct current** (**d.c.**).

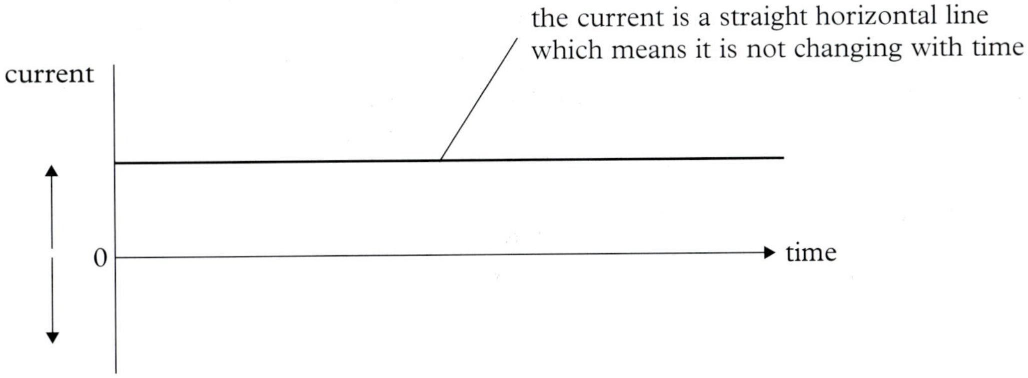
the current is a straight horizontal line which means it is not changing with time
current
0
time

Questions

1 a) Draw a labelled diagram of a bicycle dynamo.
b) Explain how the dynamo generates electricity. Explain what happens to the lights when the bicycle is
i) stationary
ii) travelling more quickly.

2 Explain the difference between alternating current and direct current.

3 What name is given to a generator which produces alternating current?

4 The mains supply in Britain has a frequency of 50 Hz. What does this statement mean?

Transformers

A portable radio needs a 9 V supply to operate correctly.

When the radio is used outside the home we are likely to use a 9 V battery for our supply. In the home however we may decide to use the electricity supplied from sockets in the house. Although this mains electricity has a voltage of approximately 230 V we can reduce this to 9 V using a device called a **transformer**.

A transformer is a device which is used to change the voltage of an a.c. supply. If the transformer decreases the voltage, as in the above example, it is called a **step-down transformer**. If the transformer increases the voltage, it is called a **step-up transformer**.

A transformer cannot change d.c. voltages. It can only change a.c. voltages.

The National Grid

The National Grid is a network of cables and wires which carries electrical energy from a power station to our homes. The National Grid uses both step-up and step-down transformers.

Transmission of electricity from the power station to our homes

The electricity we use in our homes is generated at a power station. Here coal, oil, gas or a nuclear reactor is used to produce steam. This steam drives steam turbines which turn a.c. generators to produce electricity. The electricity is produced by electromagnetic induction.

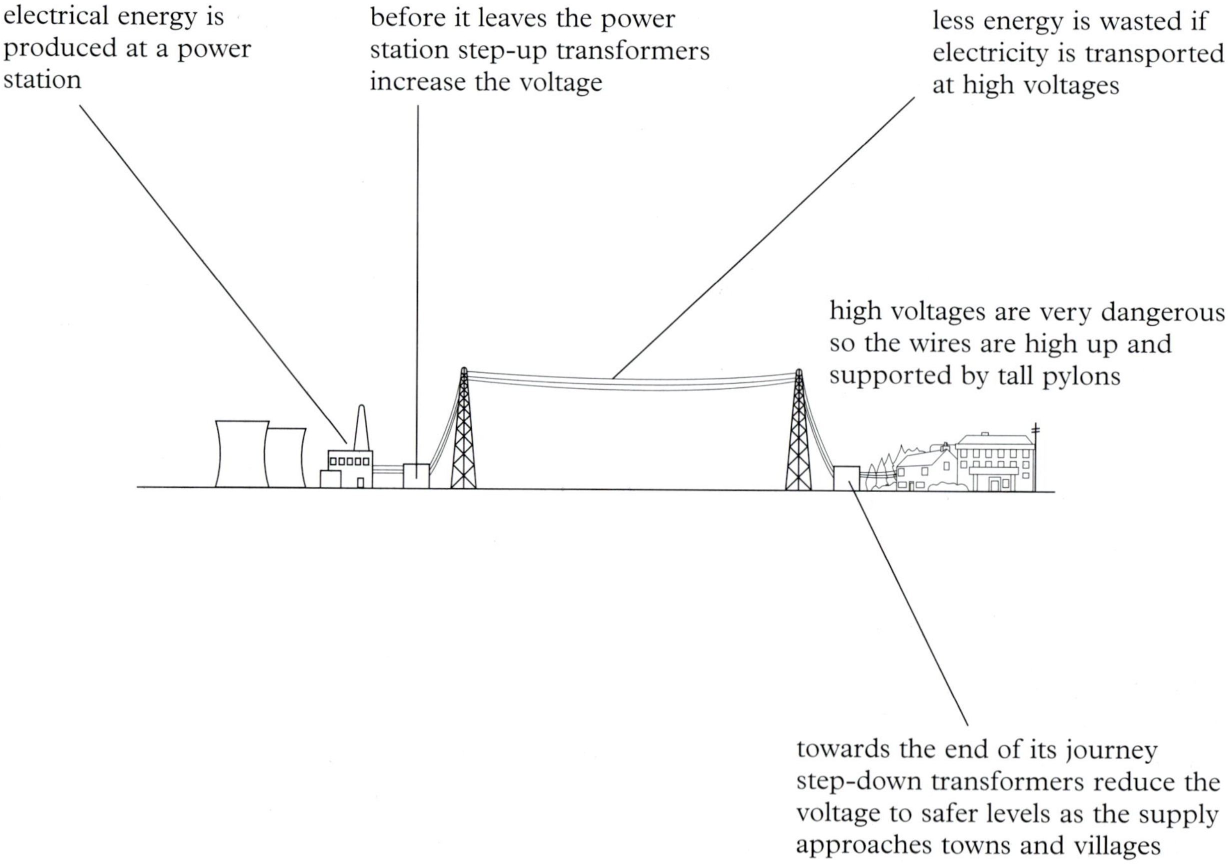
electrical energy is produced at a power station
before it leaves the power station step-up transformers increase the voltage
less energy is wasted if electricity is transported at high voltages
high voltages are very dangerous so the wires are high up and supported by tall pylons
towards the end of its journey step-down transformers reduce the voltage to safer levels as the supply approaches towns and villages

Questions

1 What is a step-up transformer?

2 What is a step-down transformer?

3 Draw a labelled diagram to show where step-up and step-down transformers are used in the transmission of electricity from the power station to the home.

Key words

alternating current	Current which is regularly changing direction.
alternator	A generator which produces alternating currents and voltages.
dynamo	A simple device for generating currents by spinning a magnet inside a coil.
electromagnetic induction	The production of a current or voltage due to the cutting of magnetic lines of force.
generator	A device for generating currents, usually by spinning a coil between the poles of a magnet.
step-down transformer	A device which is used to decrease the voltage of an alternating current.
step-up transformer	A device which is used to increase the voltage of an alternating current.
transformer	A device which is used to change the voltage of alternating currents.

Summary

If a wire cuts through a magnetic field or a magnetic field moves across a wire, current will flow through the wire. This phenomenon is called electromagnetic induction and is used in dynamos and generators to produce electricity. The size of the induced current depends upon the strength of the magnetic field, the number of turns of wire on the coil and the speed of the movement.

Current supplied from a cell or battery flows in just one direction. It is called direct current (d.c.). The current supplied from an alternator is continually changing its direction. It is called alternating current (a.c.). The mains supply in Britain is a.c. It has a frequency of 50 Hz. Alternating currents and voltages can be changed using transformers.

End of Chapter 19 Questions

1 The diagram below shows a long wire placed just above the gap between the poles of a horseshoe magnet. A very sensitive meter is connected to the wire to indicate if any current flows in the wire.

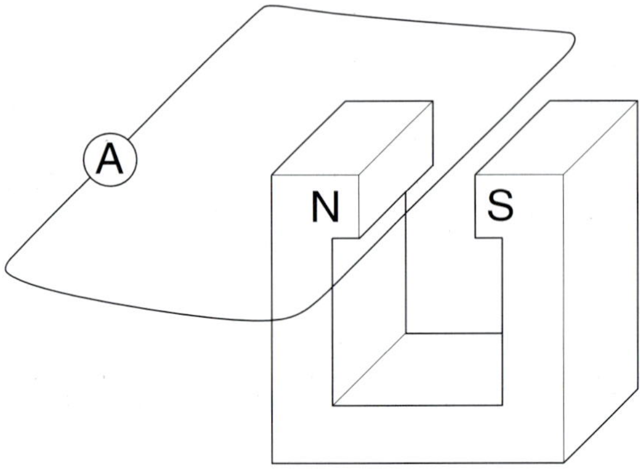

State what happens to the meter when:
a) the wire is moved quickly downwards,
b) the wire is moved slowly upwards,
c) the wire is held stationary between the poles of the magnet,
d) the wire is made into a double loop and the two pieces of wire are moved downwards quickly.

2 a) A current is produced by moving a wire across a magnetic field. What is this effect called?
b) Name one practical device which uses this effect.

3 The diagram below shows a bicycle dynamo.

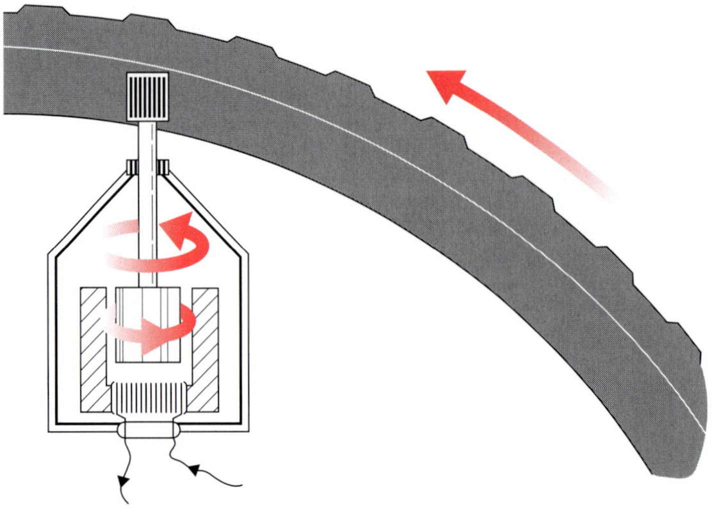

a) Explain as fully as you can why current is generated by the dynamo when the wheel is turning.
b) Why is no electricity generated when the wheel is stationary?

4 a) Explain with diagrams the difference between alternating current (a.c.) and direct current (d.c.).

b) Name one source of electrical energy which supplies d.c. and one source which supplies a.c.

5 Electrical energy generated at a power station is transmitted to our homes along the power lines of the National Grid.

a) Name one fuel which the power station could burn.

b) What device produces the electricity?

c) Before the electricity leaves the power station it passes through step-up transformers. What do these step-up transformers do?

d) Why is it economically important to use step-up transformers here?

e) Suggest one material that might be used for the power lines.

f) Why are the power lines of the transmission grid suspended high in the air on tall pylons?

g) As the power lines approach a town or village the electricity passes through step-down transformers. What do these transformers do and why?

20 Electricity In The Home

The electrical energy generated at a power station usually enters our homes through an underground cable. As soon as the cable comes into the house it is connected to an electricity meter which measures how much energy we use. From here the cable enters the **consumer box** or **fuse box**. Inside this box there are a set of **fuses**. Fuses are safety devices which break the electrical circuit if the current becomes too large. The consumer box or fuse box also contains connections for all the different circuits in the house. These circuits include lighting circuits, circuits that are downstairs and circuits that are upstairs. The box also contains a switch which allows us to turn off all the electricity in the house.

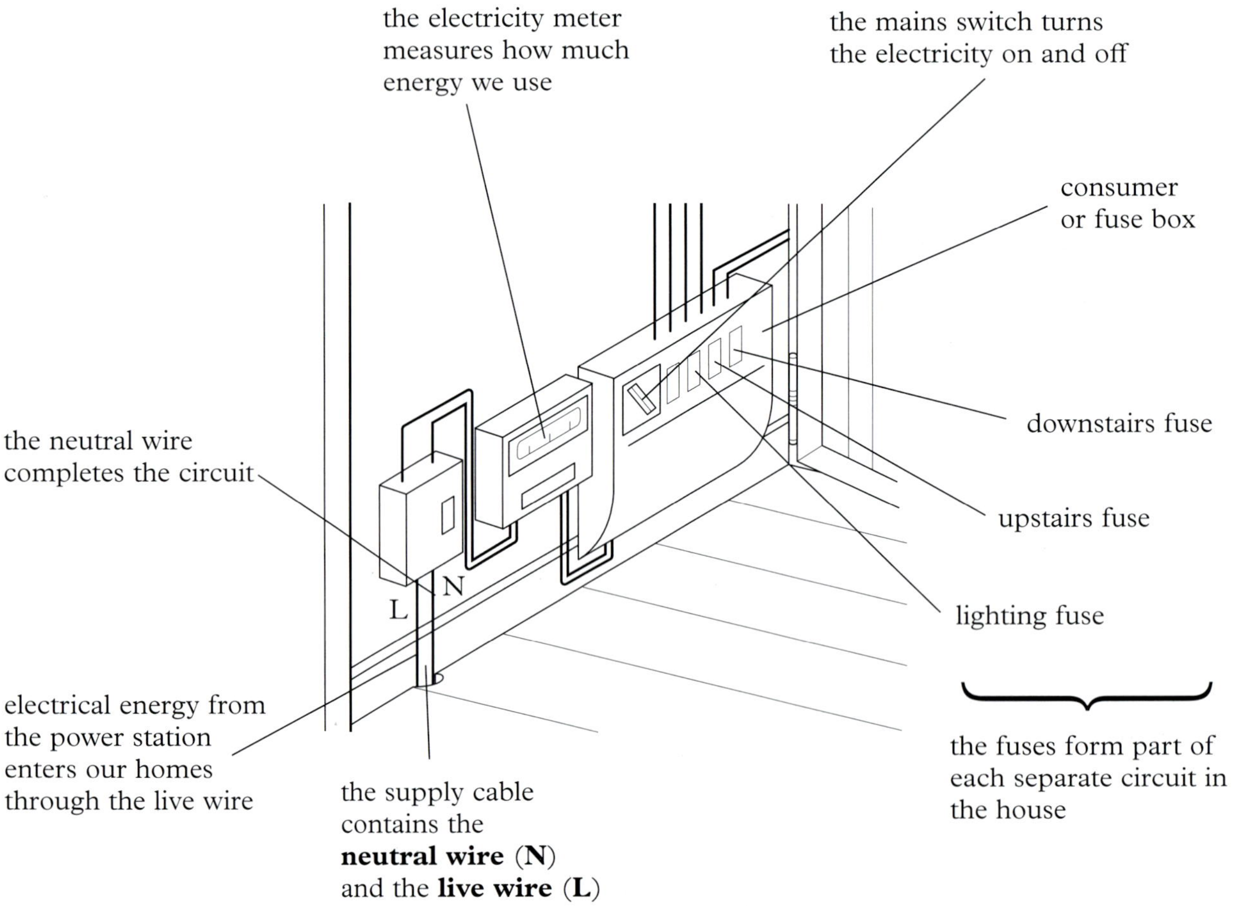

the electricity meter measures how much energy we use
the mains switch turns the electricity on and off
consumer or fuse box
the neutral wire completes the circuit
downstairs fuse
upstairs fuse
N
L
lighting fuse
electrical energy from the power station enters our homes through the live wire
the fuses form part of each separate circuit in the house
the supply cable contains the **neutral wire** (**N**) and the **live wire** (**L**)

Ring circuits

Most of the wires which leave the consumer box are connected to **ring circuits** which are usually hidden in the wall. These circuits like the one drawn over the page usually contain three wires, the live, the neutral and the earth.

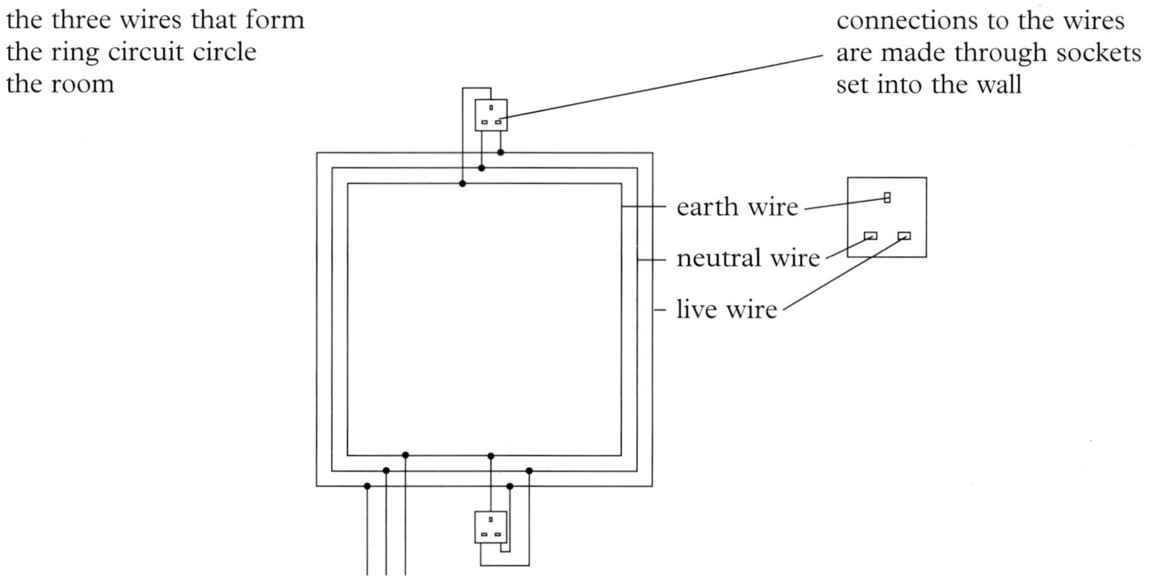
the three wires that form
the ring circuit circle
the room
connections to the wires
are made through sockets
set into the wall
earth wire
neutral wire
live wire

Questions

1 Through which wire does electrical energy enter an appliance?

2 Which wire completes the circuit?

3 What is an electricity meter?

4 Where would you find an on/off switch for all the electricity in your house?

5 What kinds of circuits carry electrical energy to the sockets in your house?

The electric plug

Mains electricity has a voltage of about 230 V. This voltage is large enough to kill if it is not used safely. To avoid accidents most electrical appliances are connected to the mains using a 3-pin plug.

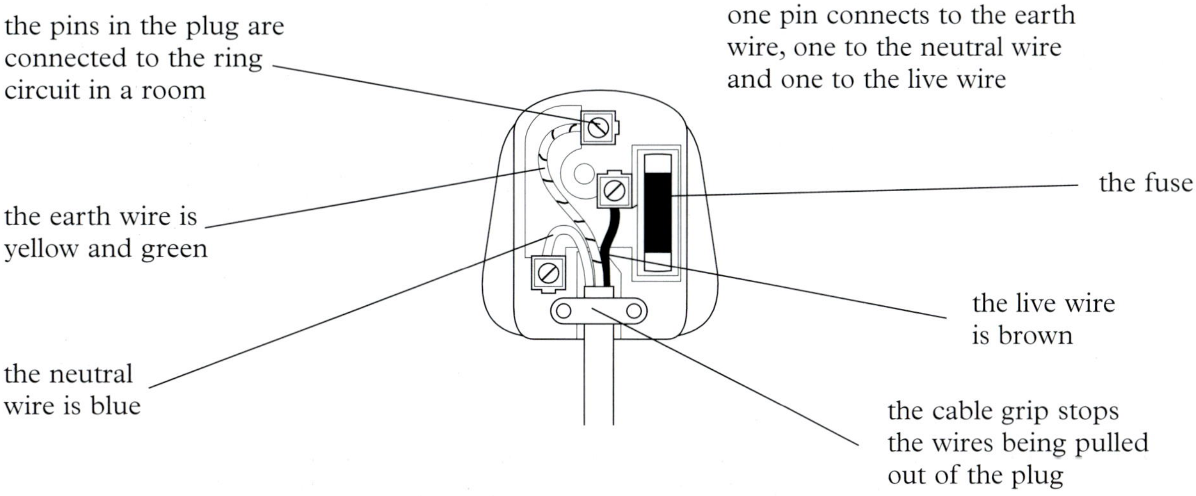
the pins in the plug are
connected to the ring
circuit in a room
one pin connects to the earth
wire, one to the neutral wire
and one to the live wire
the fuse
the earth wire is
yellow and green
the live wire
is brown
the neutral
wire is blue
the cable grip stops
the wires being pulled
out of the plug

Great care must be taken to ensure that the wires in the plug are connected to the correct pins.

Questions

1 a) Name the three wires we see in a mains plug.
b) What colours are each of these wires?

2 Draw a labelled diagram of a plug showing the correct connections for the three wires.

3 What is the purpose of the cable grip?

4 Why are the pins of a plug made of metal?

5 Why are the wires of a plug covered with plastic?

6 Write down all the faults you can see in these two plugs.

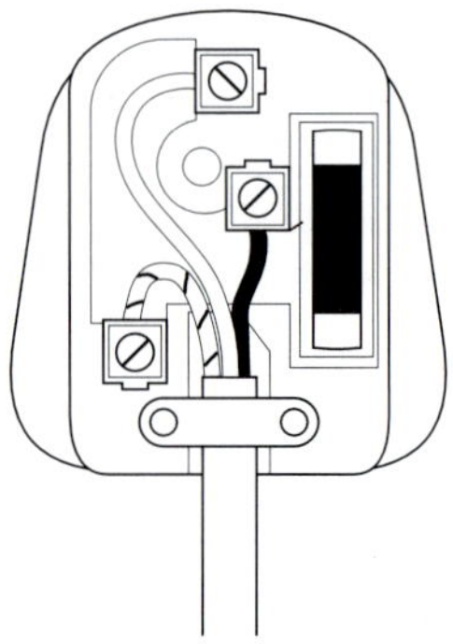

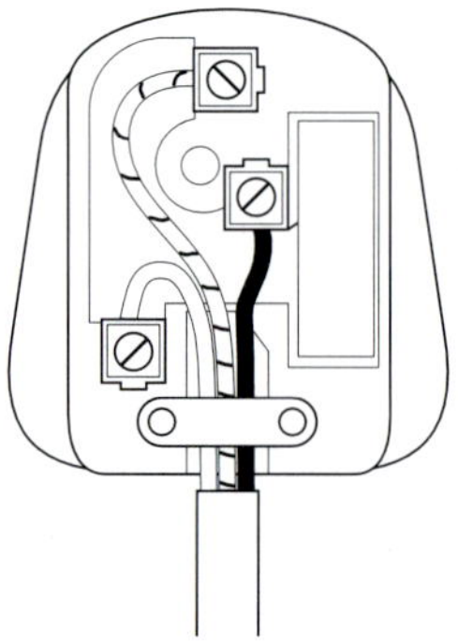

The importance of the earth wire

The **earth wire** in a circuit is there to protect us. It is there to prevent us from receiving an electric shock should an electrical appliance become faulty. The kettle in the following diagram is connected to a circuit which has no earth wire. The kettle is not faulty and will work normally when turned on.

electrical energy enters through the live wire and goes through the heating element

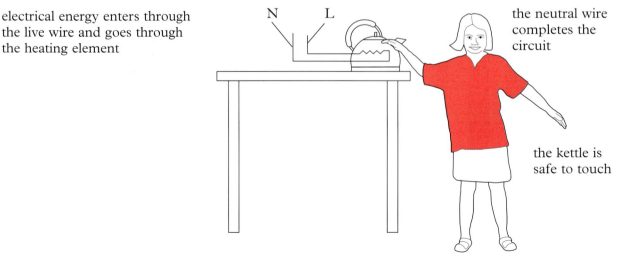

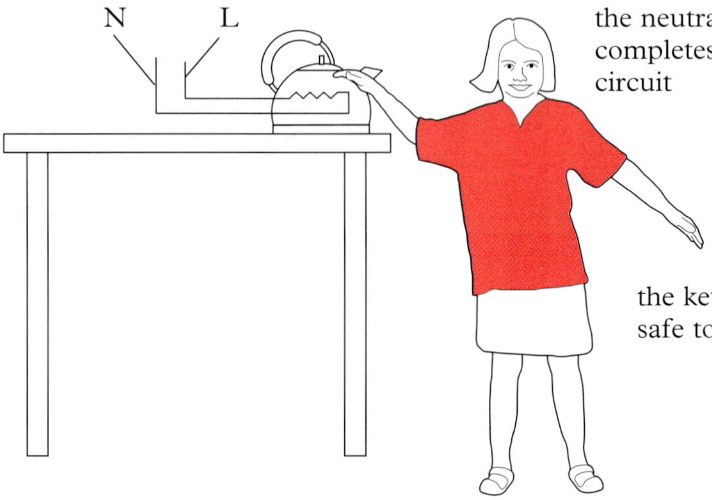

the neutral wire completes the circuit

the kettle is safe to touch

This electric kettle is also connected to a circuit which has no earth wire but this kettle is not in good working order. The heating element is broken and is touching the metal body of the kettle. If someone touches the casing whilst the kettle is turned on, electrical energy will flow through them rather than the neutral wire. There is therefore a real danger of the user receiving a severe electric shock.

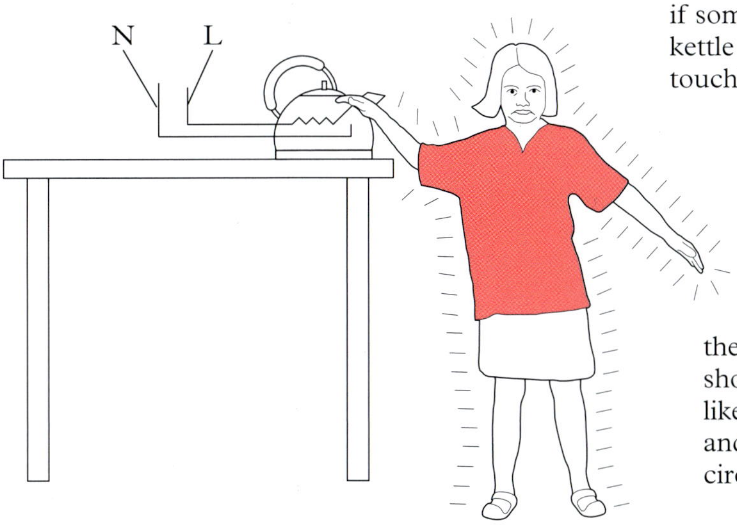
electrical energy enters through the live wire and flows through the heating element to the kettle casing
N
L
if someone touches the kettle they are effectively touching the live wire
the user receives a shock as her body acts like the neutral wire and completes the circuit

If the same kettle is connected to a circuit which has an earth wire the user will not receive an electric shock. The earth wire is connected to the metal body of the kettle. When the kettle is turned on the electrical energy from the broken live wire passes harmlessly along the earth wire rather than through the body of the user.

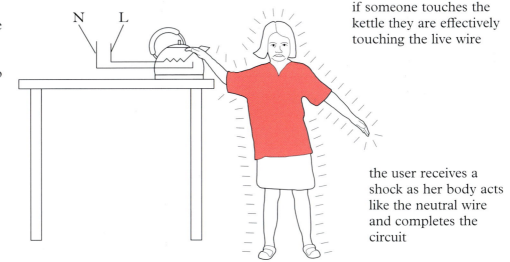

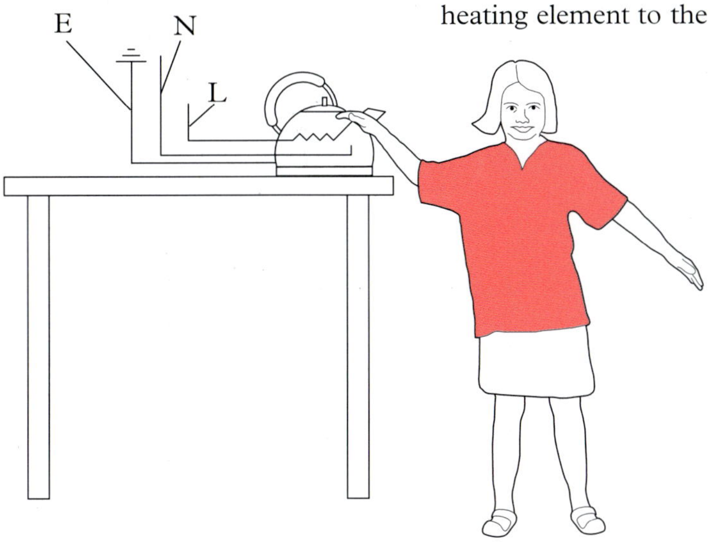
electrical energy enters through the live wire and flows through the heating element to the kettle casing
the earth wire is connected to the metal body of the kettle
E
N
L
the electrical energy of the kettle escapes through the earth wire
this time it is safe for the user to touch the kettle

Switches

Switches are used to turn appliances on and off. A switch is always connected into the live wire so that when it is is open electrical energy is unable to enter an appliance.

If a switch is mistakenly connected into the neutral wire the appliance will be turned on and off by the switch but it is the return path which is being broken. Electrical energy can still enter the appliance. Anyone coming into contact with any of the circuits of the appliances could still receive an electric shock as they then become part of the return circuit.

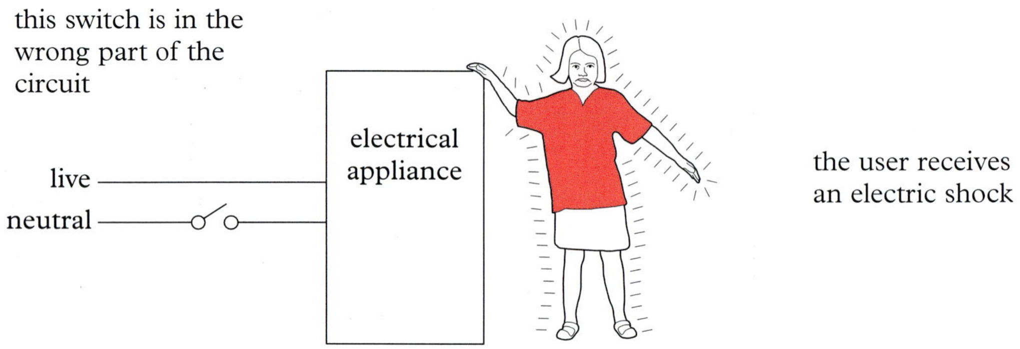
this switch is in the wrong part of the circuit
live
neutral
electrical appliance
the user receives an electric shock

the switch must always be connected in series with the live wire

the switch is in the right part of the circuit
live
neutral
electrical appliance
with the switch turned off no electrical energy can enter the appliance and it is safe to touch

Questions

1 Explain with diagrams why it is important to have an earth connection for an appliance such as an electric fire.

2 What is the main purpose of a switch in an electrical circuit?

3 Where must a switch be placed in a mains circuit?

4 Draw a labelled mains circuit diagram which includes an appliance such as a TV set, and a switch.

Fuses

If an appliance or one of its circuits becomes faulty there is a danger that too large a current will flow through the wires. This could result in the appliance being damaged. It could even cause an electrical fire. To avoid these problems **fuses** are included in all household circuits.

The most common type of household fuse consists of a thin piece of wire enclosed in a plastic cylinder. These are called cartridge fuses or cylinder fuses. If too much current passes through the wire it melts, so breaking the circuit.

a 13A fuse will blow (melt) if a current larger than 13A passes through it

a 3A fuse will blow (melt) if a current larger than 3A passes through it

the most common household fuses are rated at 3A and 13A

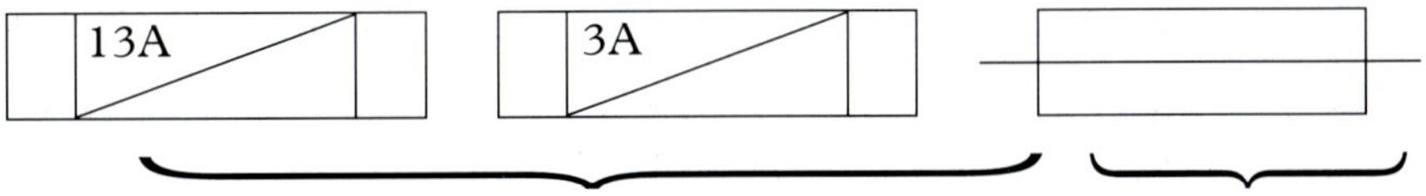

the maximum current that can flow through a fuse without melting it is called the **current rating** of the fuse

this is the circuit symbol for a fuse

Some modern fuses are in the form of a **trip switch** or **circuit breaker**. When the current exceeds a certain amount the switch automatically opens and breaks the circuit. Once the fault in the circuit has been corrected the fuse can be reset. This is usually done by pushing a button. There is no need to replace the whole fuse. A trip switch is shown below.

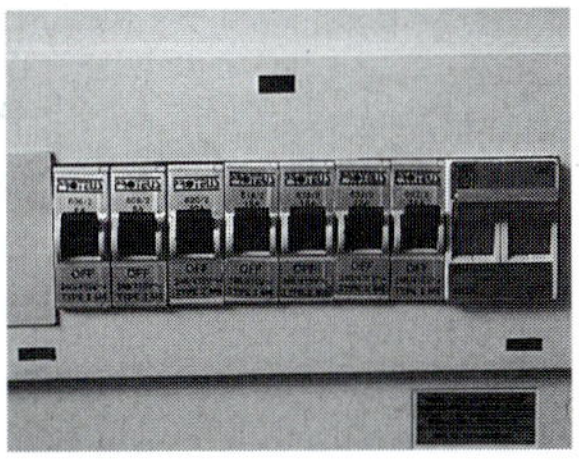

The correct fuse to fit into a circuit is one which melts or trips when a current slightly larger than the one required flows through it.

Appliance	Typical current	Correct fuse
Electric fire	12A	13A
Electric kettle	8A	13A
Hairdrier	4A	13A
Colour television	2.5A	3A
Table lamp	0.25A	3A

Earth-leakage circuit breakers and double insulation

a careless gardener might accidentally cut through the electric cable of their lawnmower
if the gardener does cut through the cable she could receive a severe electric shock

There are two ways of reducing the possibility of an electric shock.

1 Use an earth-leakage circuit breaker (ELCB). The plug from the mains socket is inserted into the ELCB, shown below, which is then pushed into the mains socket.

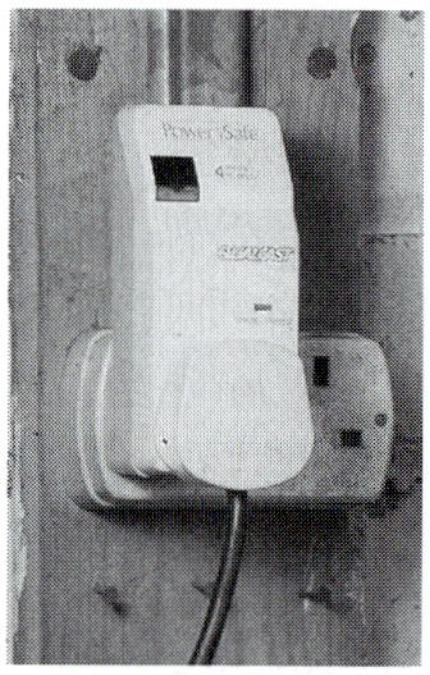

ELCBs compare the current flowing into a circuit through the live wires and the current flowing out of the circuit through the neutral wires. The two currents should be the same size. If there is a fault, such as a break in the circuit, the currents will not be the same and the ELCB immediately switches off the current to the appliance before the user is injured.

2 The second way to reduce the possibility of an electric shock is to protect the user with **double insulation**.

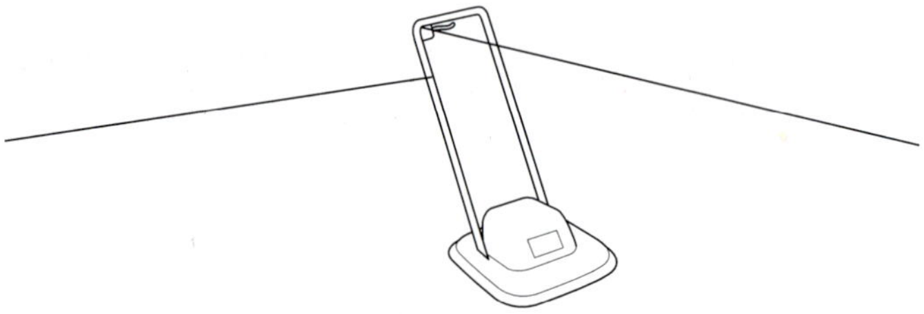
the handle of the lawnmower is made of plastic
the switch has a plastic lever and a plastic cover
when using the mower the gardener is totally insulated from any metal or electrical parts

Questions

1 Draw a diagram of a simple household fuse and explain how it is used to protect an appliance.

2 Explain what is meant by the phrase 'a fuse has a current rating of 3 A'.

3 What are the current ratings of the two most common fuses found in the home?

4 Which of the two most common fuses would you use for the following devices?
a) A spotlight which requires a current of 1 A.
b) A tumble drier which requires a current of 8 A.
c) A washing machine that requires a current of 6 A.

5 Explain the advantage of using a trip switch in a circuit rather than a cartridge or cylinder fuse.

6 Give one example of a circuit where it is important to use an earth-leakage circuit breaker.

Electrical Power

The two light bulbs in the diagram below have the figures 10 W (ten watts) 230 V and 60 W (sixty watts) 230 V printed on them. When the two bulbs are connected to the mains (230 V) the 60 W bulb glows much brighter than the 10 W bulb.

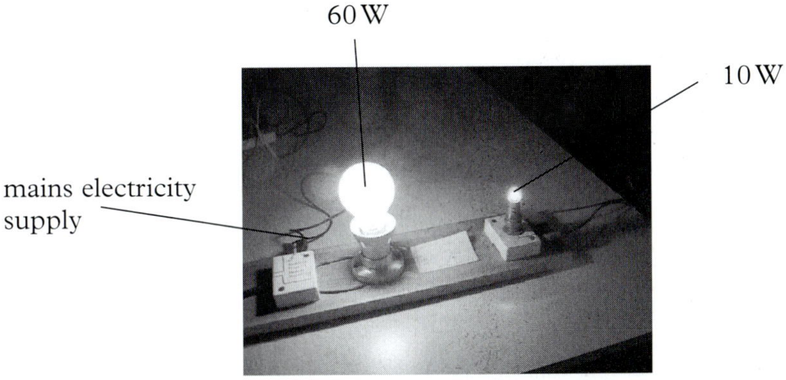

Light bulbs change electrical energy into heat and light energy. The numbers on the bulbs tell us how quickly they do this. The brighter bulb is converting electrical energy into heat and light energy at the rate of 60 J (joules) every second. The duller bulb is converting the energy at the rate of 10 J every second.

The rate at which the energy is being changed is the **power** or **power rating** of the bulb.

The bright bulb has a power rating of 60 W because it converts energy at the rate of 60 J every second. The duller bulb has a power rating of 10 W because it converts energy at the rate of 10 J every second.

If an electrical appliance is turned on for more than one second the amount of electrical energy it uses can be calculated using the following equation

energy used (E) = power in watts (P) × time in seconds (t)

The energy calculated using this equation is measured in joules.

Example
Calculate how much energy is used when a TV set rated at 600 W is turned on for 30 minutes.

Using $E = P \times t$

$$E = 600 \times (30 \times 60)$$

Remember that the time must be measured in seconds.

$$\mathbf{E = 1\,080\,000\,J} \text{ or } \mathbf{1080\,kJ}$$

Questions

1 a) Calculate how much electrical energy is used if each of the machines below is turned on for 5 s.

i)	TV set	600 W
ii)	Stereo	80 W
iii)	Kettle	1 kW
iv)	Fire	3 kW
v)	Lift (motor)	10 kW

b) How much electrical energy is used if each of the above is turned on for 5 minutes?

c) Write a sentence for each of these machines explaining the energy change that takes place in one second.

Paying for the electricity we use

electricity meter
the number on the meter tells us how much electrical energy we have used

Most homes receive an electricity bill every three months similar to the one shown below. It tells us how much electrical energy we have used and its cost. The readings are measured in Units. The Units used are the difference between the present reading and the previous reading.

Present reading on electricity meter	71073
Previous reading on electricity meter	69518
Units used	1555
Cost per Unit	6.3 p
Cost of electricity used (1555 Units × 6.3p)	£97.96
Standing charge	£8.43
Total bill	£106.39
VAT @ 5%	£5.32
Final total	£111.71

The electric meter measures this energy in kilowatt hours (kWh) or Units rather than joules.

the power rating of this fire is 1 kW

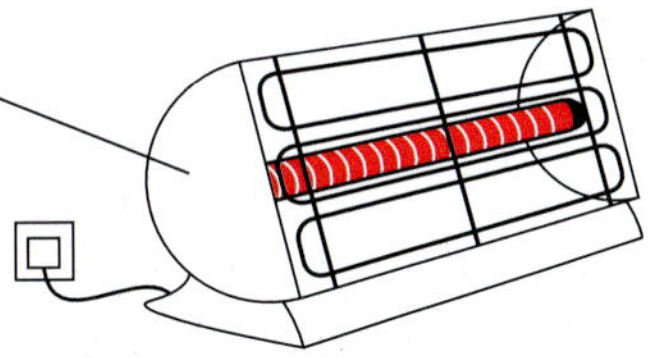

a 1 kW fire turned on for 1 hour will use 1 kilowatt-hour (kWh) of electrical energy

1 kWh is the same as 1 unit of electrical energy

the power rating of this fire is 1 kW

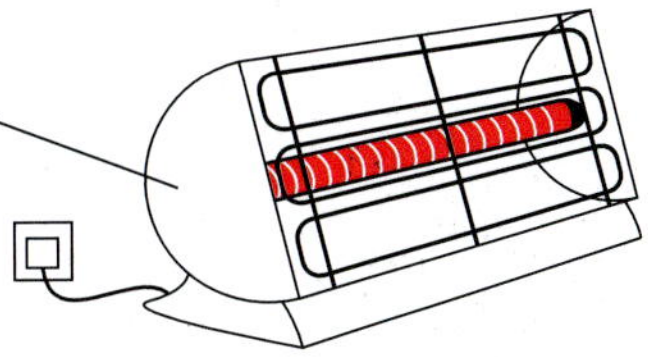

a 1 kW fire turned on for 2 hours will use 2 kWh or 2 units of electrical energy

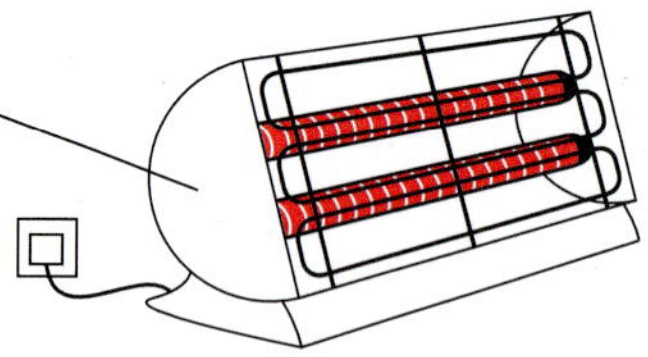
the power rating of this fire is 2kW

a 2kW fire turned on for 2 hours will use 4kWh or 4 units of electrical energy

We can summarise all the above in an equation

Number of Units used = power in kW × time in hours

Remember 1 kW = 1000 W.

Questions

1 a) Calculate the number of units of electricity used when a 3kW fire is turned on for 5 hours.

b) Calculate the number of units of electricity used when a 2kW tumble drier is turned on for 2 hours.

c) Calculate the number of units of electricity used when a 500W spotlight is turned on for 24 hours.

d) Calculate the number of units of electricity used when a 2kW washing machine is turned on for 30 mins.

Now that we have worked out the number of Units we have used we can calculate the cost of the electricity using the equation

cost of electrical energy = number of Units × cost per Unit

Example

Calculate the cost of having a 3kW electric fire turned on for one full day if the cost of one Unit is 7p.

Using the equation $E = P \times t$

$E = 3 \times 24$

$\mathbf{E = 72\ units}$

cost of electrical energy = number of Units × cost per Unit

$\text{cost} = 72 \times 7$

cost = 504 p or £5.04

Questions

1 If the cost of one Unit is 8p calculate the cost of each of the following:
a) a 2kW fire turned on for 6 hours,
b) a 1kW convector heater turned on for 3 hours,
c) a 1.5kW tumble drier turned on for 30mins,
d) a 3kW fire turned on for 20mins,
e) a 100W bulb left on for a full week.

Summary

Electrical appliances in our homes need energy to work. They obtain this energy from the mains supply. The energy enters along the live wire. The neutral wire is the return path for the electricity. The mains supply can kill. It is important that a user takes care. Most appliances are connected to the mains by a cable and three-pin plug which contains a fuse. Should an appliance become faulty a user may be protected by either the earth wire, a fuse, a circuit breaker or double insulation of the casing. The on/off switch for all appliances must be connected on the live side of the circuit so that when the switch is turned off no electrical energy can enter the appliance.

The total amount of electrical energy an appliance uses depends upon its power rating (wattage) and the length of time it is turned on. The equation describing energy use is: $\mathbf{E = P \times t}$. If the power is in watts and the time in seconds the energy used is in joules. If the power is in kilowatts and the time in hours the energy used is in kilowatt-hours or Units.

Key words

consumer box / fuse box A box which connects the incoming mains cable with all the electrical circuits in the house.

current rating The maximum current that can flow through a fuse without it breaking.

double insulation Insulation such as a plastic handle which isolates a user from the electrical parts of an appliance. This prevents any possibility of an electric shock.

earth wire The wire which is connected to the case of an appliance to protect the user when the appliance is faulty.

ELCB A safety device which turns a circuit off immediately if there is a fault.

fuse A safety device which breaks a circuit if the current is too large.

live wire The wire through which electrical energy enters an appliance.

neutral wire The return wire which completes the circuit.

power or power rating The amount of energy which is being changed or transferred by the appliance every second.

Key words (continued)

ring circuit A circuit which delivers electrical energy to all parts of a room.

trip switch / circuit breaker A special kind of fuse which can be reset.

End of Chapter 20 Questions

1 The diagram below shows the inside of a 3-pin mains plug.

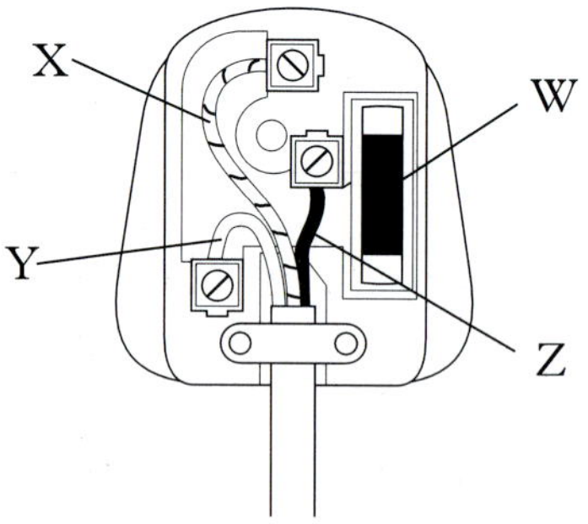

a) What is wire X called and what is its colour?
b) What is wire Y called and what is its colour?
c) What is wire Z called and what is its colour?
d) What is W?
e) Why are the pins of a plug made from metal?
f) Why are the wires of a plug covered with plastic?

2 A man is using a tumble drier which has no earth connection. The live wire inside the drier has become loose and is touching the metal casing.
a) What will happen to this man if he touches the metal casing? Explain your answer.
b) Why would this not happen if the drier had an earth connection?

The power rating for the tumble drier is 2 kW.
c) i) If it was in good working order how many Units of energy would it use if it was turned on for 2 hours?
ii) If the cost of 1 Unit of electricity is 7 p calculate the cost of using the drier for two hours.

When the drier is working properly a current of 8 A flows through it.
d) What would happen if a 3 A fuse was put into the drier's plug?
e) What size fuse should be put into the drier's plug?

3 When you use an electric hedge trimmer or a electric lawn mower it is recommended that you use an ELCB (Earth Leakage Circuit Breaker).
a) Explain why you are advised to use an ELCB with these two appliances.

Most electric lawn mowers do not have an earth wire but their handles do have double insulation.
b) Explain the phrase 'double insulation'.

21 Atomic Structure and Radioactivity

Structure of atoms

As we have already seen on page 140 we believe that an atom consists of two parts:

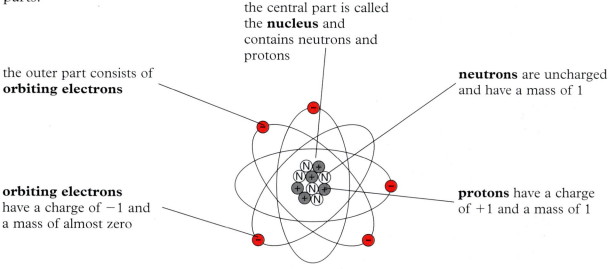

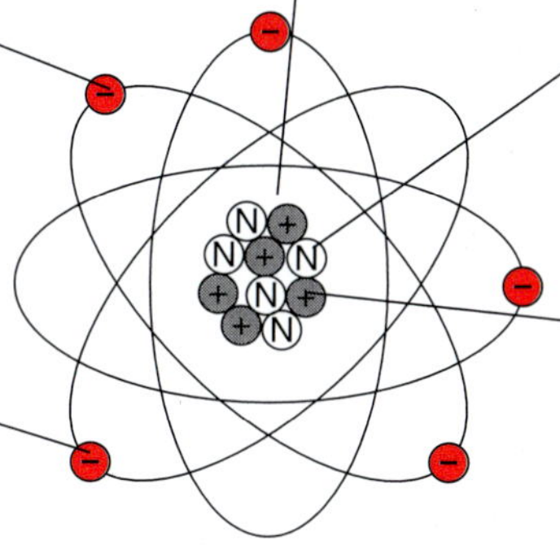

the central part is called the **nucleus** and contains neutrons and protons
the outer part consists of **orbiting electrons**
neutrons are uncharged and have a mass of 1
orbiting electrons have a charge of −1 and a mass of almost zero
protons have a charge of +1 and a mass of 1

The atom above is a neutral atom. In neutral atoms the number of protons in the nucleus is equal to the number of electrons orbiting the nucleus.

Questions

1 a) Where in an atom are the protons?
b) What charge does a proton carry?

2 a) Where in an atom are the neutrons?
b) What charge does a neutron carry?

3 a Where in an atom are the electrons?
b) What charge does an electron carry?

4 A hydrogen atom has one proton, one electron and no neutrons. Draw a diagram of a hydrogen atom.

5 A helium atom has two protons, two neutrons and two electrons. Draw a diagram of a helium atom.

6 How many electrons and protons are there in a neutral atom?

Unstable atoms

Most atoms have a structure which is stable and the number of protons and neutrons in the nucleus remain unchanged. Some atoms however have a structure which is unstable. They try to become more stable by emitting particles and waves of energy from their nuclei.

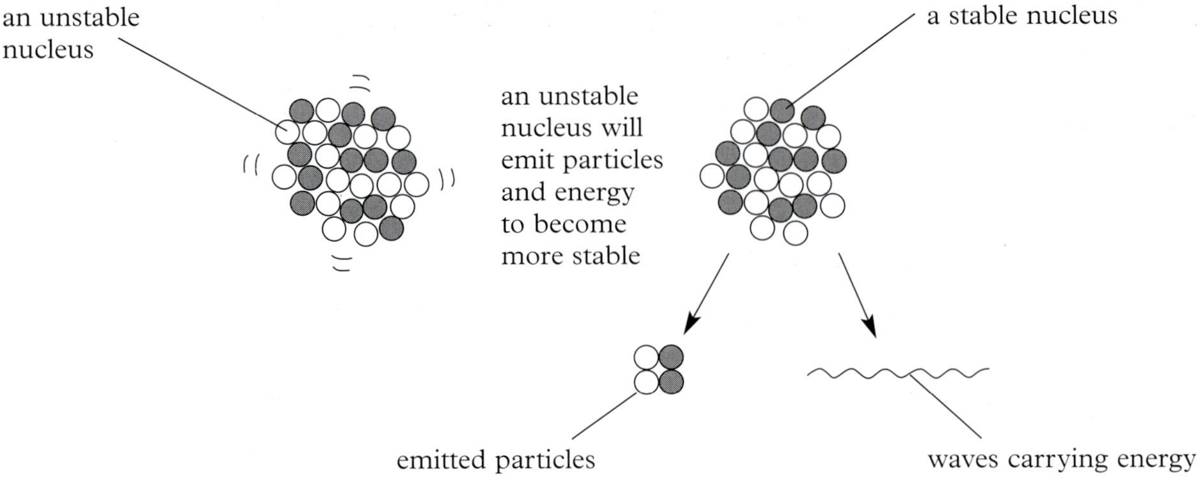
an unstable nucleus
a stable nucleus
an unstable nucleus will emit particles and energy to become more stable
emitted particles
waves carrying energy

The emission of these particles and waves from the nucleus of an atom is known as **radioactivity**. Atoms which emit these particles and waves are **radioactive**. When an atom has emitted some radiation it is said to have **decayed**.

The rate at which unstable atoms decay is totally unaffected by any physical or chemical condition.

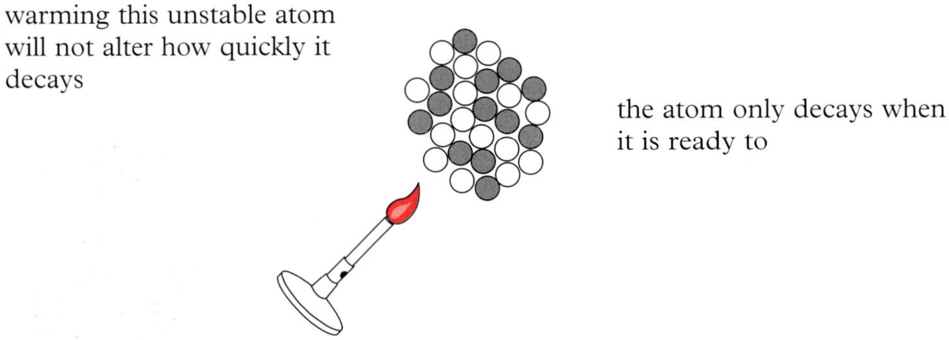
warming this unstable atom will not alter how quickly it decays
the atom only decays when it is ready to

Questions

1 a) Describe how an unstable nucleus changes into a stable nucleus.
b) What is this process called?

2 a) What is a radioactive atom?
b) How does a radioactive atom decay?

3 How can we increase the rate at which an atom decays?

Alpha, beta and gamma radiation

There are three different sorts of radiation which can be emitted from radioactive materials. These are called **alpha (α) radiation**, **beta (β) radiation** and **gamma (γ) radiation**.

All three types of radiation can be detected using a **Geiger-Muller tube and counter** or **photographic film**.

Geiger-Muller tube and counter

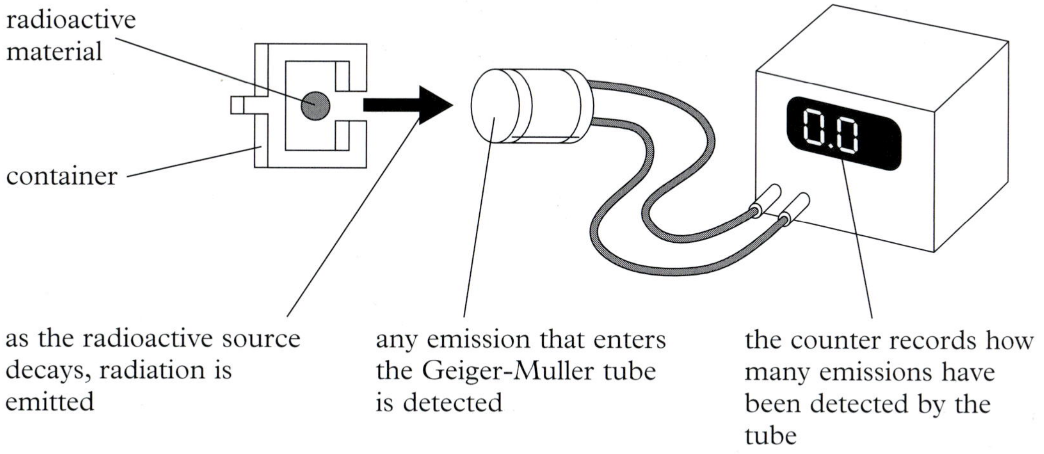
radioactive material
container
0.0
as the radioactive source decays, radiation is emitted
any emission that enters the Geiger-Muller tube is detected
the counter records how many emissions have been detected by the tube

Photographic film

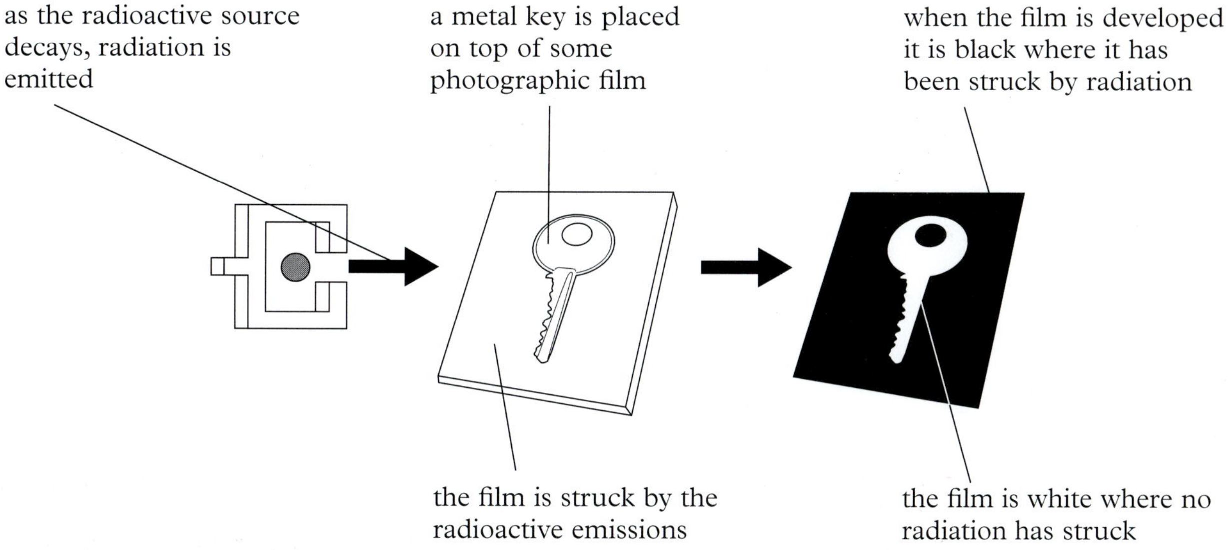
as the radioactive source decays, radiation is emitted
a metal key is placed on top of some photographic film
when the film is developed it is black where it has been struck by radiation
the film is struck by the radioactive emissions
the film is white where no radiation has struck

Properties of alpha, beta and gamma radiations

The nature of radiation

alpha radiation consists of fast moving particles

alpha particles have a charge of +2

these particles are made of 2 neutrons and 2 protons

alpha particles have a mass of 4

beta radiation consists of very fast moving electrons

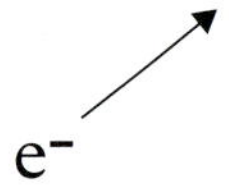

beta particles have almost no mass

beta particles have a charge of −1

gamma radiation consists of a wave, just like an X-ray

gamma radiation has no mass

gamma radiation has no charge

gamma radiation carries a lot of energy

The penetrating power of radiation

alpha radiation is the least penetrating

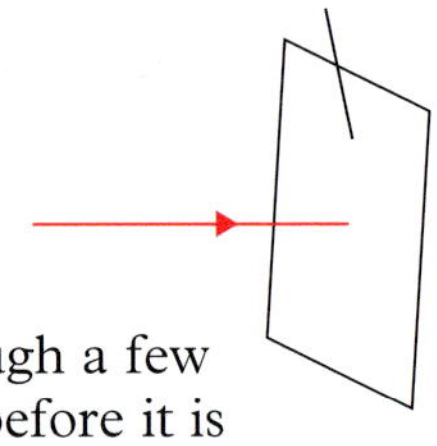

a thin piece of card will stop alpha radiation

it only travels through a few centimetres of air before it is absorbed

beta radiation is more penetrating than alpha radiation

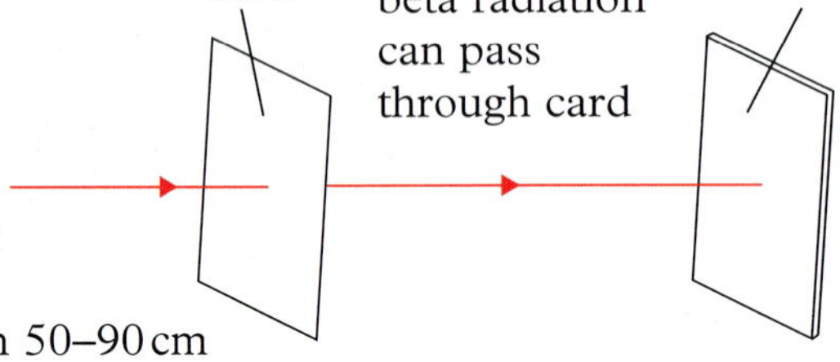

beta radiation can pass through card

several mm of aluminium will stop beta radiation

it can travel between 50–90 cm in air

gamma radiation is the most penetrating

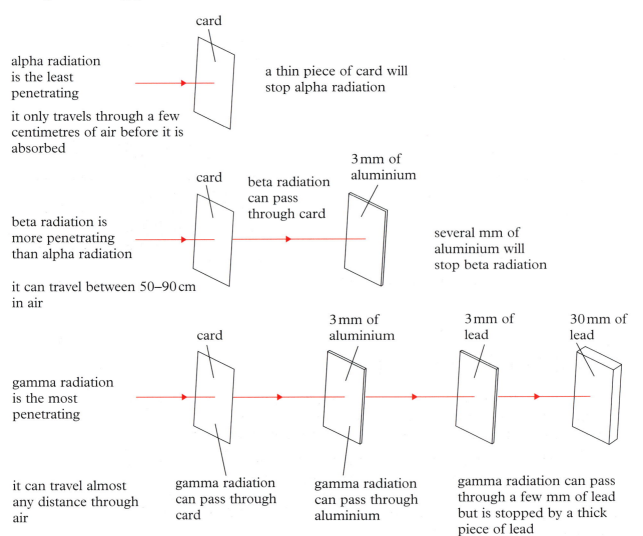

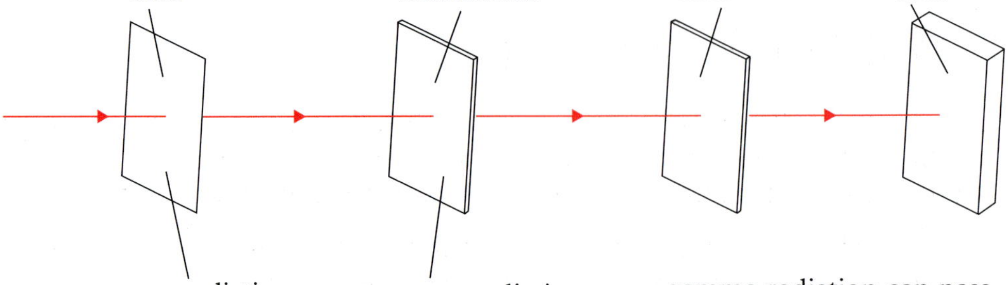

it can travel almost any distance through air

gamma radiation can pass through card

gamma radiation can pass through aluminium

gamma radiation can pass through a few mm of lead but is stopped by a thick piece of lead

Deflection of radiation

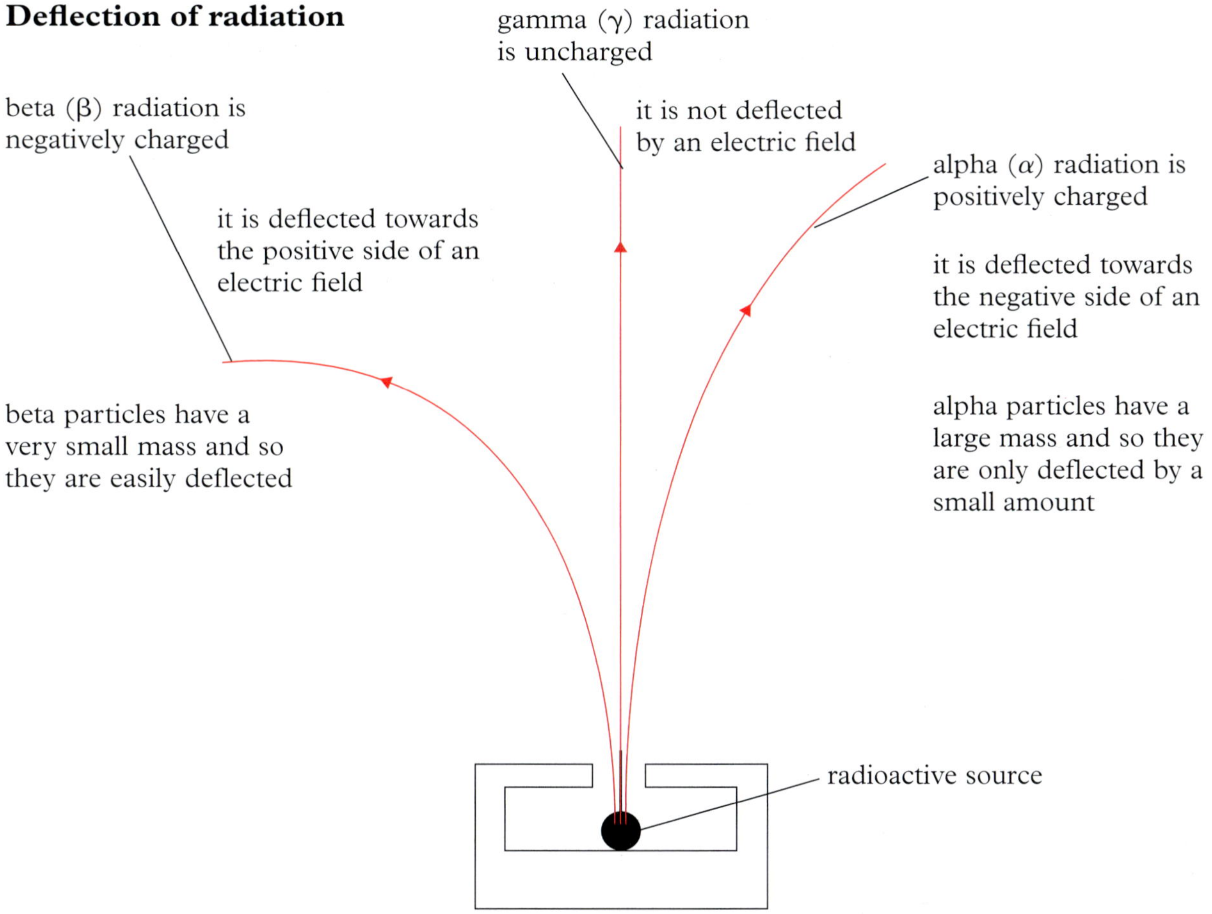
gamma (γ) radiation
is uncharged
it is not deflected
by an electric field
beta (β) radiation is
negatively charged
it is deflected towards
the positive side of an
electric field
alpha (α) radiation is
positively charged
it is deflected towards
the negative side of an
electric field
beta particles have a
very small mass and so
they are easily deflected
alpha particles have a
large mass and so they
are only deflected by a
small amount
radioactive source

Questions

1 Which radiation can pass through a sheet of paper but not through a thick sheet of aluminium?

2 Which radiation can pass through sheets of paper and aluminium?

3 Which radiation can not pass through the skin?

4 In an electric field the alpha radiation is deflected to the right.
  a) In the same field which way will beta radiation be deflected?
  b) Which way will gamma radiation be deflected?

Background radiation

Radioactivity is a natural process. There are radioactive materials all around us. They are in the ground, in the air, they are even in the food we eat. The radiation from all these sources is called **background radiation**.

Many people believe that most of the radioactivity around us is 'man-made' and has escaped from the nuclear industry. They believe it has leaked from nuclear power stations and is the result of using nuclear weapons, but as the diagram on the next page shows this is not true. Most of the background radiation comes from natural sources.

Sources of background radiation

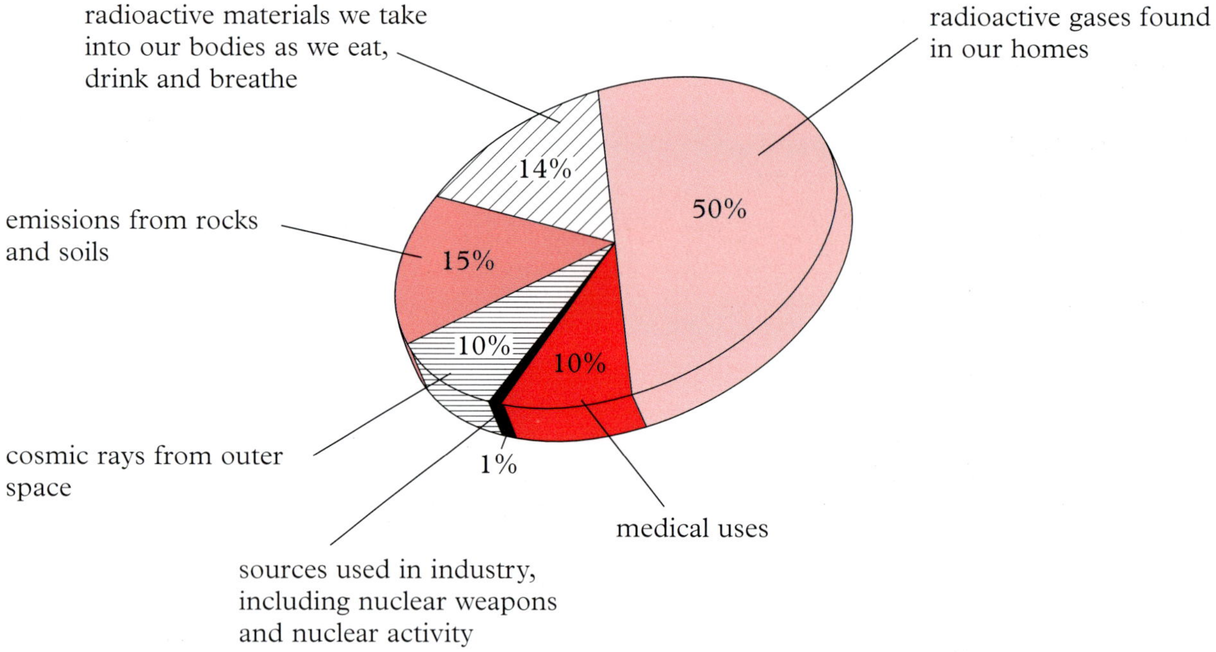

The dangers of radioactivity

Radioactivity can be harmful to us and to all living things. When radiation collides with atoms or molecules it can alter their structure. In living things this damage can cause cells to become cancerous. It may cause leukaemia, which is cancer of the blood. It can also damage reproductive organs and cause infertility. The larger the dose of radiation received the greater the risk of cancer.

Because of these dangers radioactive materials must be handled with great care. They should always be kept at arms length and where possible surrounded by lead shields.

Workers who are at risk from exposure wear badges called **dosimeters**. These record how much radiation each worker has been exposed to.

Questions

1 Answer the following questions using the information shown in the pie chart above.
- a) What is the main source of background radiation in our homes?
- b) How much of the background radiation comes from outer space?
- c) How much of the background radiation comes from medical uses?

2
- a) What is a dosimeter and what does it measure?
- b) Who might wear a dosimeter?

3
- a) Which of the three types of radiation is most easily absorbed by the skin and would therefore cause most damage to cells here?
- b) Name one illness this damage may cause.

Uses of radioactivity

Although the emissions from radioactive materials can be dangerous they can, if used correctly, be very useful.

Radiotherapy

High doses of radiation can kill cells that make up all living things. This idea is used in **radiotherapy**. The gamma radiation emitted by radioactive cobalt is used to kill cancerous cells in the body. Because the amount of radiation is carefully controlled little or no damage is done to healthy cells.

Killing germs

It is essential that the surgical instruments used in hospitals are sterile, which means they are germ-free. This used to be done by putting the instruments into boiling water. Nowadays the sterilising is carried out using gamma radiation.

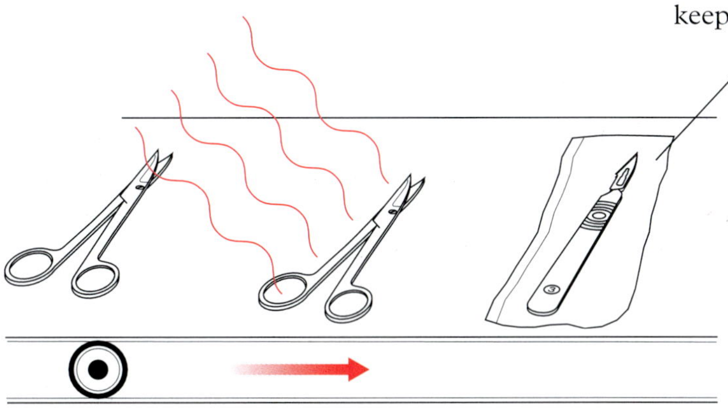
surgical instruments are exposed to gamma rays
the gamma rays kill any harmful micro-organisms on the instruments
the instruments can then be wrapped in plastic to keep them germ-free

Gamma rays can be used to keep food fresh for longer.

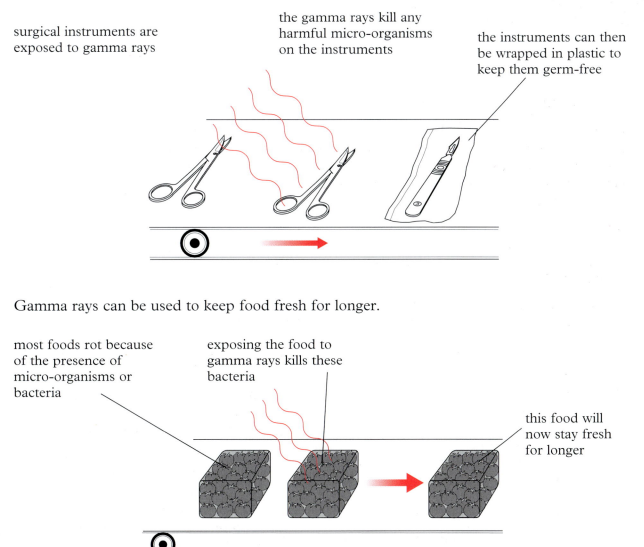

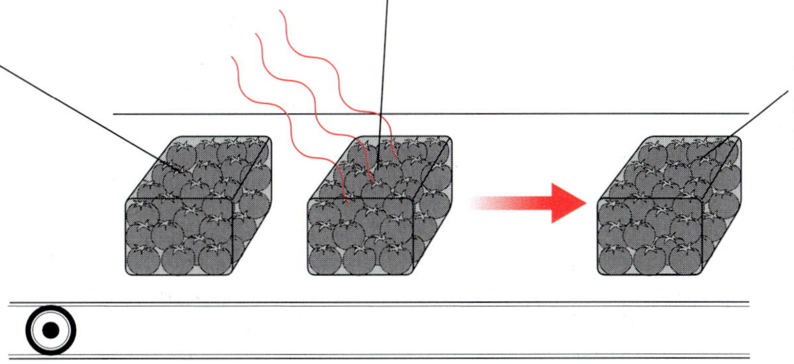
most foods rot because of the presence of micro-organisms or bacteria
exposing the food to gamma rays kills these bacteria
this food will now stay fresh for longer

Quality control

Paper manufacturers use radioactivity to measure and control the thickness of the paper they produce.

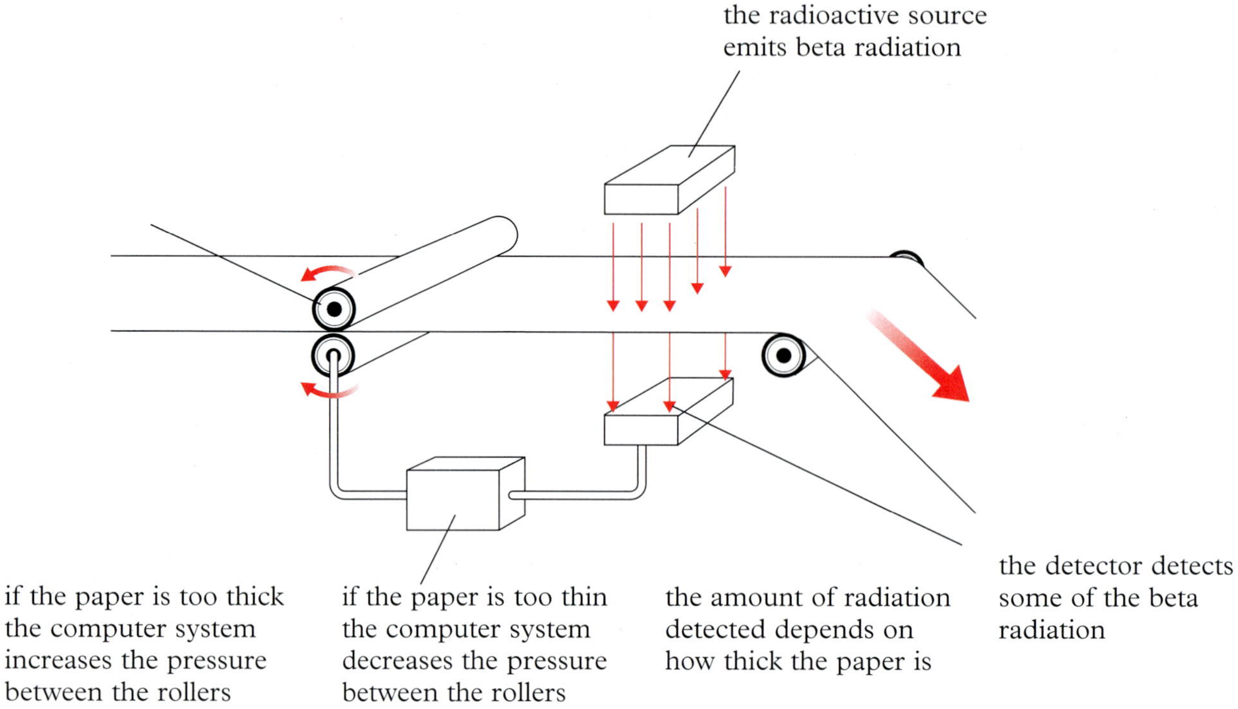
the radioactive source emits beta radiation
if the paper is too thick the computer system increases the pressure between the rollers
if the paper is too thin the computer system decreases the pressure between the rollers
the amount of radiation detected depends on how thick the paper is
the detector detects some of the beta radiation

Radioactive tracers

Radioactive tracers can be used to monitor movement and flow of substances through pipes. They can be used to detect blockages and leaks.

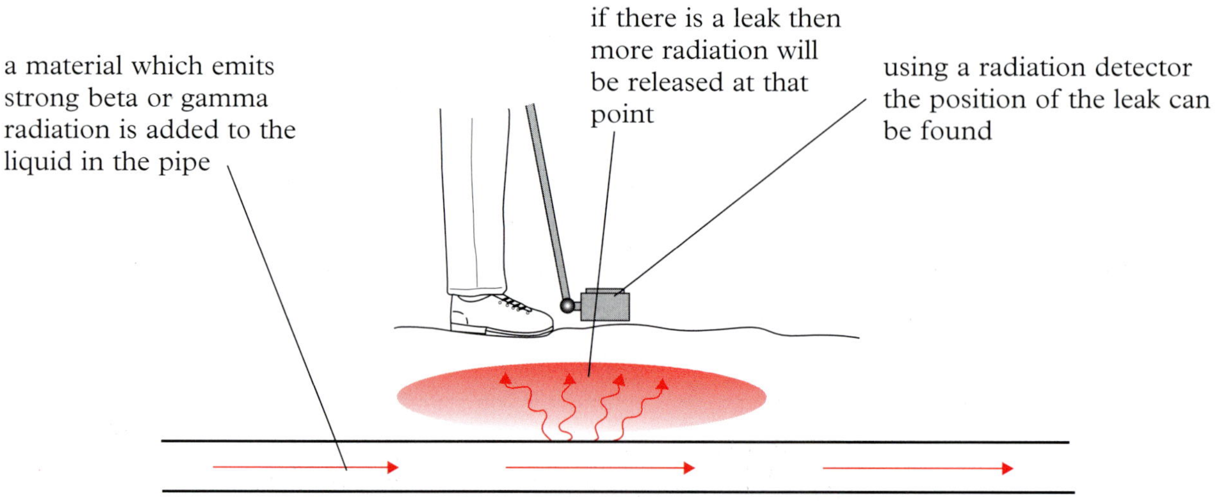
a material which emits strong beta or gamma radiation is added to the liquid in the pipe
if there is a leak then more radiation will be released at that point
using a radiation detector the position of the leak can be found

In a similar way weak radioactive tracers such as sodium-24 can be injected into the bloodstream of a patient to check his or her circulation.

Summary

Unstable atoms try to become more stable by emitting radiation from their nuclei. These atoms are described as being radioactive. There are three different types of radiation that can be emitted. These are alpha (α) radiation, beta (β) radiation and gamma (γ) radiation. Alpha radiation is the least penetrating, and carries a charge of +2. Gamma radiation is the most penetrating and carries no charge. Beta radiation has penetrating powers which are between those of alpha and gamma and carries a charge of −1.

There are many uses for radioactivity in industry, in hospitals and even in our homes. It can however damage our health and so must be used with great care.

Key words

alpha radiation Fast moving particles consisting of 2 protons and 2 neutrons, emitted by an unstable nucleus.

beta radiation Very fast electrons emitted by an unstable nucleus.

dosimeter A piece of equipment often in the form of a badge which monitors how much radioactivity a worker has been exposed to.

electron Very small, negatively charged particle which orbits the nucleus.

gamma radiation Waves similar to X-rays emitted by an unstable nucleus.

Geiger-Muller tube A piece of equipment which can detect and measure radioactivity.

neutron Uncharged particle found in the nucleus of an atom.

nucleus Centre of an atom.

proton Positively charged particle found in the nucleus of an atom.

radioactive tracer A radioactive material which is used to monitor the movement of material, like blood in a vein.

radioactivity/radioactive decay The emission of radiation by an unstable nucleus.

radiotherapy A treatment for cancer which uses gamma radiation.

End of Chapter 21 Questions

1 There are three types of radiation which could be emitted by an unstable nuclei. These are alpha radiation, beta radiation and gamma radiation. Complete the table below which shows some of their properties. The properties of alpha radiation have already been done for you.

radiation	consists of	charge	mass	penetrating power
alpha	2 protons 2 neutrons	+2	4	very weak
beta				
gamma				

2 The diagram below shows a smoke alarm which contains some radioactive material. This material emits alpha radiation as it decays.

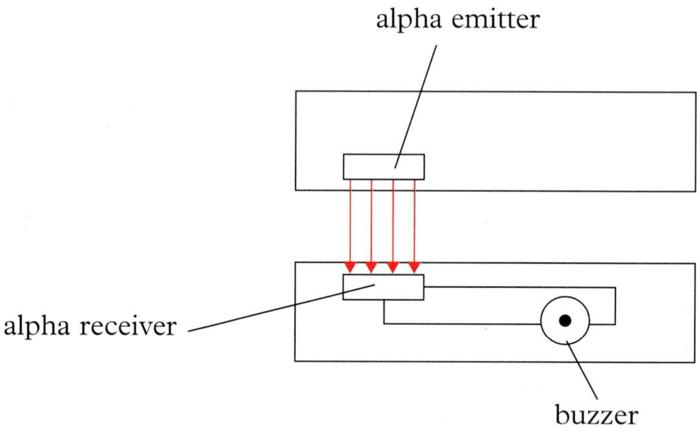

Explain why the alarm sounds if there is smoke in the room.

3 A scientist used a Geiger-Muller tube and counter to measure the radioactive count for 1 minute in each of the experiments shown on the next page. The distance between the Geiger-Muller tube and the source was kept at 5 cm. After each experiment the counter was reset to zero.

a) What was the background count for 1 minute?

b) Name two possible sources of the background radiation.

c) i) What effect does placing a piece of card between the source and the Geiger-Muller tube have on the amount of radiation detected?

ii) Explain your answer.

d) i) What effect does placing a thick piece of aluminium between the source and the Geiger-Muller tube have on the amount of radiation detected?

ii) Explain your answer.

e) i) What effect does placing a thick piece of lead between the source and the Geiger-Muller tube have on the amount of radiation detected?
ii) Explain your answer.

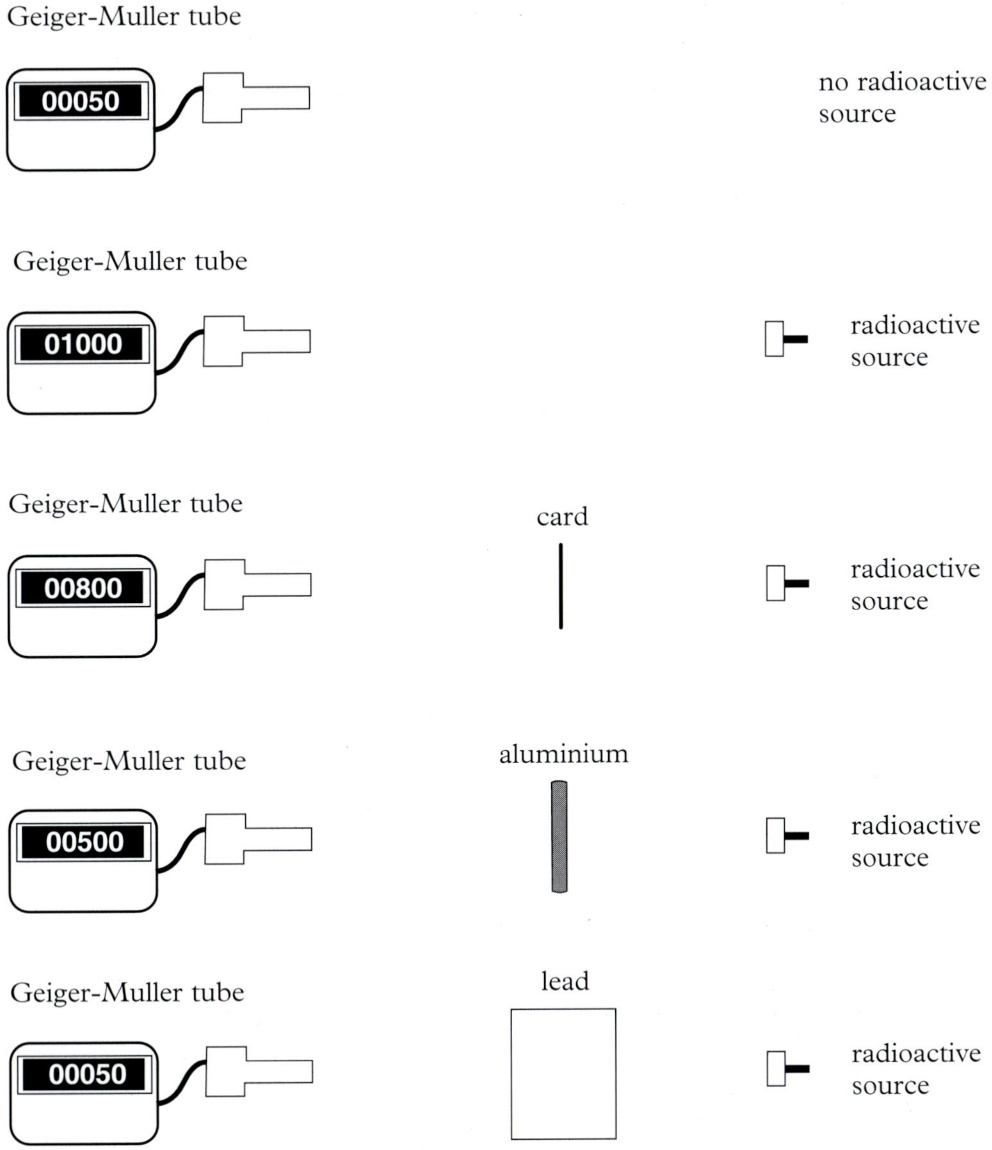

4 Nowadays many liquids and gases are transported from place to place via underground pipelines. If however, a pipe becomes faulty and leaks, it is important to find and fix the problem as quickly as possible. One way to do this is by using a **radioactive tracer**.

a) What is a radioactive tracer?
b) Explain how a radioactive tracer could be used to find a leak in an underground pipe.
c) i) What kind of radiation should the tracer be emitting if the pipe is several metres underground?
ii) Explain your answer.
d) Where might a radioactive tracer be used in a hospital?

22 Plate Tectonics

Plate tectonics tells us about the structure of the crust of our planet and how it behaves.

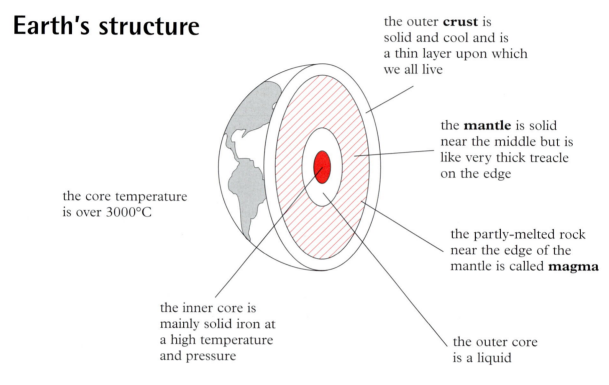

Earth's structure

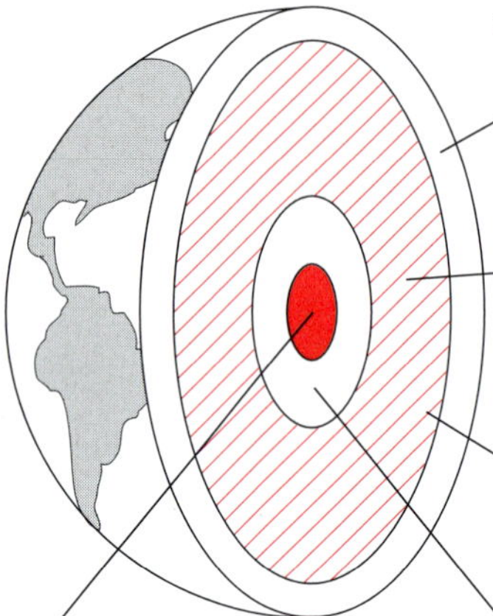
the outer **crust** is solid and cool and is a thin layer upon which we all live
the **mantle** is solid near the middle but is like very thick treacle on the edge
the core temperature is over 3000°C
the partly-melted rock near the edge of the mantle is called **magma**
the inner core is mainly solid iron at a high temperature and pressure
the outer core is a liquid

Plates

The Earth's crust is a bit like the shell of a hard boiled egg that has been cracked. The pieces of crust are called **plates**. These plates have been moving very slowly over the surface of the Earth for millions of years. Some of the plates are underwater and form the sea bed.

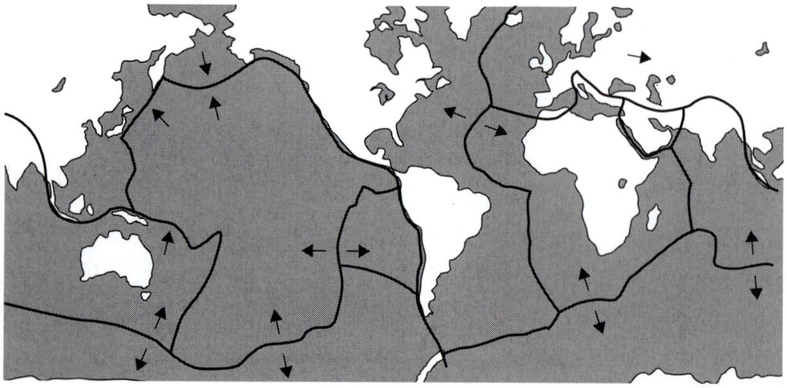
the black lines show the edges of the plates
the arrows show the direction in which the plates are moving

The plates are floating on top of the hot treacle-like mantle rock. This, like any hot liquid, is moving because of convection currents. The plates above it move at about 5 cm a year, which is about the speed at which fingernails grow.

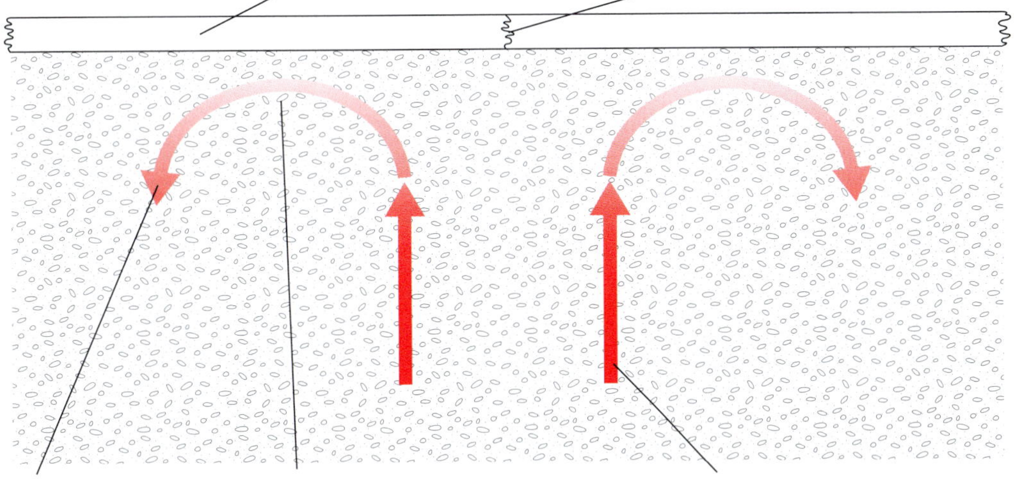
plate
this is a plate margin
the rock sinks back down again.
the rock cools as it reaches the crust
the semi-liquid mantle rock rises (like a hot air balloon in air) because it is hotter and so less dense than the rock at the top.

Earthquake areas

Areas where different plates meet are called **plate margins**. Earthquakes are likely to happen around plate margins. Earthquakes are rare in the centre of plates. Geologists have plotted the earthquake zones and built up a picture of where the plate margins are.

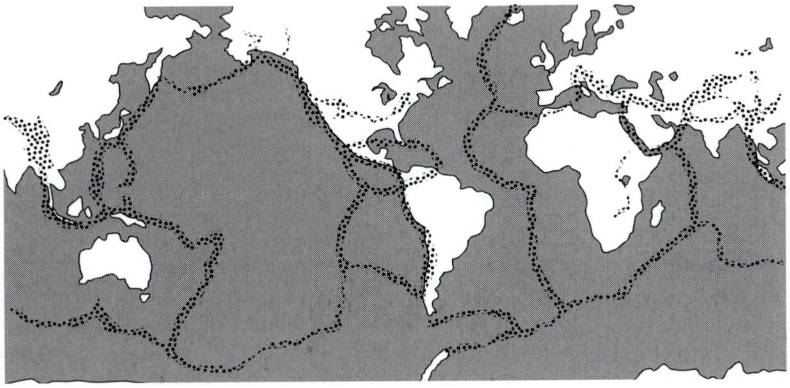

the dots show where earthquakes have taken place in the past

Questions

1 Compare the map of earthquake zones with the map of plate margins. How are they alike?

2 Find Australia on the second map. Use an atlas if you don't know where it is. Explain why you would not expect earthquakes to happen there.

When the plates move

The plates may move apart, crash into each other or slide past each other. Earthquakes and volcanoes will happen as a result of this motion.

When the plates part, new rocks are formed

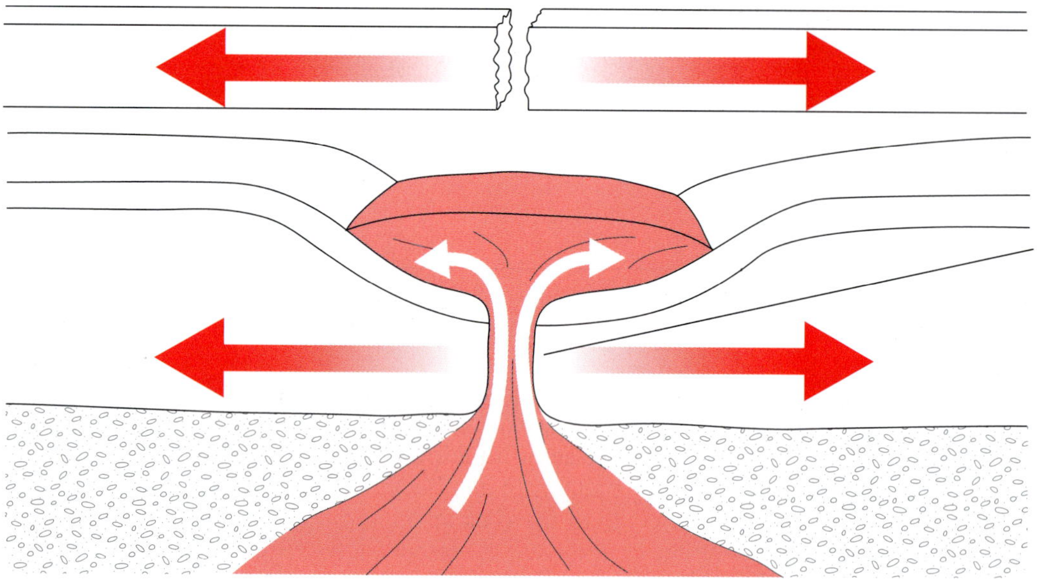
when plates move apart they leave a gap
magma from the mantle can flow up through this gap
this magma forms new rock and a volcano is created

Many plate margins are under the oceans. The areas of some sea floors, like the Atlantic Ocean, are gradually getting bigger as the plates move apart. More and more rock from volcanoes fills in the gap. This is called **sea-floor spreading**.

When the plates crash together

Rocks are lost when plates collide.

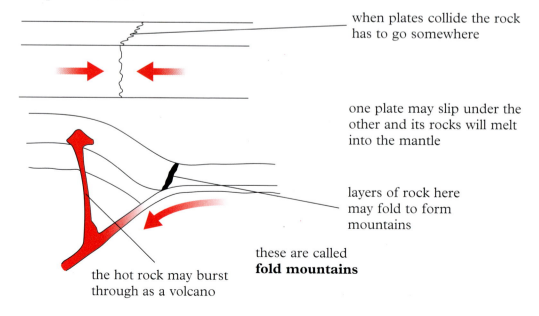

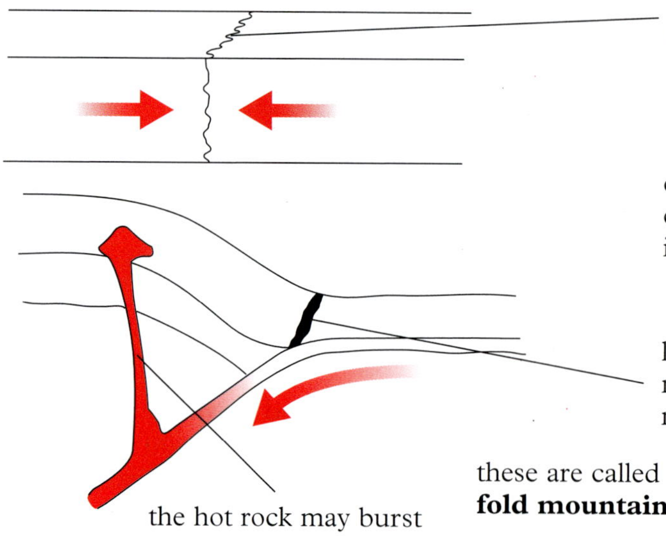
when plates collide the rock has to go somewhere
one plate may slip under the other and its rocks will melt into the mantle
layers of rock here may fold to form mountains
these are called **fold mountains**
the hot rock may burst through as a volcano

When the land is folded, rocks that were once at sea level are lifted.

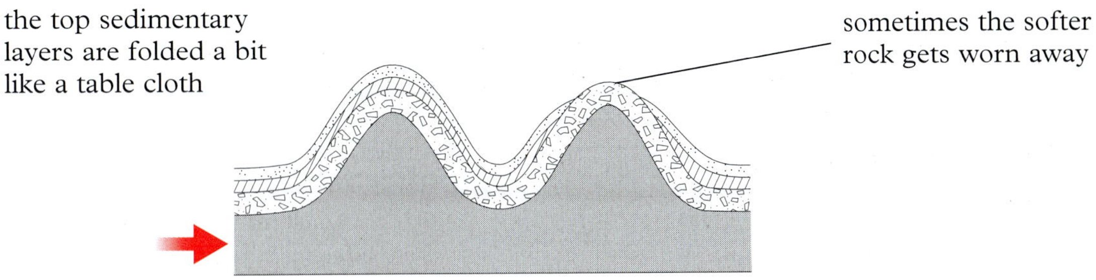
the top sedimentary layers are folded a bit like a table cloth
sometimes the softer rock gets worn away

Questions

1 a) Explain why there is a ridge all the way along the plate margins under the Atlantic ocean.

b) Iceland, shown in the diagram below, is an island made from volcanoes. It lies along the plate margin under the Atlantic ocean. Explain why Iceland is getting wider at the rate of 2 cm every year.

2 The Alps were formed when the African plate crashed into the European plate. Fossils of sea creatures have been found at the top of the Alps. Look at the diagram above and explain how this must have happened.

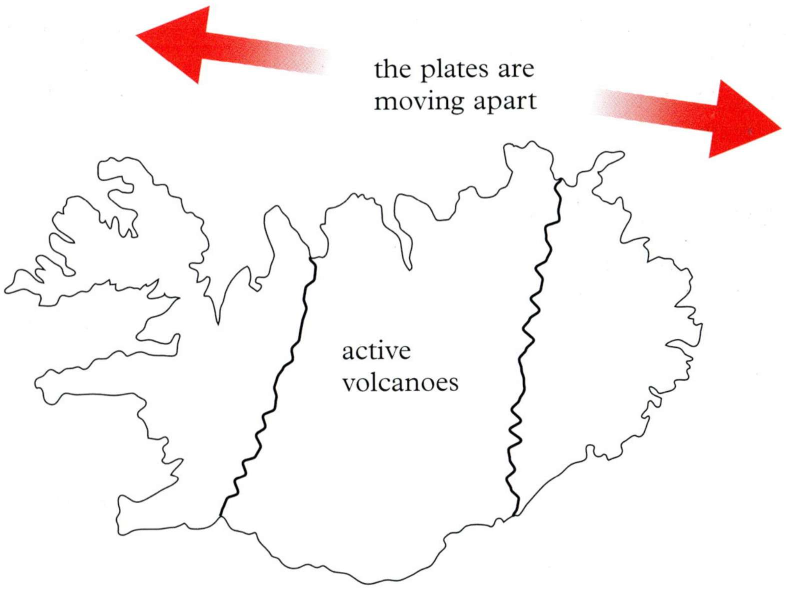
the plates are moving apart
active volcanoes

When the plates slide against each other

plates don't slide smoothly against each other

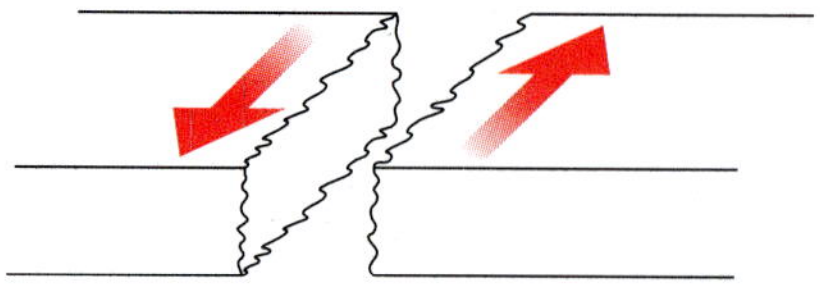

it is usually a jerky movement

This can cause large earthquakes. Sometimes the plates have been trying to slide past one another for millions of years, but friction between the plates holds them together. When the rocks finally move, it is usually sudden and violent so that the surface is shaken fiercely.

Summary

Plate tectonics describes the way the Earth's crust is constructed and the way it behaves. The crust is divided into plates which float on treacle-like rock in the mantle. Plates move at about 5 cm a year. Along the plate margins new rocks may form, rocks may be recycled and volcanoes, earthquakes and fold mountains may be created. The evidence for all this is found from the rocks themselves.

Key words

earthquake A sudden movement of the Earth's crust caused by movement of the rocks below.

fold mountains Mountains formed by the folding of the surface rock.

plates The pieces of the crust that make up the Earth's surface.

volcano A place where molten rock, ashes and gas are forced through the Earth's crust.

End of Chapter 22 Questions

1 Fill in the spaces with words from the following list. Some of the words will not be needed.

mantle planes one trees margins fingernails earthquakes five plates

The Earth's crust is divided into a series of segments which we call _________ . These are floating on semi-liquid rock in the _________ . They are moving at the rate of about _________ cm a year, which is as fast as growing _________ . There are likely to be _________ where they meet. The places where they meet are called plate _________ .

2

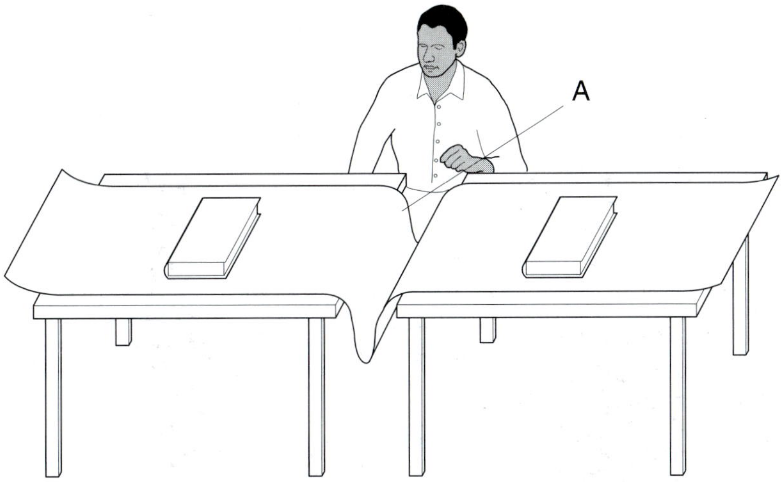

You are using a model like the one above, where the books are continents (land) and the sheet is the crust.

a) What does **A** represent?
b) What fills the space between the two continents?
c) Explain how you can show what happens when plates
   i) move apart
   ii) move together.

3 From a copy of this word square, find the following words:

mantle sea floor spreading magma convection margin core volcano treacle earthquake plate land

C	S	E	A	F	L	O	O	R	S
O	M	A	N	T	L	E	N	E	P
N	A	R	T	A	E	R	A	A	R
V	R	T	T	R	E	A	C	L	E
E	G	H	S	T	A	R	L	C	A
C	I	Q	E	I	F	L	O	O	D
T	N	U	T	N	E	W	V	R	I
I	O	A	A	H	S	H	E	E	N
O	N	K	L	A	N	D	L	O	G
N	E	E	P	M	A	G	M	A	S

23 Inside the Solar System

The Earth

We live on a **planet** called the Earth. If we could see our planet from space it would look like the photograph seen below.

This photograph, taken from space, clearly shows some of the land and sea areas which make up the surface of the Earth. High above the surface are the white wisps of cloud.

The daily journey of the Sun

Although we cannot feel it, the Earth is spinning around on its **axis** like a top. The axis is an imaginary line running through the centre of the Earth.

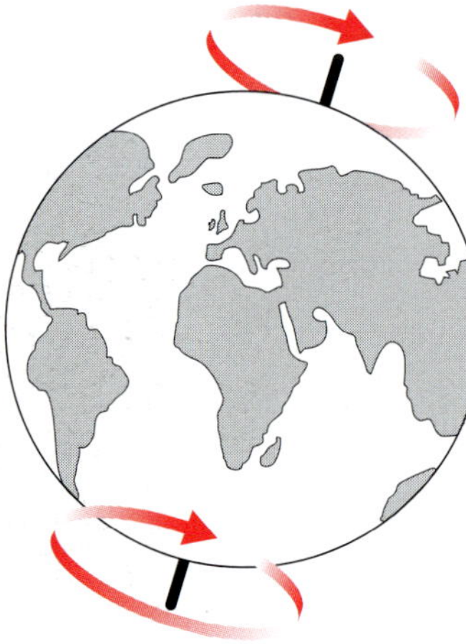
the Earth completes one full turn or rotation every 24 hours
this is **one day**
this line is the axis

This turning motion makes the Sun appear to rise in the east, travel high overhead and then set in the west.

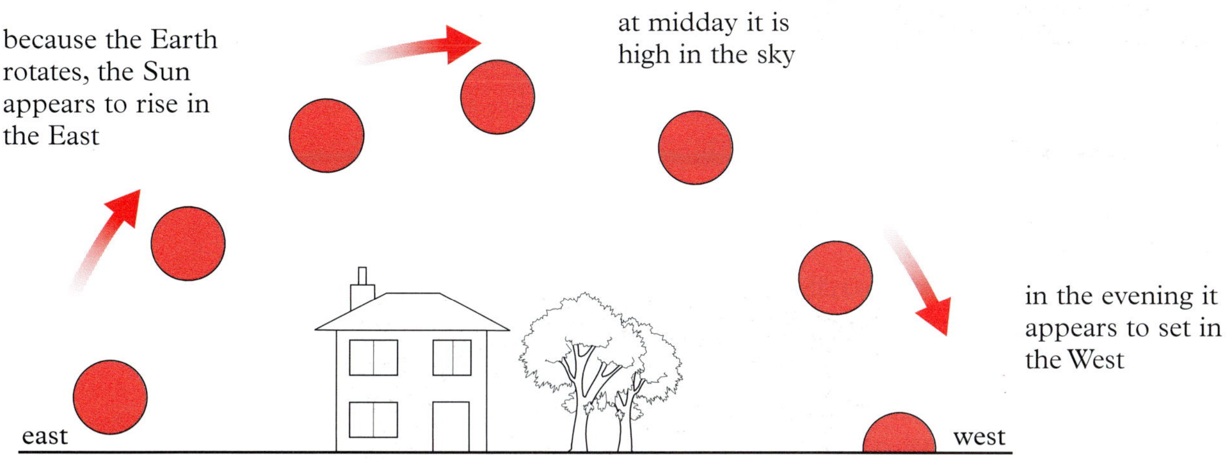
because the Earth rotates, the Sun appears to rise in the East
at midday it is high in the sky
in the evening it appears to set in the West
east
west

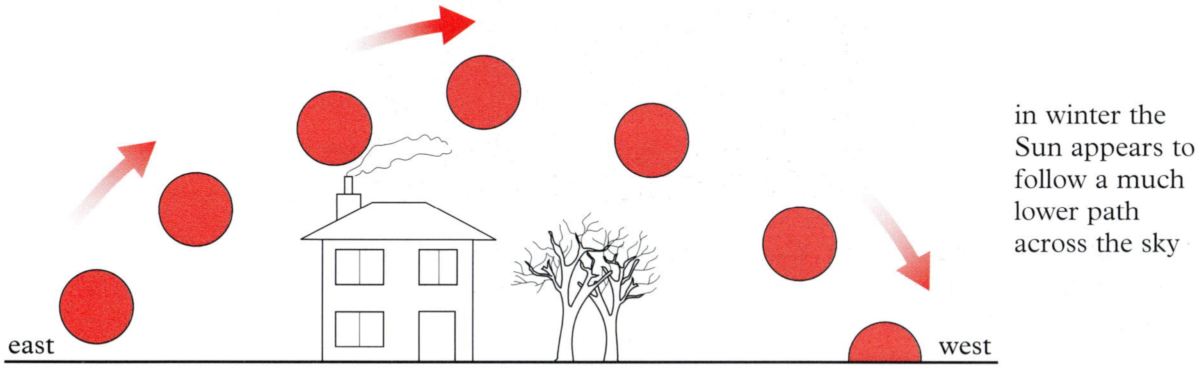
in winter the Sun appears to follow a much lower path across the sky
east
west

Day and night

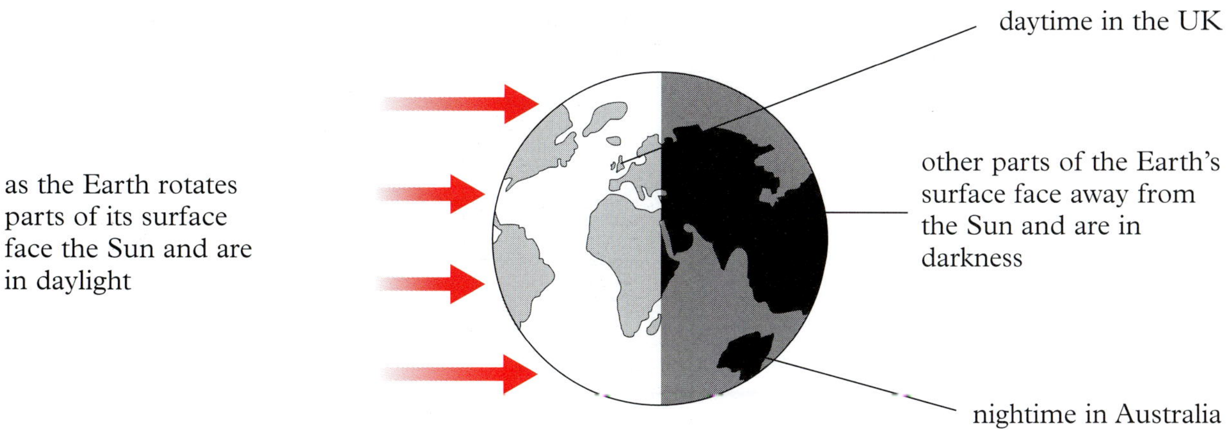
daytime in the UK
as the Earth rotates parts of its surface face the Sun and are in daylight
other parts of the Earth's surface face away from the Sun and are in darkness
nightime in Australia

Questions

1 How long does it take the Earth to complete one full turn on its axis?

2 Draw diagrams to explain why parts of the Earth are in daylight whilst other parts are in darkness (night).

3 Draw diagrams to show the difference between the path taken by the Sun across the summer sky and the path it takes in the winter.

A Year and the Seasons

A **star** is a large ball of gas that gives out heat and light. The nearest star to the Earth is the Sun. It is 150 million kilometres away. The Earth goes round or **orbits** the Sun once every year following a path called an **ellipse**. An ellipse is a slightly squashed circle.

circular orbit

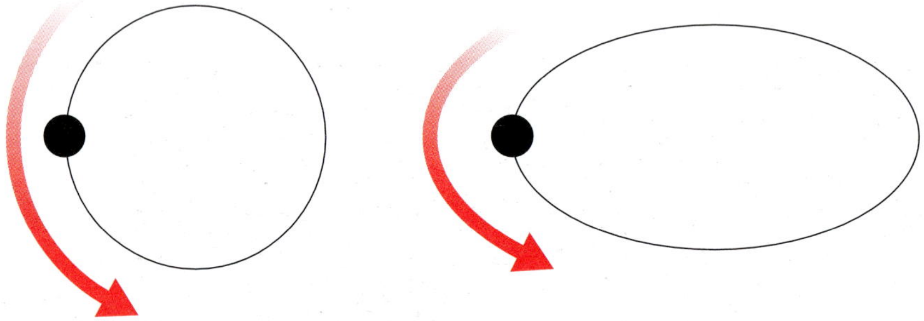

elliptical orbit

As the Earth travels around the Sun we experience the different **seasons**. These changes to our weather, climate and length of daylight happen because the axis around which the Earth spins is slightly tilted.

in spring the Earth is neither tilted towards or away from the Sun

when it is summer in Britain, our part of the Earth is tilted towards the Sun

when it is winter in Britain our part of the Earth is tilted away from the Sun

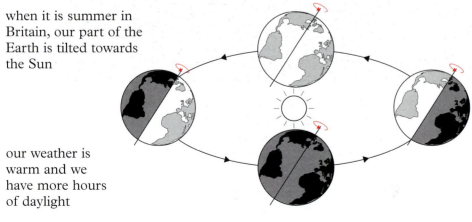

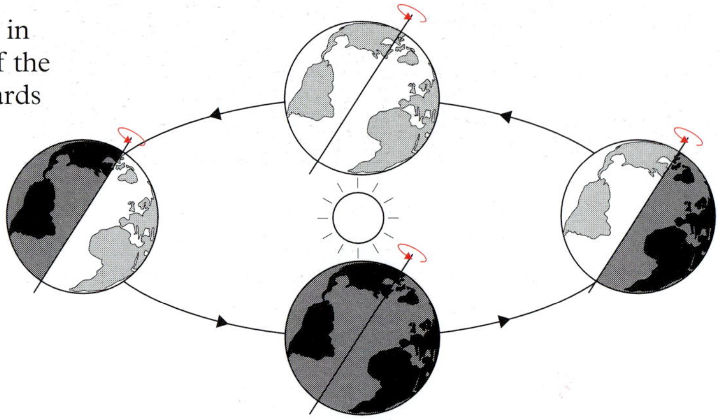

our weather is warm and we have more hours of daylight

our weather is cold and we have few hours of sunlight

in autumn the Earth is neither tilted towards or away from the Sun

Questions

1 What is the name of our nearest star?

2 How long does it take for the Earth to complete one full orbit of the Sun?

3 What is an elliptical orbit?

4 Draw diagrams to explain why in winter our weather is colder than it is in summer.

The Solar System

The Earth is one of several planets to orbit the Sun. Altogether there are nine planets. Starting with the planet nearest the Sun they are Mercury, Venus, Earth, Mars, Jupiter, Saturn, Uranus, Neptune and Pluto. An easy way to remember the names and order of the planets is to use the sentence: **M**any **V**ery **E**nergetic **M**en **J**og **S**lowly **U**pto **N**ewport **P**agnell.

The Sun, the planets and their moons make up the **solar system**.

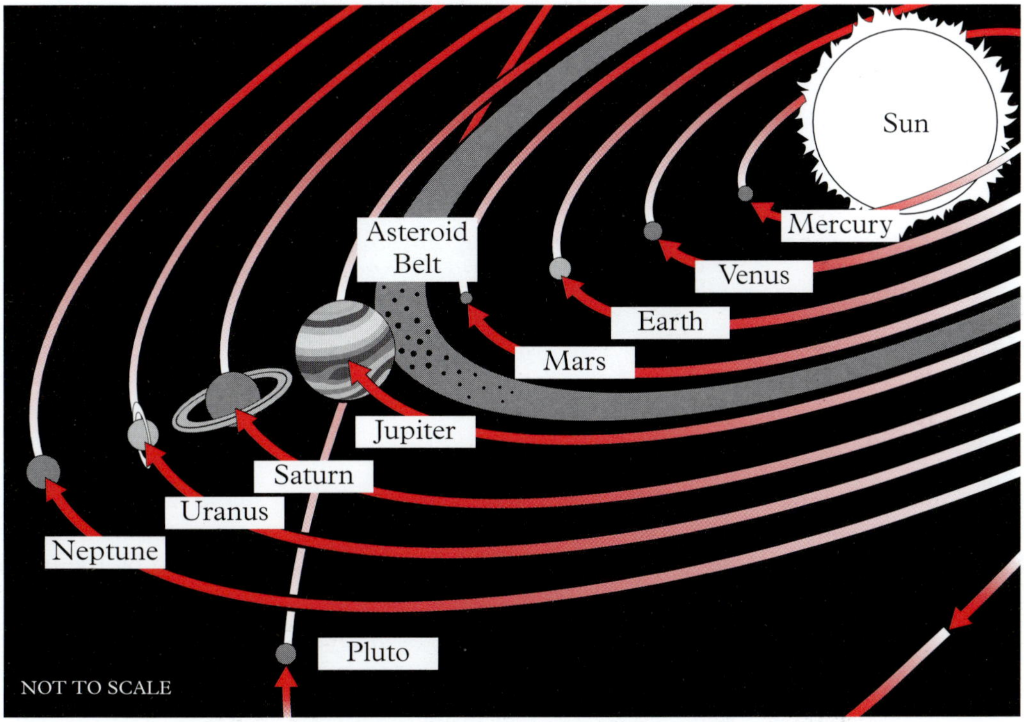

All the planets in the solar system move in elliptical orbits with the Sun near the centre. All of these planets orbit the Sun in the same plane, except for Pluto. Pluto's orbit is at an angle to this plane. The planets which are visible to the naked eye look like stars, but over a period of several weeks their positions against the distant stars change.

Asteroids

Between the orbits of Mars and Jupiter there is a belt of rock debris. These chunks of rock are called asteroids and vary in size from just a few metres to several hundreds of kilometres across.

Comets

Comets are made of dust and ice. They also circle the Sun, but their orbits tend to take them both very close to the Sun and to the outer edges of the solar system.

As a comet approaches the Sun some of its frozen gases evaporate creating a spectacular tail of dust and ice which can range from thousands to millions of kilometres long. Perhaps the most famous comet is Halley's Comet. Its orbit makes it visible from the Earth every 76 years. Halley's Comet was last seen in 1986.

comet's tail
comet's head

Moons

the Moon is a non-luminous object which means it does not produce light itself
we can see the Moon because it reflects light from the Sun
the Moon's surface is covered with craters, caused by the impact of large pieces of rock called asteroids
unlike the Earth, the Moon has no atmosphere which can protect it from meteorites

Moons are natural objects which orbit a planet. The Earth has just one moon but Jupiter has 17 moons and Saturn has 23. Mars and Venus have no moons. Our moon orbits the Earth once every 28 days. This is called a **lunar month**.

Information about the planets in our solar system

Planet	Approximate distance from the Sun compared with the Earth	Approximate diameter compared with the Earth	Approximate time to orbit the Sun in years
Mercury	½	½	¼
Venus	¾	1	½
Earth	1	1	1
Mars	1½	½	2
Jupiter	5	11	12
Saturn	10	10	30
Uranus	20	4	85
Neptune	30	3½	165
Pluto	40	¼	250

Questions

1 Draw a labelled diagram of our solar system including the Sun and the nine planets.

2 What is an elliptical orbit?

3 What is an asteroid?

4 What is a comet?

5 What is a moon?

6 Which is the smallest planet in our solar system?

7 Which is the largest planet in our solar system?

8 Which planet has the longest year?

9 Which planet is 20 times further from the Sun than the Earth is?

10 Which two planets are just half the size of the Earth?

Summary

The Earth is one of nine planets which orbit a star we call the Sun. These, together with other bodies such as moons, asteroids and comets, make up our solar system.

The Earth spins on its axis once every day. It is daytime in the half of the Earth's surface which is receiving sunlight. It is nightime in the half which is receiving no sunlight. The rotation of the Earth makes the Sun appear to rise in the East and set in the West.

The Earth orbits the Sun once every year. Because the axis of the Earth is tilted we experience seasonal changes in temperature and length of day as the Earth moves around the Sun.

Key words

asteroids Large pieces of rock usually found in a belt between Mars and Jupiter.

axis Imaginary line running through the centre of the Earth.

comets Lumps of dust and ice which follow elongated orbits around the Sun.

daytime When the Earth's surface is receiving sunlight.

ellipse A slightly squashed circle.

moons Natural bodies which orbit planets.

nightime When the Earth's surface is receiving no sunlight (when it is facing away from the Sun).

one day The time it takes the Earth to make one full rotation.

orbit The path a planet follows around the Sun.

planet A non-luminous body which orbits a star.

star A luminous body around which planets may travel.

seasons Periods of time when the Earth is in different parts of its orbit around the Sun consisting of spring, summer, winter and autumn.

solar system A family of planets, asteroids, comets etc. orbiting the Sun.

End of Chapter 23 Questions

1 a) What causes the Sun to appear each day in the east, travel across the sky and set in the west?

b) Fill in the gaps.
   i) A day is the time it takes the Earth to _________ .
   ii) A year is the time it takes the Earth to _________ .
   iii) Explain with diagrams, why parts of the Earth are in daylight whilst other parts are in darkness.

2 Below are the names of seven planets in our Solar System.

Neptune Venus Mercury Saturn Uranus Earth Pluto

a) Name the two planets in our Solar System whose names are not here.
b) Which planet has the largest orbit?
c) Which planet has the shortest orbit?
d) Which planet completes one orbit of the Sun in 365 days?
e) Which star do all these planets orbit?
f) What is the shape of all these orbits?
g) What is an asteroid and where in the Solar System would we find them?
h) i) What is a comet?
   ii) Name one famous comet sometimes visible from the Earth.
i) i) What is a moon?
   ii) Name one planet which has a moon.

3 a) Explain why we are able to see stars.
b) Explain why we are able to see the Moon.
c) Explain why we are unable to see stars in the daytime.

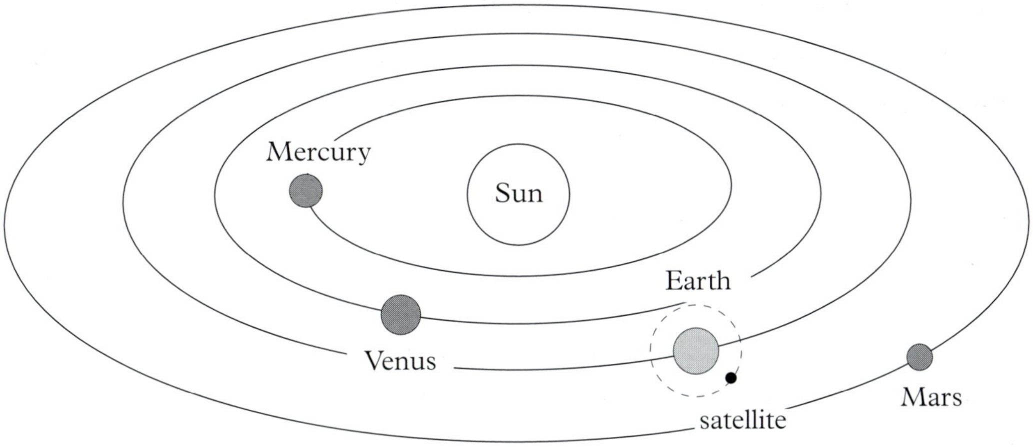

4 The diagram above shows the orbits of some of the planets.
a) What shape are these orbits?
b) Name one planet whose year is longer than the Earth's.
c) What is the name of the natural satellite which is orbiting the Earth?
d) Comets also orbit the Sun. Copy the above diagram and add to it the path of a typical comet.

5 The diagram below shows the Earth in four different positions as it orbits the Sun.

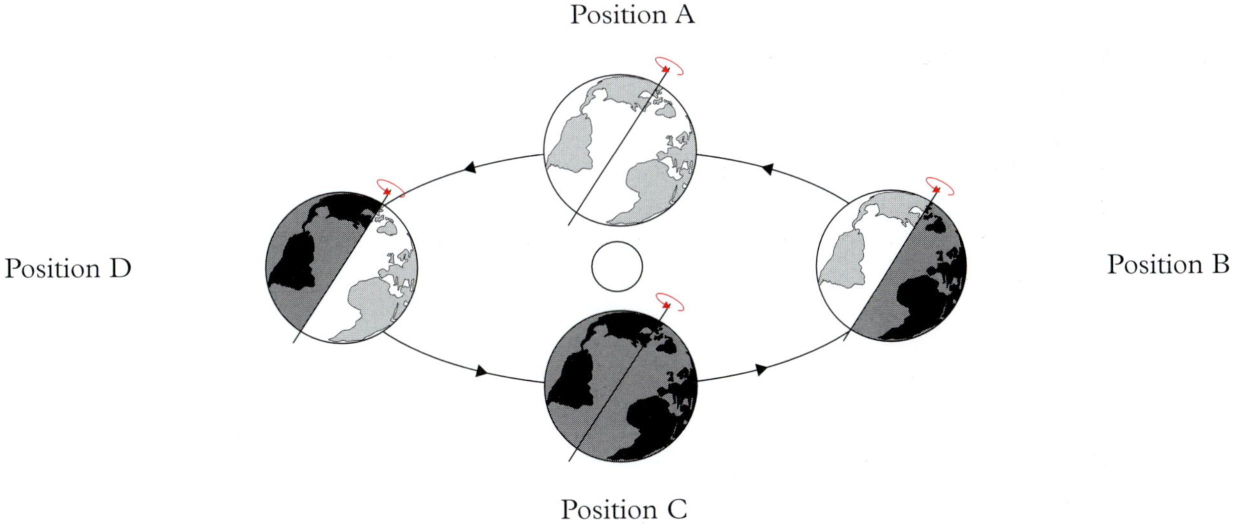

a) How long does it take for the Earth to complete one full orbit of the Sun?
b) i) In which position is it winter in the Northern hemisphere?
ii) Explain your answer.
c) i) In which position is it spring in the Northern hemisphere?
ii) Explain your answer.
d) Give two changes that take place on Earth as the seasons move from summer to winter.

24 The Solar System and Beyond

Gravitational forces

For many many years mankind observed the motions of the stars and planets but was unable to explain why they moved as they do. In 1687 a scientist called Isaac Newton put forward a theory which not only explained the motions of all the bodies in the sky but eventually explained how the stars and their planets were first formed.

Newton suggested that all objects are attracted to other objects by forces he called **gravitational forces**. The size of these forces depends upon the mass of the objects and their separation.

the gravitational attraction between small objects is extremely weak

the gravitational attraction between large objects is very strong

The greater the masses of the objects the greater the forces of attraction.

the gravitational attraction between the Earth and the Moon is strong

if the Moon were further away however . . .

. . . the gravitational attraction between the Earth and the Moon would be much less

The closer the objects are the greater the forces of attraction.

Moving in a circle

Any object which is travelling along a curved path is experiencing a force.

to make this ball travel around in a circle you have to pull hard on the string

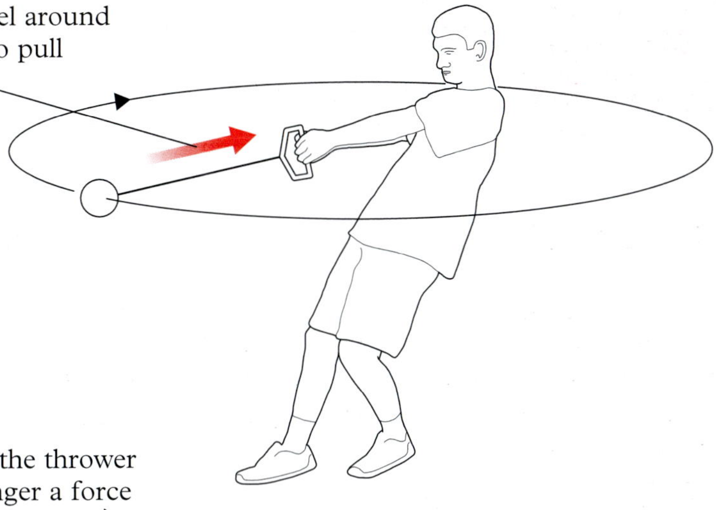

if the chain snaps or the thrower lets go there is no longer a force pulling the ball around

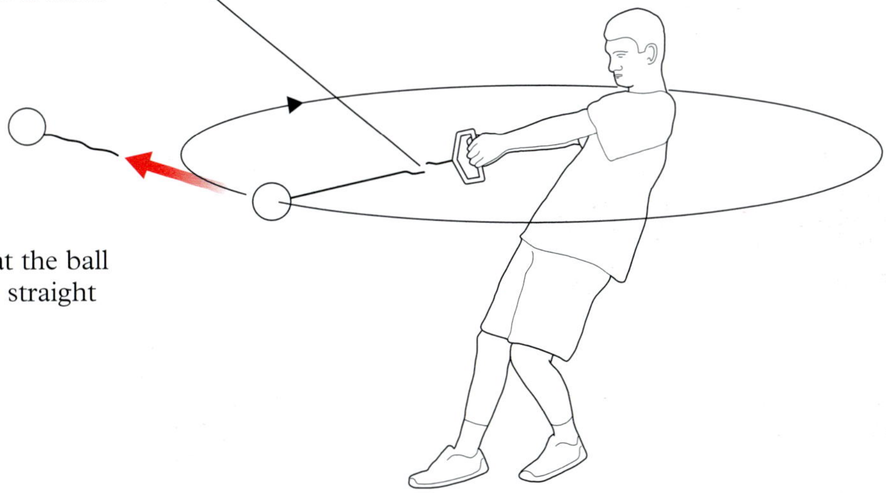

this means that the ball flies away in a straight line

Planets and moons also travel in curved paths.

a force causes the planets and moons to follow curved paths

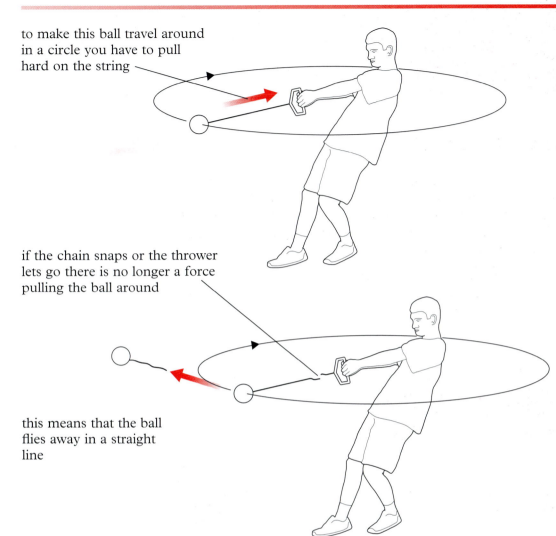

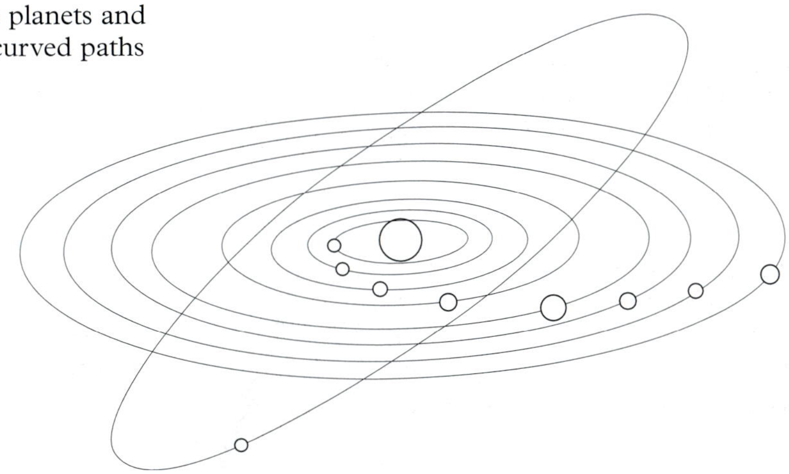

this force is the attractive force between objects due to gravity

Our Sun contains 99% of the mass of the solar system. It is the gravitational attraction between it and the planets which holds our solar system together. The planets nearest the Sun experience the largest forces and follow the most curved paths. Planets furthest from the Sun experience the smallest gravitational forces. Their paths are much less curved.

Questions

1 What must be acting upon an object to make it travel along a curved path?

2 Name two bodies which travel along curved paths.

3 Describe what happens to the gravitational attraction between two bodies when the folowing happens:
a) the distance between them increases
b) the mass of both bodies increases.

Stars

Stars are luminous objects, which means that they give out their own light. The star at the centre of our solar system is the Sun. The Sun is a very average star. It is not particularly large or hot compared with other stars. The Sun's surface temperature is about 6000 °C whilst the temperature at its centre is about 15 million °C. These high temperatures are the result of **nuclear reactions** involving the gas hydrogen.

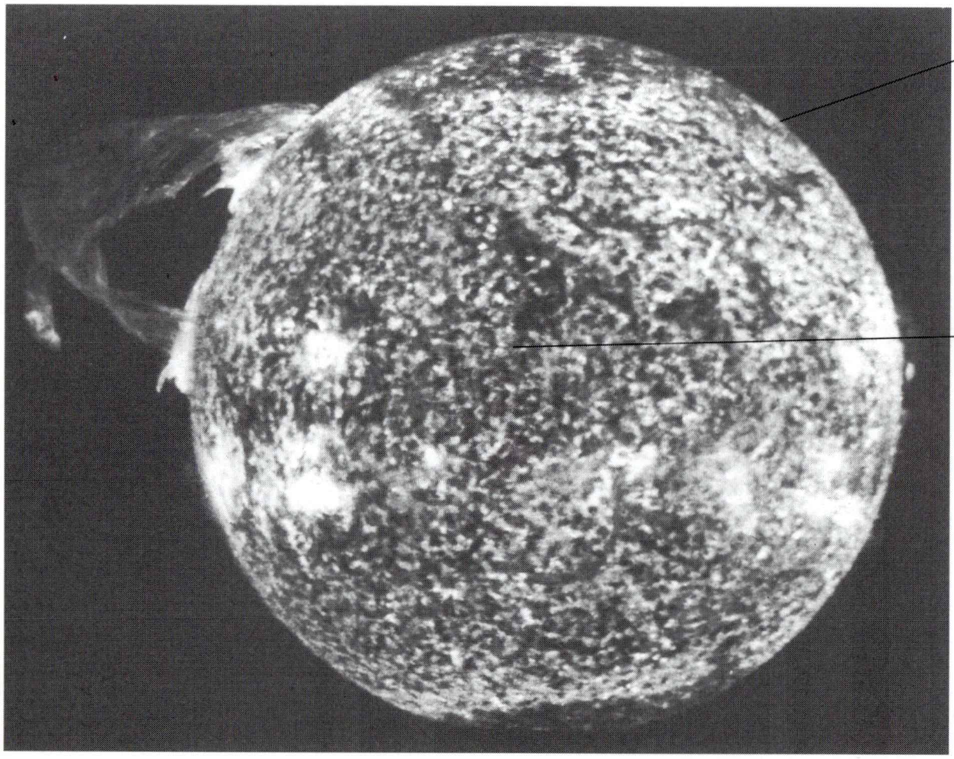
the Sun's surface temperature is about 6000 °C
at its centre the temperature of the Sun is about 15 million °C

All the heat and light energy emitted by stars are a result of nuclear reactions within the gases.

The Birth of the Solar System

Stars like our Sun are formed from giant clouds of dust and gas. Gravitational forces draw the clouds together to form the hot ball of gas we call a star. Some of the cloud may then form planets around the star.

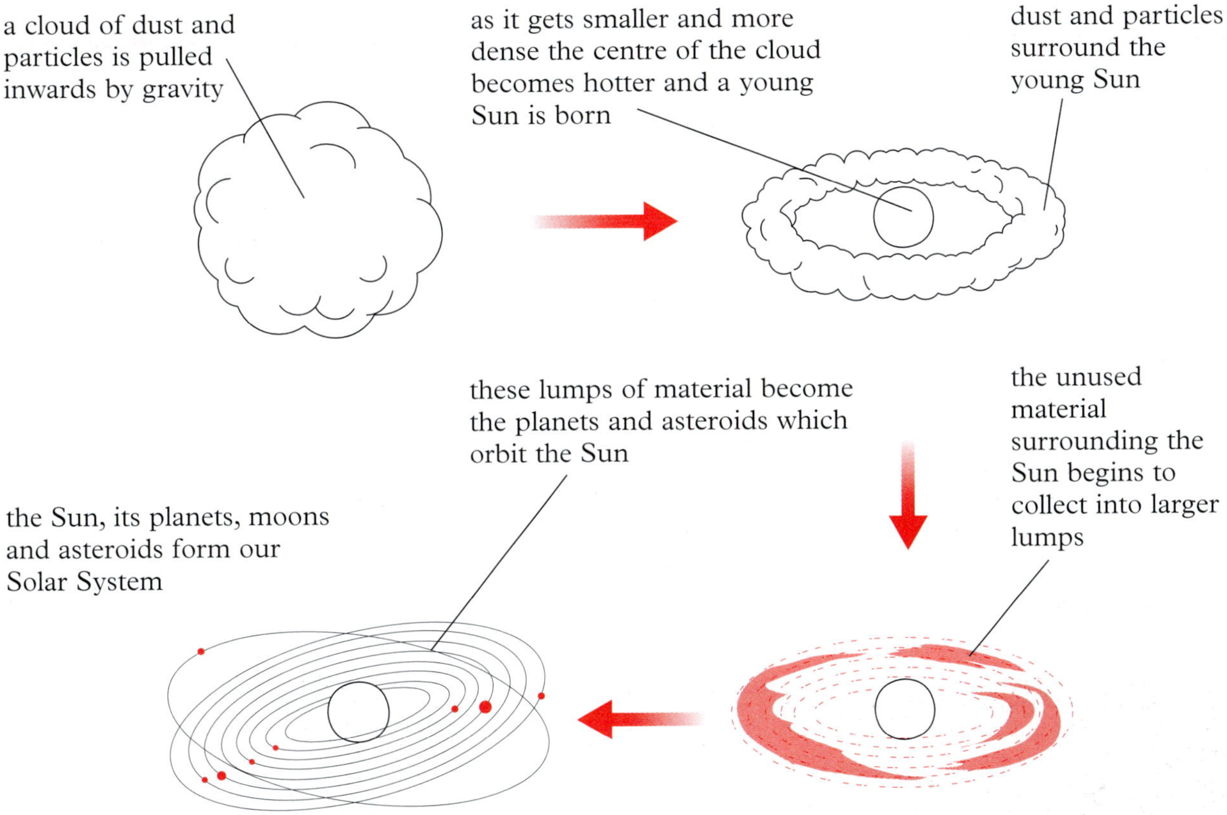
a cloud of dust and particles is pulled inwards by gravity
as it gets smaller and more dense the centre of the cloud becomes hotter and a young Sun is born
dust and particles surround the young Sun
the unused material surrounding the Sun begins to collect into larger lumps
these lumps of material become the planets and asteroids which orbit the Sun
the Sun, its planets, moons and asteroids form our Solar System

The smaller bodies orbiting the Sun are called planets. Like moons, planets are not luminous. They do not emit any light of their own.

The Birth and Death of a Star

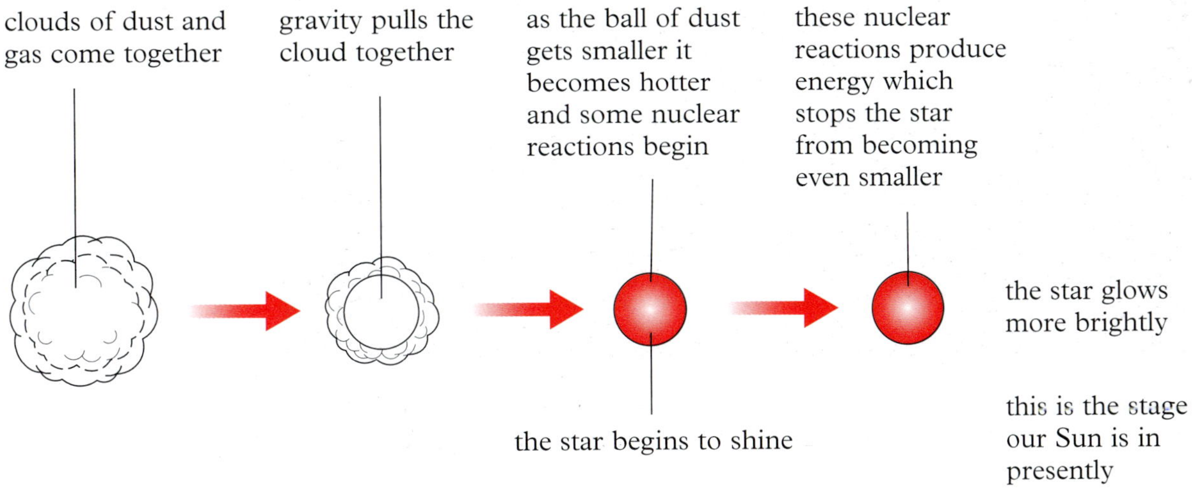
clouds of dust and gas come together
gravity pulls the cloud together
as the ball of dust gets smaller it becomes hotter and some nuclear reactions begin
these nuclear reactions produce energy which stops the star from becoming even smaller
the star begins to shine
the star glows more brightly
this is the stage our Sun is in presently

After millions of years the gases in stars like our Sun will be used up.

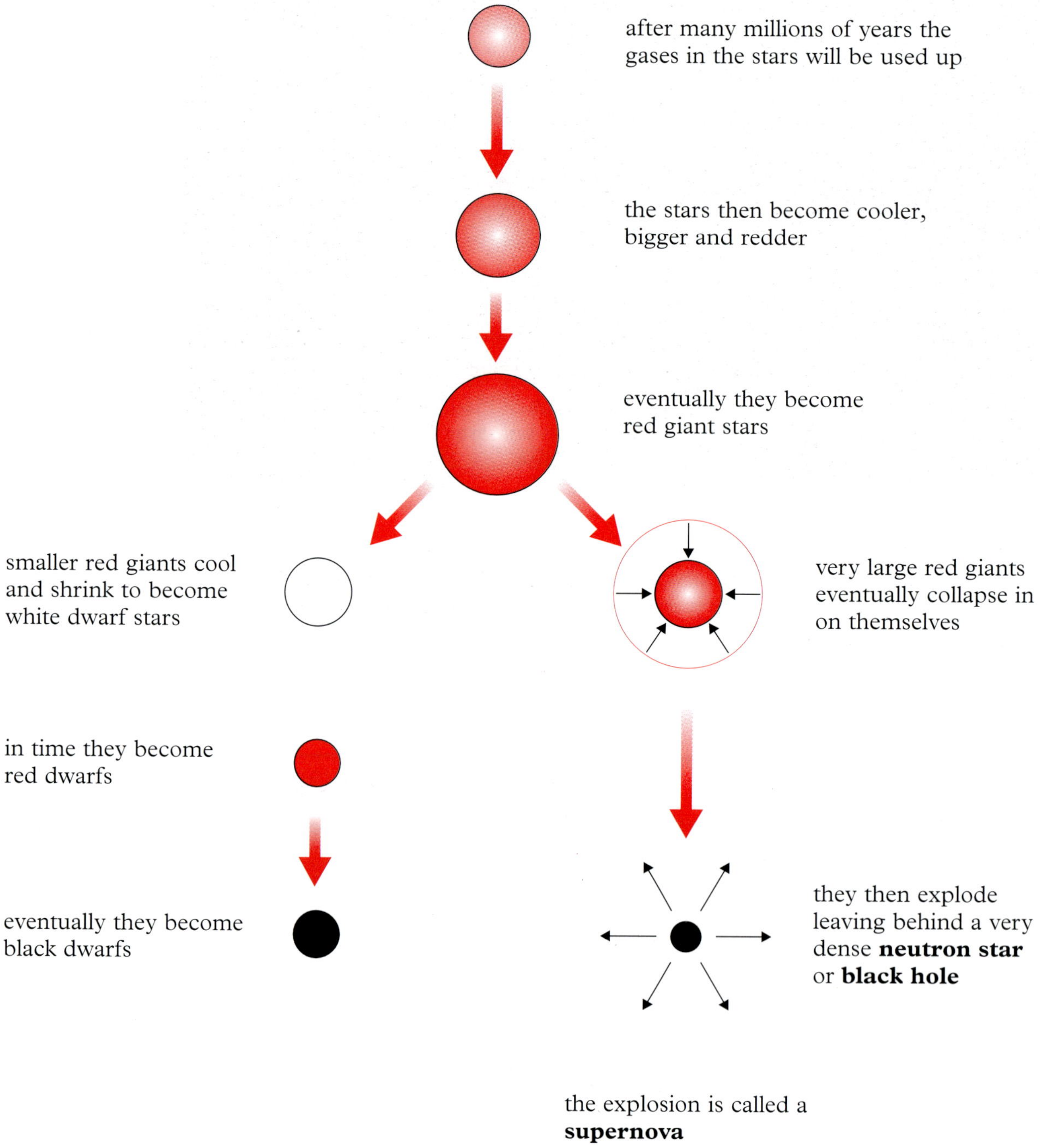
after many millions of years the gases in the stars will be used up
the stars then become cooler, bigger and redder
eventually they become red giant stars
smaller red giants cool and shrink to become white dwarf stars
very large red giants eventually collapse in on themselves
in time they become red dwarfs
eventually they become black dwarfs
they then explode leaving behind a very dense **neutron star** or **black hole**
the explosion is called a **supernova**

Constellations

During the daytime the light from the Sun is so bright it is impossible to see any other stars. They only become visible as the Sun sets. Even then these other stars seem small and dim compared with our Sun. This is because they are much much further away.

In the night sky there are groups of stars which may appear to be close together. These groups are called **constellations**. You will probably have seen and heard of some of them. Astronomers often join up the stars with imaginary lines to make them easier to see.

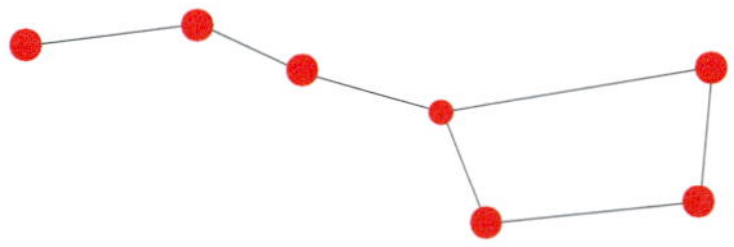

The Plough

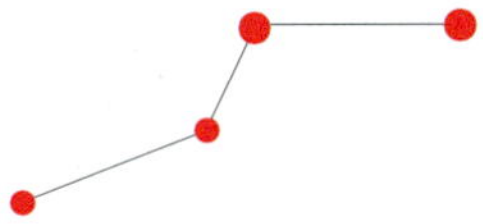

Pisces

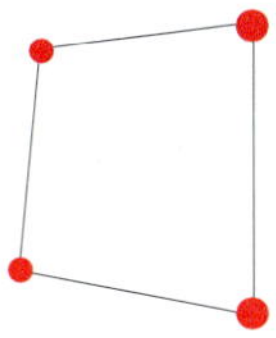

Pegasus

Galaxies

Stars and constellations cluster together in enormous groups called **galaxies**. Our galaxy is called the Milky Way and contains about 200 000 million stars.

the Milky Way is a spiral galaxy

seen from above it looks just like a spiral

seen from the side, the Milky Way looks like a flattened disc

There are billions of galaxies spread throughout the universe.

an elliptical galaxy

a spiral galaxy

Questions

1 Describe how a star is born.

2 What is the source of all the energy which is emitted by a star?

3 a) What is a constellation?
b) Draw two constellations.

4 What is a galaxy?

5 What is the name of the galaxy where we live?

The Light Year

Distances in space are so great that our normal units of measurement, the metre and the kilometre, are far too small. The Sun is 150 million km from Earth. But it takes just 8.3 minutes for a ray of light to travel this distance.

We use this idea of how long it will take light to travel between two points as our new distance measuring unit. It is called the **light year**. A light year is the distance a ray of light will travel in one year.

the light from this star takes one year to reach the Earth

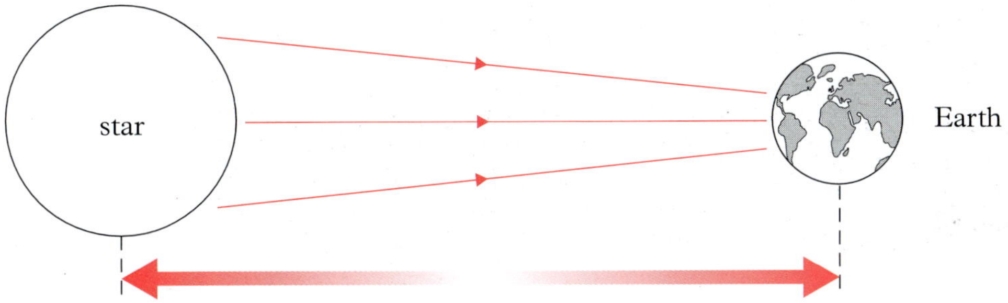

the distance between the star and the Earth is 1 light year

Our next nearest star is so far away that its light takes 4.3 years to reach the Earth. It is 4.3 light years from Earth.

Light from our nearest galaxy, the Andromeda galaxy, takes 2 million years to reach us.

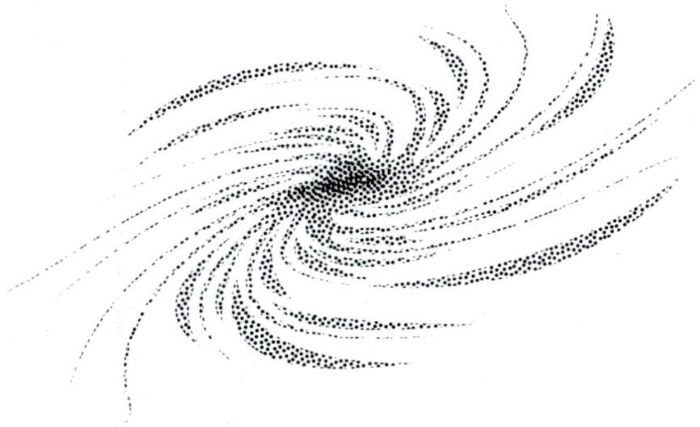

the Andromeda galaxy is 2 million light years from Earth

Some galaxies are thousands of light years across and millions of light years away.

The moving night sky

If we were to take a photograph of the stars in the sky every hour, as shown in the photograph on the next page, we would discover that they appear to revolve and change position. This happens because the Earth is rotating. Only one star the Pole Star, appears not to move, as it is directly above the axis of rotation.

this photograph was taken over a few hours

it shows that the stars appear to move in circles

this happens because the Earth spins on its axis

Questions

1 What is a light year?

2 The distance from one end of our galaxy to the other is 100 000 light years. Explain what this means.

3 A camera on a stand takes one picture of the northern night sky every hour. The photographs show that all the stars except one appear to be moving.
a) What is causing the stars to appear to move?
b) Which star does not appear to be moving?

Artificial satellites

the moon orbits the Earth – it is a **natural satellite**

there are many man-made objects which also orbit the Earth

these are called **artificial satellites**

These artificial satellites have many uses, as we can see below.

Observing the Earth from above

This photograph was taken by a weather satellite. From high above the Earth it is able to provide information which makes weather forecasting more reliable.

Communications

Communication satellites allow people to send and receive messages from all over the world. Radio, television, the Internet, even mobile phones make use of these artificial satellites.

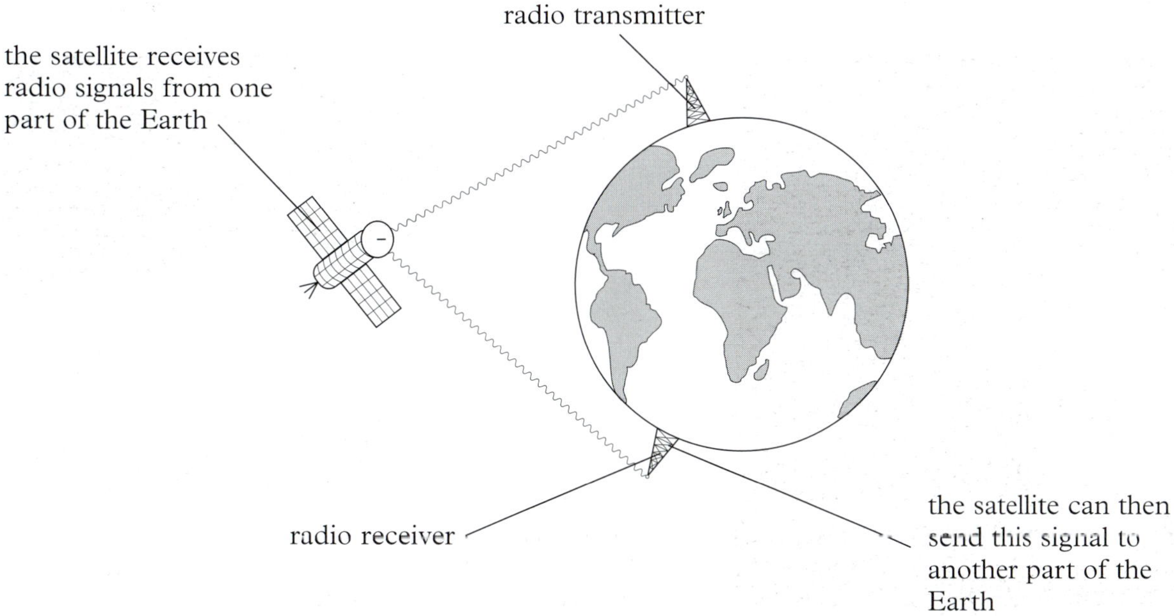
radio transmitter
the satellite receives radio signals from one part of the Earth
radio receiver
the satellite can then send this signal to another part of the Earth

Observing space from above the Earth

When scientists use telescopes to observe objects in space they have until recently had to look through the Earth's atmosphere. Under certain circumstances this can prevent them from seeing objects clearly. They are now overcoming this problem by mounting telescopes on satellites which are orbiting above the Earth's atmosphere. One such telescope is the Hubble telescope. It is orbiting 600 km above the Earth's surface.

Exploring the solar system

Although man has shown with his visits to the Moon that he is able to travel in space, artificial satellites offer a cheaper and safer way of investigating our solar system and beyond.

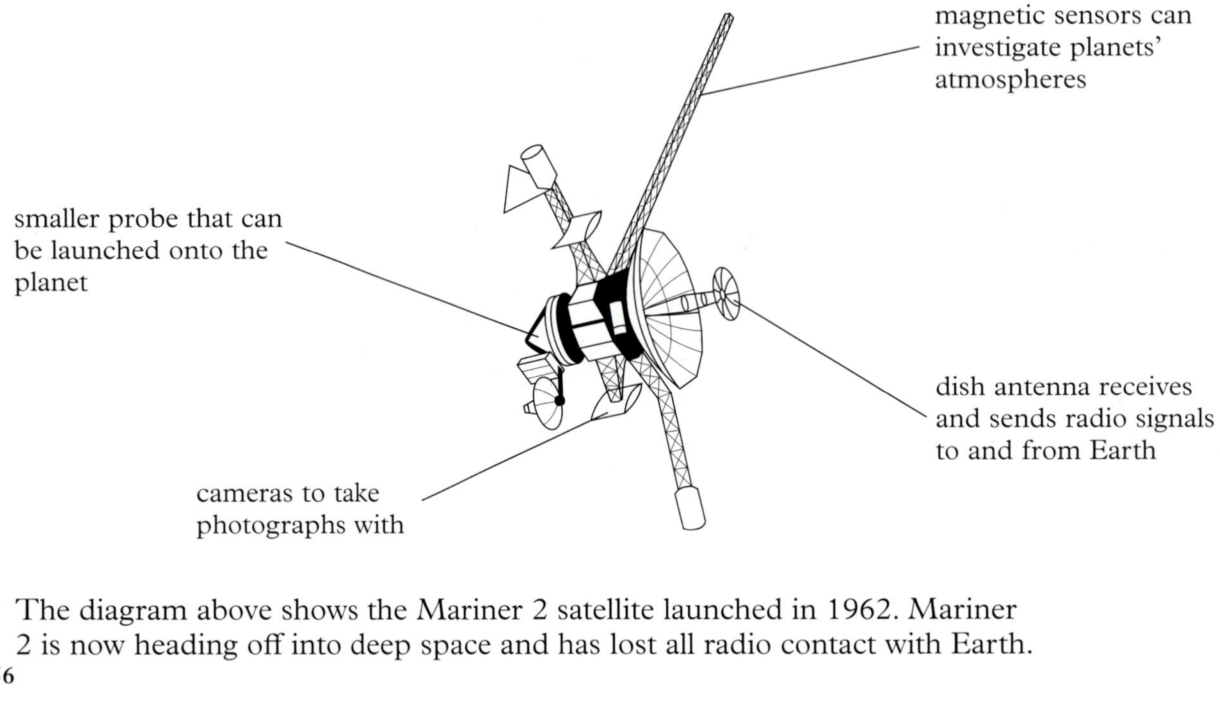
magnetic sensors can investigate planets' atmospheres
smaller probe that can be launched onto the planet
dish antenna receives and sends radio signals to and from Earth
cameras to take photographs with

The diagram above shows the Mariner 2 satellite launched in 1962. Mariner 2 is now heading off into deep space and has lost all radio contact with Earth.

Questions

1 Name one natural satellite.

2 What is an artificial satellite?

3 Describe four uses for an artificial satellite.

Satellite Orbits

Satellites can orbit the Earth at different heights and at different speeds.

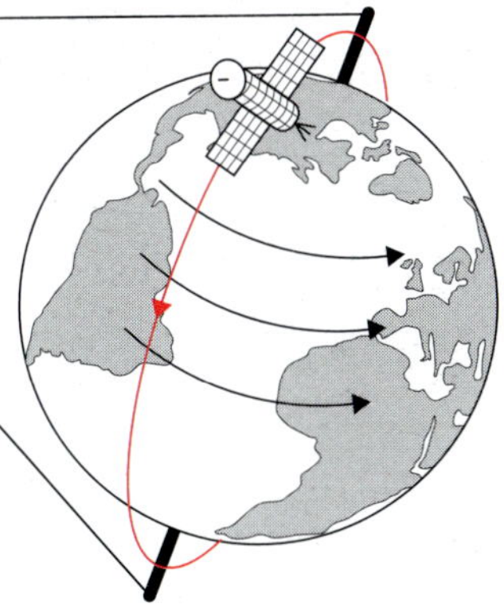
satellites which monitor the environment are put into fast low orbits which take them over the North and South poles
using these orbits they can scan the whole of the Earth's surface in a very short time

A satellite which orbits high above the equator and circles the Earth once every 24 hours will remain above the same point on the Earth's surface. A satellite which does this is called a **geostationary satellite**. Geostationary satellites are used for communications purposes so that they are always in the correct position to receive and redirect messages.

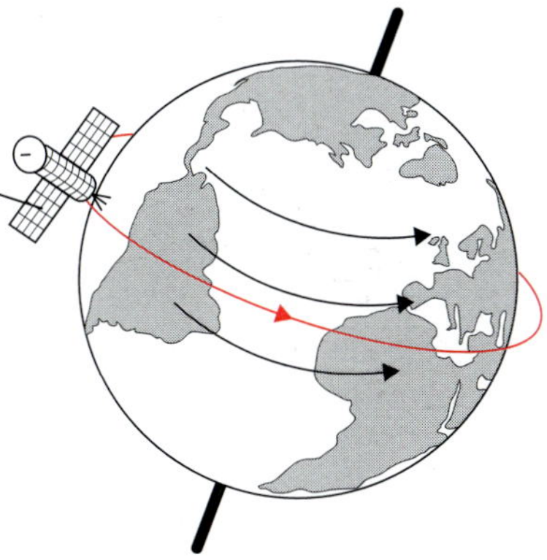
as the Earth rotates the satellite moves with it
the satellite stays above the same part of the Earth
this is called a **geostationary orbit**

The Origins and the Future of the Universe

Nobody knows exactly how the universe began but there is evidence which suggests it started thousands of millions of years ago with a **Big Bang**. This explosion sent matter spreading out in all directions. Observations by astronomers confirm that even now the universe and everything in it is still spreading out. Some scientists believe that the universe will continue to expand for ever. This theory is called the **Continuous Expanding Universe**. Others believe that the forces of gravity will slow down this expansion and eventually pull all the matter back together. At which point the whole process will begin again. This theory is called the **Pulsating Universe**.

Questions

1 What is a geostationary satellite?

2 Suggest two uses of a geostationary satellite

3 What kind of orbit would you use for a satellite if you wanted to monitor the whole of the Earth's surface each day?

4 Explain how most scientists now believe the universe began.

Summary

Our Sun is just one star in many millions in a galaxy we call the Milky Way. Stars are massive luminous bodies which are formed when large amounts of dust and gas are pulled together by gravitational forces. As the matter is drawn together, very high temperatures are created which begin nuclear reactions within the gases. The star is now in the main stable period of its life. Our Sun is at this stage in its life. After a very long period of time these gases are used up, new reactions begin and the appearance and temperature of the star changes. Groups of stars which from the Earth appear to be close together are called constellations.

Smaller amounts of gas and dust encircling a star may be pulled together to form less massive, non-luminous bodies called planets. The motions of stars, planets, moons, constellations and galaxies are controlled by gravitational forces.

Artificial satellites can be put into orbit around the Earth and used to observe the Earth's surface, monitor the weather, send messages between places which are a long way apart on the Earth and observe the universe without the Earth's atmosphere getting in the way. They can also be launched by rocket to travel to other planets or moons where they are used to gather and send back valuable information.

Key words

artificial satellites	Man-made objects which orbit a planet.
Big Bang	The explosion which began the expanding universe.
constellation	A small group of stars, like the plough.
galaxy	A very large cluster of stars (thousands of millions of stars).
geostationary orbit	An orbit which keeps a satellite over the same point on the Earth's surface.
geostationary satellite	A satellite which remains over the same point on the Earth's surface.
gravitational forces	Attractive forces between bodies.
light year	The unit used to measure large distances in space. It is the distance light travels in one year.
natural satellites	Natural bodies which orbit a planet e.g. a moon.
nuclear reactions	Reactions between the nuclei of atoms which release vast quantities of energy.
star	Hot ball of gas inside which nuclear reactions are taking place.
supernova	Spectacular explosion signalling the death of a large star.

End of Chapter 24 Questions

1 Look at the list below.

Earth Moon Sun constellation Universe Solar System

Put them in order of size beginning with the smallest and finishing with the largest.

2 The diagram below shows a communications satellite in geostationary orbit high over the equator.

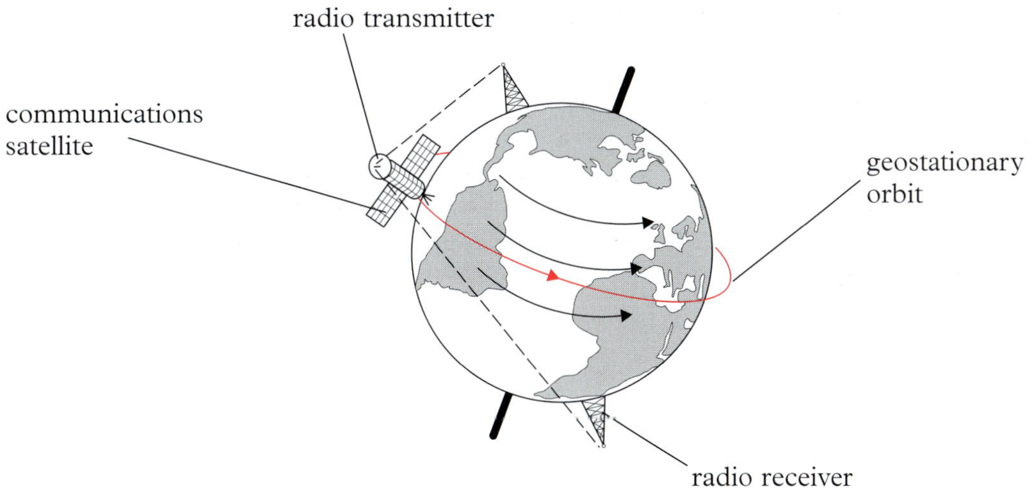

a) Explain how the people on opposite sides of the Earth can use the satellite to communicate.
b) Why is it important that the satellite is in a geostationary orbit?
c) What forces keep a satellite in orbit around the Earth?

3 Astronomers observed one part of the sky for several weeks. Apart from the Moon they observed several bright objects. Choose words from below to correctly name the following objects.
a) Object A is one of many bright objects which was in the same position each night.
b) Object B is a bright object which over several weeks moved slowly across the fixed pattern.
c) Object C is also a bright object slowly moving across the fixed pattern but it appears to have a long bright tail.
d) Object D is a bright object which crosses the sky several times in one night.

satellite star planet comet

4 The diagram below shows part of the solar system

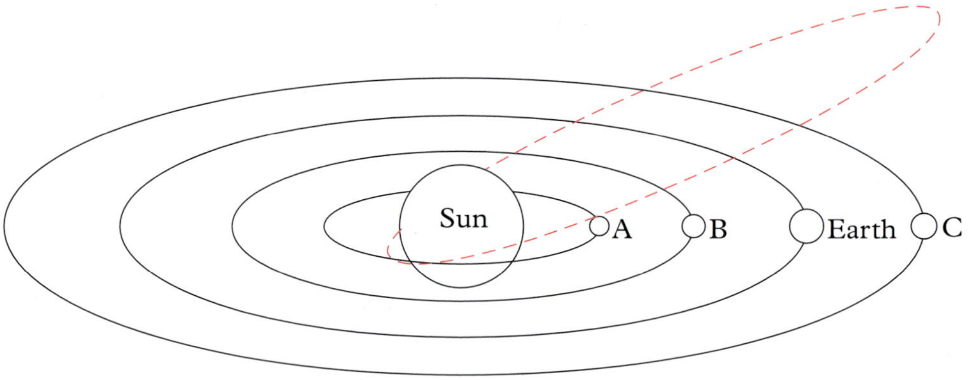

a) i) Name planet A.
ii) Name planet B.
iii) Name planet C.
b) What kind of body might follow an ellipse similar to that shown by the dotted line?
c) What forces keep the planets in elliptical orbits?
d) Which planet in the diagram experiences the largest forces?
e) i) Name one planet which has a natural satellite.
ii) What is the name of this satellite?
f) What is an asteroid?
g) Where in the solar system is there a belt of asteroids?

5 Complete each of the sentences below by choosing the correct words from the list. Each word may be used once, more than once or not at all.

Sun Universe Milky Way constellation planet moon

The Earth is a _________ . It orbits the Sun. The Sun is a _________ . Large clusters of stars are called galaxies. Our galaxy is called the _________ . There are billions of galaxies in our _________ .

Index